Lecture Notes in Civil Engineering

Volume 566

Lecture Notes in Civil Engineering (LNCE) publishes the latest developments in Civil Engineering—quickly, informally and in top quality. Though original research reported in proceedings and post-proceedings represents the core of LNCE, edited volumes of exceptionally high quality and interest may also be considered for publication. Volumes published in LNCE embrace all aspects and subfields of, as well as new challenges in, Civil Engineering. Topics in the series include:

- Construction and Structural Mechanics
- Building Materials
- Concrete, Steel and Timber Structures
- Geotechnical Engineering
- Earthquake Engineering
- Coastal Engineering
- Ocean and Offshore Engineering; Ships and Floating Structures
- Hydraulics, Hydrology and Water Resources Engineering
- Environmental Engineering and Sustainability
- Structural Health and Monitoring
- Surveying and Geographical Information Systems
- Indoor Environments
- Transportation and Traffic
- Risk Analysis
- Safety and Security

To submit a proposal or request further information, please contact the appropriate Springer Editor:

- Pierpaolo Riva at pierpaolo.riva@springer.com (Europe and Americas);
- Swati Meherishi at swati.meherishi@springer.com (Asia—except China, Australia, and New Zealand);
- Wayne Hu at wayne.hu@springer.com (China).

All books in the series now indexed by Scopus and EI Compendex database!

Y. K. Guruprasad · G. R. Dodagoudar ·
K. S. Nanjunda Rao · P. C. Pandey · H. B. Nagaraj
Editors

Recent Advances in Infrastructure Development and Management—Volume 2

Select Proceedings of MTCEIDM 2023

Springer

Editors
Y. K. Guruprasad
Department of Civil Engineering
BMS College of Engineering
Bengaluru, Karnataka, India

K. S. Nanjunda Rao
Department of Civil Engineering
Indian Institute of Science Bangalore
Bengaluru, Karnataka, India

H. B. Nagaraj
Department of Civil Engineering
BMS College of Engineering
Bengaluru, Karnataka, India

G. R. Dodagoudar
Department of Civil Engineering
Indian Institute of Technology Madras
Chennai, Tamil Nadu, India

P. C. Pandey
Formerly Professor (Retd.)
Department of Civil Engineering
Indian Institute of Science Bangalore
Bengaluru, Karnataka, India

Emeritus Professor
School of Civil Engineering
REVA University
Bengaluru, Karnataka, India

ISSN 2366-2557 ISSN 2366-2565 (electronic)
Lecture Notes in Civil Engineering
ISBN 978-981-96-1557-5 ISBN 978-981-96-1558-2 (eBook)
https://doi.org/10.1007/978-981-96-1558-2

This Springer imprint is published by the registered company Springer Nature Singapore Pte Ltd.
The registered company address is: 152 Beach Road, #21-01/04 Gateway East, Singapore 189721, Singapore

Preface

Advancement and modernization of construction practices in the field of Civil Engineering have led into rapid development in infrastructure and in construction management. The second volume of this book aims to serve as a comprehensive resource for academicians and professionals. The chapters in this volume cover a wide range of topics, and each chapter presents research contributions that address current challenges and throws light on future developments in the respective areas.

This volume contains select proceedings of the International Conference on 'Modern Trends in Civil Engineering Infrastructure Development and Management', held from 8th to 10th December 2023. This gathering brings together leading researchers, practitioners, and innovators to discuss and share advancements in Civil Engineering that shape the future of infrastructure. Civil Engineering has always been at the forefront of transforming society, and today, it faces a unique challenge: to develop sustainable, resilient, and technologically advanced solutions that cater to the growing needs of our world.

The themes of this conference cover a broad range of pressing topics and advancements in Civil Engineering. **IoT Applications in Monitoring of Civil Engineering Infrastructure** highlights the role of the Internet of Things (IoT) in enhancing infrastructure safety and efficiency through real-time monitoring and predictive maintenance. **Induction of Recycled Waste in Construction** emphasizes sustainable construction practices that reduce environmental impact by incorporating recycled materials, fostering a more circular economy in the construction industry. Parallel to this, **Futuristic Building Materials** explores newer materials that promise greater durability, flexibility, and sustainability in modern structures.

The theme of **Innovation in Foundation Design and Soil Stabilization** is crucial as engineers face the complexities of diverse soil types and challenging terrains. Novel approaches in foundation design and soil stabilization ensure stability and safety, meeting the demands of various environments. Additionally, **Modalities in Design and Behaviour of Mega Structures** delves into the unique challenges associated with large-scale structures, which require sophisticated engineering approaches to withstand extreme loads and environmental stresses. **Effective Construction Management Practices** addresses the importance of efficient project management, emphasizing strategies to minimize delays, reduce costs, and maximize resource utilization. Infrastructure for **Water Resources and Transportation Engineering** is essential to support growing populations, and this theme explores sustainable water resource management and advanced transportation solutions that are crucial for economic and social development. Finally, **GIS & Remote Sensing Applications to Civil Engineering** sheds light on how spatial data technologies are revolutionizing planning, design, and infrastructure management by enabling more precise and informed decision-making.

Each chapter in this volume has been selected after a thorough check and peer review to ensure that the quality and academic rigour are maintained. The editors appreciate the contributions made by the authors and thank the reviewers for their invaluable feedback. This volume would not have been possible without their dedication and expertise. The variety of topics covered in this volume shows the multifaceted aspects in Civil Engineering. From theoretical research to practical applications, each chapter contributes to a deeper understanding of the individual domain and addresses towards the current challenges. The interdisciplinary nature of the research also underscores the importance of collaboration across different domains of expertise, fostering innovation.

We hope that this volume serves as a valuable reference for researchers, engineers, and students, providing them with the inspiration leading them towards application of this knowledge into practice.

Bengaluru, India Dr. Y. K. Guruprasad, Ph.D. (IISc)

Contents

About the Editors

Dr. Y. K. Guruprasad is working as an Associate Professor in the Department of Civil Engineering at BMS College of Engineering, Bengaluru. He received his Ph.D. from the Indian Institute of Science (IISc), Bengaluru. He carried out part of his Ph.D. research at the University of Edinburgh, Scotland, United Kingdom in 2009 as part of the UK-India Education and Research Initiative (UKIERI) programme on the repair of structures adopting Fibre Reinforced Polymer (FRP) and the behaviour of structures exposed to fire. Dr. Guruprasad has also completed his post-doctoral research at IISc. He completed M. Tech. in structural engineering from M.S. Ramaiah Institute of Technology (MSRIT), Bengaluru, and his B.E. in Civil Engineering from BMS College of Engineering, Bengaluru. His areas of research include but are not limited to, repair and retrofit of structures, retrofit of structures adopting FRP composites, adoption of IoT in structural health monitoring, non-destructive testing evaluation, earthquake engineering and seismic resistant analysis and design of structures, structural dynamics and vibration analysis, fracture mechanics, microanalysis of concrete exposed to high temperature. He has multiple publications in refereed international and national journals, conferences and book chapters. He has guided seven M. Tech. students in structural engineering. He has also guided many final-year B.E. projects, which have been recognized and funded by the Karnataka State Council for Science and Technology (KSCST) at the IISc Campus. He has also carried out consultancy work in non-destructive evaluation of structures and structural analysis and design of structures. He is a reviewer for international journals. He is a life member of the Institution of Engineers (India), Chartered Engineer (CE) and Institution of Engineers (India), Indian Concrete Institute(ICI), Indian Society of Non-destructive Testing (ISNT), Indian Association for Computational Mechanics (IndACM) and Institute for Research Development and Training Of Construction Trades & Management (INSTRUCT). He is also a fellow member of the Association of Consulting Civil Engineers (ACCE).

Dr. G. R. Dodagoudar is presently working as a Professor in the Department of Civil Engineering, Indian Institute of Technology (IIT) Madras, India. He obtained his Ph.D. degree from IIT Bombay. His areas of research interest include computational geomechanics, contaminant transport modelling, reliability analysis of geotechnical systems, soil dynamics and earthquake engineering, and seismic hazard and risk assessment. Dr. Dodagoudar has guided 21 Ph. D. and seven M.S. (research) thesis and many M. Tech. dissertations. He has extensive publications in national and international refereed journals. Dr. Dodagoudar has been actively involved in many continuing education programmes. He is an active member of the New Delhi Standing Appellate Committee of the All India Council for Technical Education (AICTE). Dr. Dodagoudar has completed many prestigious sponsored research and consultancy projects.

Dr. K. S. Nanjunda Rao is currently the Chief Research Scientist in the Department of Civil Engineering at IISc, Bangalore. He obtained his BE (Civil Engineering) and M.E. from BMS College of Engineering, Bangalore, and Ph.D. (structural engineering) from IISc Bangalore. His research interests include mechanics of composite plate and shell structures, finite element structural analysis, structural masonry, reinforced and prestressed concrete, earthquake-resistant design of structures, and alternative building materials. He is actively involved in experimental facilities like shock table testing and strong floor development for structural testing. The book co-authored by him titled 'Alternative Building Materials and Technologies' received the ACCE–NAGADI

AWARD in 2008 for Best Publication in Civil Engineering. He has supervised 10 Ph.D. and M.Sc. (Engineering) theses. He has published over 60 papers in refereed journals and conferences. He has successfully completed several sponsored research projects. He has served as the chair and as a member of several academic and administrative committees.

Dr. P. C. Pandey retired as a professor of structural engineering from IISc, Bengaluru, and is currently serving as the Professor Emeritus in Civil Engineering at REVA University, Bangalore. He has obtained his B.E. from BIT Mesra, ME from IISc, Bengaluru, and Ph.D. from the University of Liverpool, UK. He has previously served as faculty at Nanyang Technological University, Singapore, and at BITS Pilani. He completed his post-doctoral fellowship at the University of Wales, Swansea, and later served there as an Erasmus Mundus Visiting Professor. His broad areas of research interest are computational mechanics, finite element analysis, and damage mechanics. He has supervised six Ph.D. students and has published over 120 papers in international journals and conferences.

Dr. H. B. Nagaraj completed his bachelor's in Civil Engineering from SJCE, Mysore, in 1986 and obtained his post-graduation in geotechnical engineering from the National Institute of Technology (NIT), Surathkal. He completed his doctoral degree in soil mechanics from IISc, Bengaluru. He has been serving as a faculty in the Civil Engineering Department at BMS College of Engineering since 1997 and is currently heading the department. He has keen research interests in the engineering behaviour of fine-grained soils and alternate building materials. He has published more than 40 technical papers in international journals and conferences. He is an active reviewer for most of the leading international journals in the field of geotechnical engineering and building materials. He has been a recipient of many international awards as a reviewer from 'Publons' and has also received the Outstanding Reviewer Award from two international journals the 'Journal of Cleaner Production' and 'Applied Clay Science'. He has recently been successful in implementing his laboratory-scale research findings in the construction of a rammed earth structure in one of the residential building projects in Bengaluru, taken up in collaboration with Made in Earth-Sustainable Architecture and Building Practices, Bengaluru. In 2016, he was invited to review an international project proposal (seeking funding) submitted by a leading materials manufacturing and research organization to the United States Department of Agriculture, National Institute of Food and Agriculture under the grant programme Small Business Innovation Research (SBIR), USA. He has supervised two doctoral dissertations along with four master's of science and 40 master's degree thesis. He has filed several patents, a few of which have been published. He has handled various externally funded projects, including those of the Department of Science & Technology (DST) and AICTE. He has also reviewed the BIS Code: Review of Geotechnical Provisions in Indian Seismic Code IS 1893 (Part 1): 2002 as a part of the Gujarat State Disaster Management Authority, Gandhinagar (GSDMA)-IIT Kanpur project. He has delivered more than 20 lectures for e-learning programmes on geotechnical engineering. He has handled more than 50 major consultancy projects in the fields of geotechnical investigation and forensic geotechnics and has recently managed two major industry-sponsored research projects to use iron ore tailings from Rai Bahadur Seth Shreeram Narasingdas (RBSSN) Pvt. Ltd., Karnataka, to produce building blocks using iron ore tailings. He is presently supervising the practical implementation of the production units.

Investigating the Practicality of Incorporating Waste Expanded Polystyrene Beads and Rice Husk in Lightweight Concrete Blends

Harshith Ankanayakanahalli Mohan Kumar, Alsa Nashith, Siri Hemanth, and Vijay V. Nair

Abstract This experimental study investigates the potential of utilizing sustainable waste expanded polystyrene (EPS) beads and rice husk as a lightweight aggregate in concrete mixtures. As the construction industry seeks sustainable alternatives to traditional materials, the incorporation of EPS beads derived from recycled sources and rice husk were anticipated as a promising solution. The primary objective of this research is to assess the mechanical performance of lightweight concrete formulations containing varying proportions of waste EPS beads and rice husks. The study employs a systematic approach, beginning with the characterization of the waste EPS beads to determine their physical and chemical properties. Subsequently, concrete mixtures were designed by substituting conventional aggregates with different percentages of EPS beads and rice husk. The mechanical properties, such as compressive strength, flexural strength, and density, were evaluated to assess the structural integrity and performance of the lightweight concrete. Preliminary findings suggest that the inclusion of sustainable waste EPS beads and rice husk could significantly reduce the density of concrete, resulting in a lightweight material with acceptable mechanical properties for certain applications. The study aims to provide insights into the feasibility of implementing such sustainable practices in the construction industry, contributing to the development of sustainable building materials.

Keywords Lightweight concrete · Concrete mix design · Waste management · Resource conservation · UN SDG 11

H. A. Mohan Kumar · A. Nashith · S. Hemanth
Department of Civil Engineering, Malnad College of Engineering, Hassan 573202, India
e-mail: hs@mcehassan.ac.in

Present Address:
V. V. Nair (✉)
School of Engineering, Chanakya University, Bangalore 562165, India
e-mail: vijay.v@chanakyauniversity.edu.in

Y. K. Guruprasad et al. (eds.), *Recent Advances in Infrastructure Development and Management—Volume 2*, Lecture Notes in Civil Engineering 566,
https://doi.org/10.1007/978-981-96-1558-2_1

1 Introduction

Lightweight concrete (LWC) has become increasingly popular for various construction applications due to its unique properties and benefits which include reduced weight, improved insulation, enhanced fire resistance, better workability, and high strength-to-weight ratio [1]. Numerous research studies were reported in the realm of LWC investigating its performance across diverse applications [2, 3]. The inherent characteristics of LWC offer a promising potential for various applications. In comparison with normal concrete of density 2400 kg/m^3, LWC is reported to exhibit a density ranging between 675 and 1760 kg/m^3 [4], which is usually achieved through partial replacement of fine and/or coarse aggregates with lightweight and sustainable alternatives [3, 5]. Although LWC was known to be in use during the Roman Empire, it was Haug and Field, who scientifically demonstrated a floating concrete platform which was developed using LWC [5]. Later, researchers across the globe explored the application possibilities of floating concrete as structural elements in platforms and bridges [5–7].

Pradeep et al. [8] reported the use of rice husk ash, micro-silica, and microsphere in preparing LWC for its application in building a lightweight narrow water vessel [8]. An attempt is made in the present study to develop a LWC mix particularly focusing on the achievement of floatability through numerous trials with a huge prominence on overall density. Additionally, focus has been made to obtain considerable compressive strength. Coarse aggregate being the major contributor toward heavy weight of conventional concrete, it was hypothesized that on the replacement of the conventional coarse aggregate with lightweight and sustainable aggregates would obtain structural LWC which will in turn ensure floatability. The main objective of the study was to develop an unconventional concrete mix design for floatable concrete using lightweight and sustainable alternatives and analyze its fresh and hardened concrete properties.

2 Materials and Methods

2.1 *Raw Materials*

Cement was used as a binding element in both concrete and mortar. Ordinary Portland Cement (OPC) of 53 Grade conforming to IS 12269: 2013 was used. Fly ash of class F was procured from Karnataka Power Corporation Limited which was replaced with cement (IS 3812: 2003—Part 1). Super plasticizers used were Sika® Lightcrete and Air Entraining admixture procured from Talrak—Construction Alchemists & Solution Providers®. Further, aggregates such as expanded polystyrene (EPS) beads, lightweight expanded clay aggregates (LECA), and rice husk were used as a replacement for coarse and fine aggregates, respectively. EPS has good thermal insulation properties with stiffness and compression comparable to medium clay, whereas

Table 1 Physical properties of raw materials

Property	Fine aggregates	Coarse aggregates	Rice husk	EPS	LECA
Fineness modulus	3.39	7.06	6.00	7.00	7.58
Specific gravity	2.67	2.73	0.54	0.01	0.50
Water absorption, %	3.10	0.50	70	-	24
Surface moisture, %	1.60	1.50	–	–	–
Density, kg/m^3	834	1087	14.93	1.05	0.31

LECA and rice husk being less dense compared to conventional aggregates were hypothesized to enormously decrease the overall density of the concrete which consequently would promote floating.

All raw materials were examined and analyzed as per the standards for testing and specification of aggregates used in concrete viz., IS 2386 (Part 3): 2021, IS 383: 2016, ASTM C29, ASTM C566. The physical properties of raw materials used in the study are presented in Table 1.

The proportion of various constituents in concrete mixtures, such as cement, aggregates, water, and admixtures, to achieve the desired properties in the finished concrete was calculated as per IS 456: 2000. Several factors including the type and strength of concrete required, the characteristics of materials available, and environmental conditions were considered. M25 and LWC concrete specimen cubes were cast to test their mechanical properties in the casting yard of the Department of Civil Engineering, Malnad College of Engineering. Triplicate specimens were casted and standard curing conditions for 28 days were practiced. Specimens were tested to determine their 3-day, 7-day, 14-day, and 28-day compressive strengths using VEB Werkstoffprüfmaschinen Leipzig® 2000kN compression testing machine in the Material Testing Laboratory of the Department of Civil Engineering, Malnad College of Engineering.

3 Results and Discussion

The relative density of the LWC cube specimens was determined to be 775 $\pm$ 25 kg/m^3 fulfilling the criteria reported earlier by Kumar et al. [4]. The relative density of LWC was nearly 32% of the relative density of conventional M25 concrete and lighter than water. The LWC concrete cubes were found to be floating in water as shown in Fig. 1.

The mechanical properties of the M25 and LWC specimens were compared by determining their incremental increase in compressive strengths during 3- to 28-day period. The compressive strengths were measured after 3rd, 7th, 21st and 28th days of curing at an approximate water loading rate of 120 kg/cm^2/min. The gain in compressive strengths of the concrete cube specimens is illustrated in Fig. 2. Comparatively, it was noted that although the initial compressive strengths (after 3rd and 7th

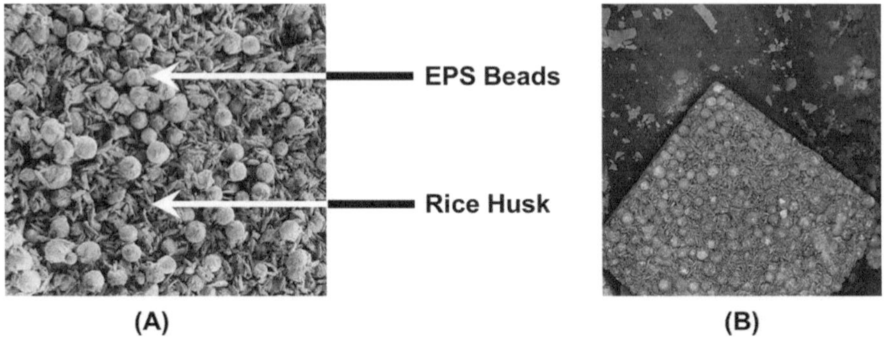

Fig. 1 Photograph of (**A**) Voids between the EPS beads in a LWC specimen cube. (**B**) LWC specimen cube floating in water

day) were insignificant for LWC specimens, an incremental increase in compressive strengths up to 16.2 MPa was recorded. The 28-day compressive strength of LWC specimens was approximately 39% of that of M25 concrete specimens. This incremental increase in compressive strengths of LWC might be due to better moisture absorption properties of rice husk and LECA in the concrete blends as presented in Table 1. The absorbed water in rice husk and LECA might have contributed to maintenance of the right moisture condition during the curing process in the bulk of the concrete cubes through a possible interphase boundary supporting the complete and effective hydration of cement, leading to the development of a strong and durable concrete specimen.

Fig. 2 Compressive strength development pattern of M25 concrete and LWC cube specimens

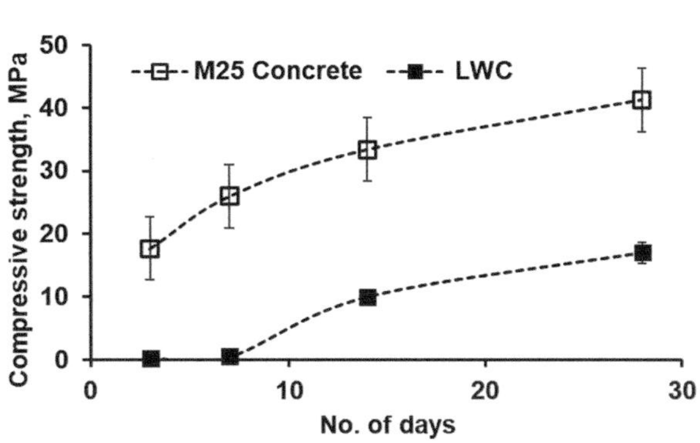

4 Conclusion

Although the relative density of the LWC cubes prepared based on the developed mix design was determined to be 775 ± 25 kg/m^3 without developing cracks or formation of fungus enabled the floatation of the cube specimens in water, the rice husk and EPS beads significantly contributed to formation of many voids. It was anticipated to witness a significant reduction in compressive strength. Following conclusions could be drawn from tests performed on conventional M25 concrete cubes and LWC cubes to determine their mechanical properties casted using rice husk, LECA, and waste EPS beads:

1. The 28-day compressive strength of LWC specimens (16.2 MPa) was approximately 39% of that of M25 concrete specimens.
2. The incremental increase in compressive strength of LWC specimens after 7th day of curing might be due to the possible development of an interphase boundary between the rice husk and LECA components of the concrete cubes offering higher specific area of contact with water in the bulk of concrete specimen.

References

1. Chandra S, Berntsson L (2002) Lightweight aggregate concrete. Elsevier Science, United States
2. Elango KS, Sanfeer J, Gopi R, Shalini A, Saravanakumar R, Prabhu L (2021) Properties of light weight concrete–a state of the art review. Mater Today Proceed 46:4059–4062
3. Rumšys D, Bačinskas D, Spudulis E, Meškėnas A (2017) Comparison of material properties of lightweight concrete with recycled polyethylene and expanded clay aggregates. Proced Eng 172:937–944
4. Kumar VK, Priya AK, Manikandan G, Naveen AS, Nitishkumar B, Pradeep P (2021) Review of materials used in light weight concrete. Mater Today Proceed 37:3538–3539
5. Adhikary SK, Ashish DK, Rudžionis Ž (2021) Expanded glass as light-weight aggregate in concrete–a review. J Clean Prod 313:127848
6. Haug AK, Fjeld S (1996) A floating concrete platform hull made of lightweight aggregate concrete. Eng Struct 18(11):831–836
7. Kim MO, Qian X, Lee MK, Park WS, Jeong ST, Oh NS (2017) Determination of structural lightweight concrete mix proportion for floating concrete structures. J Korean Soc Coast Ocean Eng 29(6):315–325
8. Pradeep S, Deepak P, Dheepak S (2020) Rice husk ash based cementitious material for concrete canoe. In: IOP Conference series: materials science and engineering 912(6), 062004, IOP Publishing

"Multi-Objective Optimization Model for Resource–Time–Cost, and Demolition Waste Reuse Optimization"

R. S. Hemanth and S. P. Sreenivas Padala

Abstract Optimization of construction activities is necessary to minimize cost, time, and resource utilization of any project. Better decisions taken will help in solving these problems, and (Yusoff et al. in Proced Eng 15, 2011) study suggests NSGA a prominent optimization tool. The main objective of this study is to explore solutions for optimization of time–cost–resource and demolition waste reuse for sustainable construction and also to develop a multi-objective oriented (MOO) model for making better sustainable decisions. To improve the scheduling, time management, resource allocation, management, design of a project and provide a better sustainable and green building design, it was found NSGA is good optimization tool for construction activities. An average of 8.5–9% decrease was observed for cost, time, and environmental impact of the project by the model.

Keywords NSGA-III · Pareto · Multi-objective optimization (MOO)

1 Introduction

Multi-objective-oriented sustainable construction refers to the practice of designing and constructing buildings and infrastructure that aim to achieve multiple goals simultaneously, including economic, environmental, and social objectives, while also promoting long-term sustainability. Sustainable construction is achieved through a range of design strategies and construction practices that prioritize energy efficiency, renewable energy, water conservation, and the use of environmentally friendly materials. These practices aim to minimize the building's whole life cycle and its effects on the environment, including construction, usage, and eventual removal or reuse. The issues that building projects confront are discussed, along with how MOO may successfully address these challenges by producing a set of Pareto-optimal solutions

R. S. Hemanth (✉) · S. P. S. Padala
Department of Civil Engineering, M. S. Ramaiah University of Applied Sciences, Bangalore, Karnataka, India
e-mail: Hemanth.36.rs@gmail.com

Y. K. Guruprasad et al. (eds.), *Recent Advances in Infrastructure Development and Management—Volume 2*, Lecture Notes in Civil Engineering 566,
https://doi.org/10.1007/978-981-96-1558-2_2

that reflect different trade-offs between goals. The concept of Pareto dominance and the distinction between convergence and diversity in solution sets are explored. Currently, MOO model is developed for optimization of time and or cost and resource together; in this paper, optimization of resource–time–cost is done altogether with demolition waste use optimization.

Literature Review: In [1] study between time, cost, and sustainability is not complete, in-depth investigation is necessary. Limitations for sustainable construction are defined without solutions for it in [2]. Operation uncertainty is considered only for set of ideal parameters but not based on location of the project in [3], and in [4] study, optimal duration for construction linking with sustainability concept has yet to be found,

In [5] limited study, lot of important objectives are to be explored; in [6], embodied carbon link with sustainability has to be done in-depth for conditions suiting our geography; in [7], management of the tool has to be discussed; in [8], detail study about the technology used has to be discussed; in [9, 10, 13], on-site validation has to be discussed; in [11, 14], detailed and large-scale implications have to be discussed; in [12], time and resource use has to be discussed; and in [15], construction phase calculations have to be discussed.

2 Proposed Concept and Methodology

The construction industry operates within a dynamic environment characterized by numerous interconnected variables and conflicting objectives. Traditional construction planning and decision-making approaches often struggle to balance diverse goals such as cost minimization, project duration reduction, resource optimization, and quality enhancement. In response, the multi-objective optimization (MOO) model emerges as a promising framework to address these complexities. This paper presents an overview of the application of multi-objective optimization in the construction domain.

Through literature survey and past data, a BIM-based model is developed for convention method of construction and sustainable construction using 30% of demolition waste (fly ash and ceramic tile). From the BIM model using Revit and Microsoft project, the required dataset is formed for time, cost, and resource, and their environmental impact is calculated. After which the following procedure is followed (Fig 1):

i. **Identify the objectives:** The first step is to identify the objectives that are important for sustainable construction. These might include minimizing environmental impact, maximizing social benefits, and minimizing construction costs.
ii. **Define the decision variables:** The next step is to define the decision variables that will be used in the model. These might include the materials used in construction, the design of the building, and the construction methods employed.

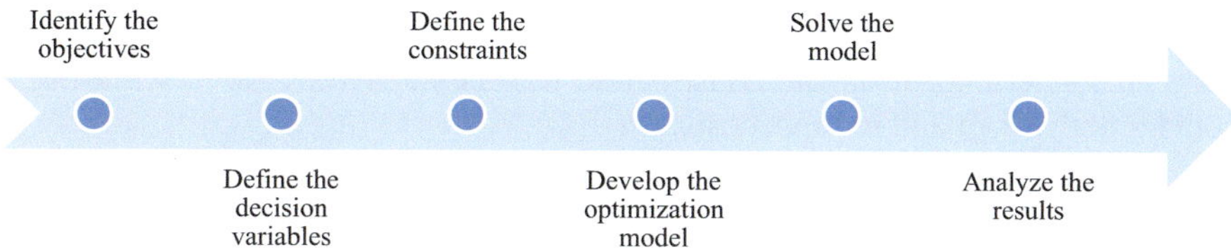

Fig. 1 Flowchart of methodology followed

iii. **Define the constraints:** There will be constraints that limit the options available for each decision variable. For example, there may be regulations on the types of materials that can be used, or limits on the amount of energy that can be consumed.

iv. **Develop the optimization model:** Using the objectives, decision variables, and constraints identified in the previous steps, develop an optimization model that can be used to find the best possible solution.

v. **Solve the model:** Once the optimization model has been developed, it can be solved using mathematical techniques such as linear programming or genetic algorithms. This will produce a collection of solutions that show the compromises made between the various goals.

Analyze the results: Final step is to analyze the results to determine the best solution to balance the various objectives. This may involve examining the trade-offs between different objectives or exploring the sensitivity of the model to different assumptions.

3 Defining Objective Functions and Variables

Objective Function 1: Minimizing construction time is of paramount importance in the construction industry due to its significant impact on project success, costs, and overall efficiency.

Minimize Construction Time: Σ Time required of the activity, cost of the activity, environmental impact of the activity

Decision Variable 1: Minimize time required for the completion of a activity: Project completion time (PCT) = The total of all the activities in a construction project’s critical path, or PCT, is equal to ACT.

$$\text{Total time} = \sum_{A \in CP} \text{CPA} \ldots\ldots \tag{1}$$

A Activity
CPA Critical path activity

Decision Variable 2: Minimize cost of the activity: project completion cost (PCC) is the algebraic total of all the construction project's activities ACC, while ACC includes both direct and indirect costs.

$$\text{Project cost: } \sum_{A} \text{CA} \ldots\ldots \quad (2)$$

A Activity
CA Cost per activity direct + indirect cost

where PCC represents the $\sum_{A}$ CA project completion, which is the total of the direct costs associated with each task, and the project's indirect costs can be calculated by multiplying the PCT by the daily indirect costs.

Decision Variable 3: Minimize environmental impact**:** The measurement of the activity environmental impact (AEI) is expressed in terms of greenhouse gas emissions comparable to CO_2. The total AEI of all project activities is equal to PEI.

$$\text{Environmental impact: } \sum_{A=1} \text{EI}_A \text{in Kg } CO_2 \ldots\ldots \quad (3)$$

EI Environmental impact of activity A

where n is the total number of activities in the project, AEI is the environmental impact of activity A, and PEI is the project's overall environmental impact.

Objective Function 2: Minimize embodied carbon of demolition waste reuse: As the world grapples with the urgent need to address climate change, the construction and demolition industries are increasingly being scrutinized for their significant contributions to carbon emissions. While the focus on sustainable building practices is growing, the environmental impact of demolition waste is often overlooked

$$\text{Minimize embodied carbon} = \Sigma \text{ demolition energy, process energy,}$$
$$\text{transportation energy}$$

Decision Variable 1: Minimize demolition energy (DE): The energy needed to demolish a building and dispose of its components is known as demolition energy. The process of calculating demolition energy involves multiplying the total fuel used to demolish the structure by the energy density of diesel fuel, which is 38.7 MJ/liter (E_d). The entire amount of fuel used to demolish a structure is calculated by multiplying the individual machine run periods for that task by the fuel consumption rates of those machines.

$$\text{DE: } \sum (v_1 \times E_d) + (v_e \times E_e) \ldots\ldots \quad (4)$$

v_l Volume of fuel consumed
v_e Volume of energy consumed
E Energy conversion factor

$$1 \text{ litter diesel} = 38.7\text{MJ/Liter (Ed)}, \ 1 \text{ kWh} = 16.30\text{MJ/kWh (Ee)}$$

Decision Variable 2: Minimize process energy**:** The energy needed to recycle and convert demolition debris into recycled aggregates is known as process energy. There are two methods for computing the process energy: a wet processing system and a dry processing system.

$$\text{Process energy (PE):} \sum \text{Pre - crushing energy use} + \text{Crushing phase energy use}$$

Decision Variable 3: Minimize transportation energy (TE): The energy required to move demolition trash, such as concrete, masonry, and tile waste, from the demolition site to the C&D recycling facility is known as transportation energy. The process of calculating the transportation energy involves several steps. Total energy used to move demolition debris to the C&D recycling facility is as follows:

$$\text{TE} = V_d \times E_d \ldots\ldots. \tag{5}$$

where

$$V_d = T \times F \ldots\ldots. \tag{6}$$

V Volume of diesel
E Energy conversion factor
T Total no of trips
F Fuel consumed

Decision Variable 4: Minimize operation Energy (OE): Energy consumption and efficiency of a building or construction project during its operational phase. It encompasses the energy required for heating, cooling, lighting, ventilation, and powering the building's systems and equipment.

$$\text{OE:} \sum \text{Energy consumed} \times E_e \ldots\ldots. \tag{7}$$

E Energy conversion factor

Objective Function 3: Minimize embodied energy of resources**:** Embodied energy refers to the total amount of energy required to produce, transport, and eventually dispose of a material or product throughout its entire life cycle. This includes all the energy inputs associated with the extraction, processing, manufacturing, packaging, transportation, and sometimes even the end-of-life disposal or recycling processes.

Decision Variable 1: Minimize embodied energy of materials (EEm):

$$EE_M = \sum_{n=1}^{r} EEn \times \int n \times V_n \ldots\ldots \quad (8)$$

where EEn = Materials embodied energy factor;

$\int$ Material's density
V_n Quantity of materials used in building;
n total number of materials

Decision Variable 2: Minimize equipment energy (EEq):

$$EEq = \sum_{i=1}^{n} H \times FC \times D_f \times EF \ldots\ldots \quad (9)$$

H Total operating hours of an equipment.
FC Average fuel consumption for an equipment per unit time (m3 /hour).
D Density of fuel (kg/m3).
EF Energy factor for certain fuel type.

Decision Variable 3: Minimize transportation energy (TE):

$$TE = \Sigma_i^n Dm * RTm * EC\mathit{f} \ldots\ldots \quad (10)$$

Dm Hauling distance of the material
RTm Round trips.
EC*f* The fuel energy coefficient.

Objective Function 4: Minimize cost: Cost minimization is a critical objective in the construction industry, driven by a combination of economic, environmental, and social considerations. Construction projects involve significant investments of resources, and optimizing costs can lead to greater financial viability, reduced environmental impact, and improved societal outcomes.

Decision Variable 1: Minimize project cost (PC):

$$PC: \sum MC \times n \ldots\ldots \quad (11)$$

MC Material cost.
n No of material units.

Decision Variable 2: Minimize equipment cost (EC):

$$EC = \sum ui * EQ \ldots\ldots. \tag{12}$$

ui Cost of Equipment per hour.
EQi Equipment type.

Decision Variable 3: Minimize transportation cost:

$$\mathrm{EC} = \sum_{i=1}^{n} \sum_{k=0}^{n} C_{ij} \times S_{ij} \ldots\ldots. \tag{13}$$

C_{ij} = Cost per unit from i and j.
S_{ij} = Amount i and j.

Decision Variable 4: Minimize operation cost(OC):

$$\begin{aligned}&\text{Operating costs} = \text{Labor, gasoline, upkeep, tire, track,}\\&\text{and other component replacement expenses,}\\&\text{plus overhead} \ldots\ldots.\end{aligned} \tag{14}$$

4 Case Study

BIM-based model: BIM is a collaborative and data-driven approach to construction and design that enables the creation, management, and sharing of comprehensive and detailed digital models throughout a project's lifecycle. In a BIM-based model, various aspects of a building, including its geometry, spatial relationships, structural systems, architectural elements, mechanical systems, and construction sequencing, are represented in a three-dimensional (3D) virtual environment.

The proposed building is composed of Basement + Ground + Three floors on a site area of 16,000 Sq. ft and has an approximate built area of 100,000 Sq. Ft. Where Option 1 refers to conventional method of construction, and Options 2 and 3 are considered by using construction and demolition waste with 30% fly ash and ceramic tile, respectively. Figures 2 and 3 show the side elevation and typical floor plan of proposed building for optimization.

MOO Model: A multi-objective optimization method called NSGA-III (Nondominated Sorting Genetic method III) is made to handle situations when there are several competing objectives. It is a member of the evolutionary algorithm family, which draws inspiration from the principles of natural selection, and is an expansion of

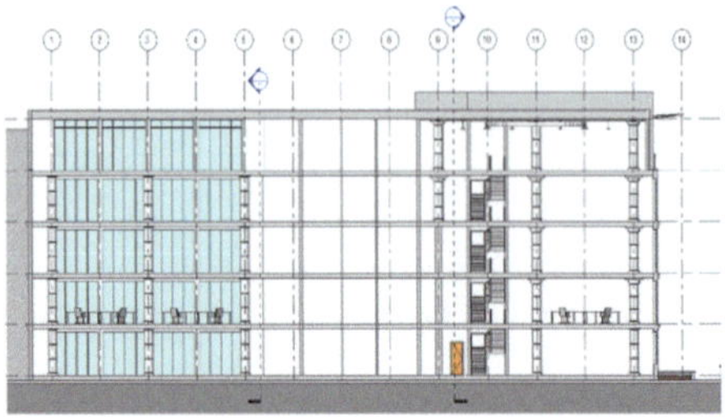

Fig. 2 Side elevation of case study

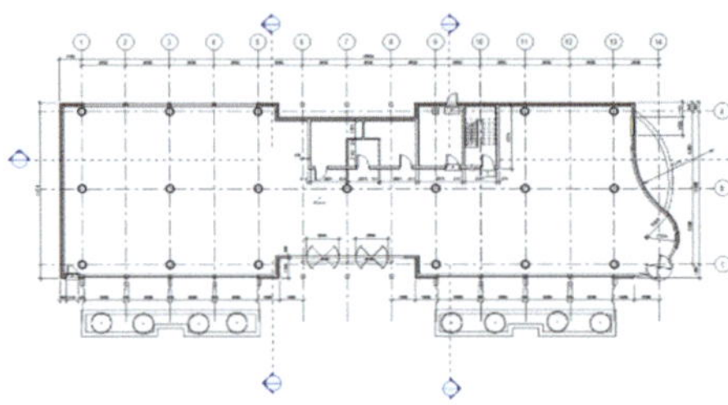

Fig. 3 Typical floor plan

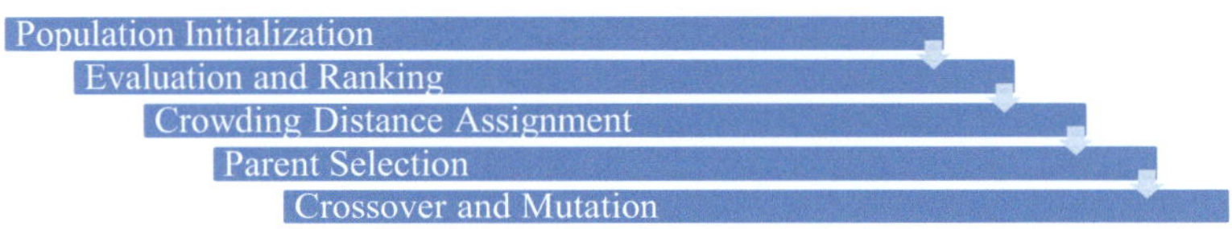

Fig. 4 Flowchart of MOO model working

the original NSGA algorithm. (Fig. 4) symbolizes the steps taken in the NSGA Optimization process.

MATLAB and Research Parameters:

MATLAB (Matrix Laboratory) is a powerful programming and computational environment commonly used for various scientific, engineering, and mathematical tasks. It provides a comprehensive set of tools and functions for optimization, making it a popular choice for solving optimization problems. (Fig. 5) Represents the research parameters considered for MOO model.

5 Results and Discussions

The results from model can help project managers, engineers, and stakeholders make informed decisions, optimize resource allocation, and balance competing objectives for construction projects (Table 1). These results are valuable for achieving more sustainable, cost-effective, and efficient construction outcomes. Table 2 shows the

```
%% NSGA-III Parameters
% Generating Reference Points
nDivision = 10;
Zr = GenerateReferencePoints(nObj, nDivision);
MaxIt =100;  % Maximum Number of Iterations
nPop =25;  % Population Size
pCrossover = 0.5;        % Crossover Percentage
nCrossover = 2*round(pCrossover*nPop/2); % Number of Parnets (Offsprings)
pMutation = 0.5;        % Mutation Percentage
nMutation = round(pMutation*nPop);  % Number of Mutants
mu = 0.02;      % Mutation Rate
sigma = 0.1*(VarMax-VarMin); % Mutation Step Size
```

Fig. 5 MATLAB NSGA-III parameters considered

Table 1 Results obtained from MOO model

Type of material	Time	Environmental impact	Cost
Conventional	322	61,74,541	1,15,000,000
30% Fly ash	374	74,54,483	1,32,610,000
30% Ceramic	339	93,68,238	1,46,410,000
Model	298	59,22,300	1,06,115,000

Table 2 Percentile comparison of results obtained with conventional

Type of material	Time percentage %	Envi impact percentage %	Cost percentage %
Conventional	–	–	–
30% Fly ash	11.61	12.0	11.53
30% Ceramic	9.06	12.5	11.04
Model	– 9.25	– 9.5	– 9.22

time, cost, and environmental impact of conventional, fly ash, ceramic aggregate, and model results. It can be observed that an average of 9% reduction was achieved from the multi-objective optimization model. It can be said the MOO model helps in making better decisions for project managers for future scheduling of projects.

6 Conclusion

For the purpose of attaining sustainable construction, the proposed work represents a multi-objective optimization model (MOOM) for optimizing embodied energy and cost. The study shows the optimization model works by finding a sustainable solution which incorporates time–cost–environmental impact by making the use of demolition waste. To evaluate the time, cost, and impact on the environment of sustainable construction, the framework is built on combining a mathematical optimization model for the best choice of materials and best cost for different building components with BIM. The results from model can help project managers, engineers, and stakeholders make informed decisions, optimize resource allocation, and balance competing objectives for construction projects. These results are valuable for achieving more sustainable, cost-effective, and efficient construction outcomes. It can be observed that an average of 9% reduction was achieved from the multi-objective optimization model. It can be said the MOO model helps in making better decisions for project managers for future scheduling of projects.

Some studies have an increase in cost of 9.8% and time of 9.3% and we can reduce the environmental impact by 7.9%. Further studies suggest with the incorporation of sustainable practices, the cost and environmental impact can be reduced by 40–45% in the next 25 years. The study shows the optimization model works by finding a sustainable solution which incorporates time–cost–environmental impact by making the use of demolition waste.

References

1. Hyari KH, El-Rayes K, El-Mashaleh M (2009) Automated trade-off between time and cost in planning repetitive construction projects. Constr Manag Econ Taylor Francis J 27(8):749–761
2. Asl MR, Stoupine A, Zarrinmehr S, Yan W (2015) Optimo: a BIM-based multi-objective optimization tool utilizing visual programming for high performance building design. https://doi.org/10.52842/conf.ecaade.2015.1.673
3. Asl, MR, Bergin M, Menter A, Yan W (2014) BIM-based parametric building energy performance multi-objective optimization. 2. https://doi.org/10.52842/conf.ecaade.2014.2.455
4. Chaturvedi S, Rajasekar E, Natarajan S (2020) Multi-objective building design optimization under operational uncertainties using the NSGA II algorithm. Buildings. https://doi.org/10.3390/buildings10050088
5. Yan W, Asl MR, Su Z, Altabtabai J (2015) Towards multi-objective optimization for sustainable buildings with both quantifiable and non-quantifiable design objectives. https://doi.org/10.1061/9780784479681.024
6. Peng J, Feng Y, Zhang Q, Liu X (2023) Multi-objective integrated optimization study of prefabricated building projects introducing sustainable levels. Sci Rep. https://doi.org/10.1038/s41598-023-29881-6
7. Wang W, Rivard H, Zmeureanu R (2003) Optimizing building design with respect to life-cycle environmental impacts. IBPSA Conference on eighth international building performance simulation association
8. Yusoff Y, Ngadiman M, Zain A (2011) Overview of NSGA-II for optimizing machining process parameters. Proced Eng. https://doi.org/10.1016/j.proeng.2011.08.745
9. Hyari K, El-Rayes K (2004) A multi-objective model for optimizing construction planning of repetitive infrastructure projects. https://doi.org/10.13140/2.1.5017.3128
10. Hyari K, El-Rayes K, Asce M (2006) Optimal planning and scheduling for repetitive construction projects. J Manag Eng. https://doi.org/10.1061/(ASCE)0742-597X(2006)22:1(11)
11. Cook AJ (2006) Structural steel framing options for mid- and high rise buildings. Massachusetts: Massachusetts Institute of Technology
12. Alsadi A, Cabrera N, Faggin M, He Y, Patel M, Trevino F, . . . Zirakian T (2019) Comparative study on the cost analysis of a green versus conventional building. ACET, 5

13. Anon (2015) An assessment of energy technologies and research opportunities. America: Department of energy
14. Baskaran K, Gopikrishna A, Thiruvaranakan P (2017) Study on filler slab construction. Reserch gate. p 9
15. Busby Perkins; Will, Consulting Stantec (2018) Roadmap for the integrated design process. GreenSpaceNCR

Structural Performance of Axially Loaded UPVC—Encased Circular Concrete Columns

Manish Sharma, Mayank Mishra, Shreya Kushwaha, Vineet Jaiswal, Rohit Sharma, and Abhishek Kumar Gupta

Abstract This study investigates the structural performance of axially loaded UPVC-encased circular concrete columns, aiming to enhance strength, durability, and deformation capacity. Various columns with UPVC tubes (diameters: 160 mm, 110 mm, 90 mm) and PPC concrete mixes (M15, M20, M25) were tested under axial compression. The yield load–displacement curve and specimen deformation patterns were analyzed, documenting shear and macrocracking failures in all specimens. The results show that UPVC encasement improves load-bearing capacity, enhances deformation capacity, and delays the onset of cracks. The specimen with a 160 mm diameter and M25 concrete mix exhibited the highest performance, achieving an ultimate load of 653 KN. These findings demonstrate that UPVC offers a promising and cost-effective alternative for structural applications by improving strength, durability, and crack resistance.

Keywords UPVC-encased columns · Structural performance · Composite structural systems · Ductility

1 Introduction

As per the numerous surveys conducted by various agencies, it has been concluded that concrete is the second largest used material in the world when it comes to construction. Observing the ever-increasing construction practices, the materials used in construction, which are known for harming our environment, are also increasing at an alarming rate. Hence, there is an obvious requirement to inspect and peep on the possible methods of replacement of these materials as well as search for new materials serving as substitutes in this regard, which are comparatively more environmentally friendly and economical for the last pockets of society as well [1–3].

M. Sharma (✉) · M. Mishra · S. Kushwaha · V. Jaiswal · R. Sharma · A. K. Gupta
Department of Civil Engineering, Rajkiya Engineering College Mainpuri UP India, Mainpuri, India
e-mail: smanish337@gmail.com

Y. K. Guruprasad et al. (eds.), *Recent Advances in Infrastructure Development and Management—Volume 2*, Lecture Notes in Civil Engineering 566,
https://doi.org/10.1007/978-981-96-1558-2_3

By the technique of combining two or more materials in a definite proportion serving as an additive, suitable changes can be observed that prove the enhancement of the properties of the material and make it much more environmentally friendly, and hence can be considered for further use and research [4]. By combining two or more components in a specific ratio, it is observed that the strength and other properties of the resultant material are enhanced. Studies have demonstrated that confined concrete exhibits superior mechanical properties compared to unconfined concrete [5]. At the same time, this paper focuses on composite columns as well [6]. Composite columns are structures that are made up of two or more different materials. Normal columns are made up of cement sand aggregate and some kind of admixture functioning as additives. A normal column is also known for containing a large quantity of concrete and some fixed amount of steel reinforcement within it with IS code-specified parameters. In this paper, an effort to make a strong compression-resisting and high-tensile-strength composite column by using UPVC is studied and designed so that the amount of concrete used can be reduced to a valuable extent while also making the finishing work easy [7–11].

UPVC is made up of PVC plastic, which is available in different structures and sizes. Due to its unique, incredible, and unbeatable properties and its light weight, it is used for various purposes, like making doors and window casing, UPVC pipes, etc. UPVC shows high resistance to corrosion; it also has high durability, fire resistance, thermal insulation, and a very low maintenance cost [12]. UPVC casing acts as a protective layer, mitigating environmental deterioration while potentially enhancing the structural integrity of the column. Moreover, UPVC's lightweight and durable nature makes it a viable option for various structural applications, particularly in corrosive environments [13, 14]. Additionally, an extensive review of plastic tube-confined concrete has highlighted improvements in axial strength and durability, with additional benefits such as corrosion resistance and reduced maintenance costs [15].

Corrosion is known as the biggest trouble in construction practices as it creates a huge amount of loss in the economic life of a user, community, or nation. UPVC seems to replenish or compensate for the same for the industries.

The thermal insulation and fire resistance make it ahead of many other compound chlorides. Due to these properties, it is a preferable replacement for other materials like timber and iron used for making accessories like furniture, goods, and pipes. These properties create the scope of usage and further research within it for its justification toward productive use for other purposes and new possible production markets with UPVC-disabled goods. Recent studies indicate that polyvinyl chloride (PVC) tubes filled with fibrous concrete exhibit enhanced axial load capacity and ductility, suggesting that similar benefits can be achieved with UPVC encasement [16, 17]. Furthermore, the use of scrap ceramic tiles as a partial replacement in concrete mixtures has demonstrated promising results in improving concrete properties, which could be relevant for UPVC-encased columns [18].

In this paper, different sizes of UPVC pipes filled with plane concrete are tested, and the effect of UPVC is examined, as shown in Fig. 1. The test load is applied only to concrete, and then its strength, flexure, cracking, failure, and other engineering parameters are calculated and observed. A varying compressive load is applied to the a  b 

Fig. 1 **a** Different size of UPVC **b** concrete fill UPVC

specimen, and its failure model is studied. Different sizes of UPVC pipes filled with their respective types of concrete are tested, and their failure models are examined [13, 19]

As per the above studies, the use of PVC as a composite material is found valuable. As PVC increases its mechanical properties, it was not observed to have an unpleasant and harmful impact on the environment. Methods of suitable appearance can also be observed, and production costs can also be significantly reduced. Based on the previous research in this field, work on the use of UPVC in a more efficient way has been done, and considerable results have been obtained.

2 Investigational Study

2.1 Sample Configuration

Utilizing various tubes and plain concrete as shown in Fig 2, specimens with various nominal diameters were produced (grades M15, M20, and M25). During specimen

Three layers of freshly prepared concrete are then poured into the mound. Each layer is applied, and then the appropriate vibration and compaction are carried out.

a  b 

Fig. 2 **a** Different samples **b** Mix Concrete preparation, pipes are inserted in a circular mound

Table 1 Specimen details

Sample ID	Grade	Outer dia (mm)	Inner Dia(mm)	Wall Thickness(mm)	Height (mm)	Slenderness Ratio
160PC15P3	M15	160	148	6	300	8.11
110PC15P3	M15	110	102	4	300	11.76
90PC15P3	M15	90	82	4	300	14.63
160PC20P3	M20	160	148	6	300	8.11
110PC20P3	M20	110	102	4	300	11.76
90PC20P3	M20	90	82	4	300	14.63
160PC25P3	M25	160	148	6	300	8.11
110PC25P3	M25	110	102	4	300	11.76
90PC25P3	M25	90	82	4	300	14.63

The specimen was covered in a thick polyethylene sheet to stop evaporation. The prepared sample is designated as 160PC25P3, where 162 denotes the outside diameter, PC25 denotes standard M25-grade concrete, and P3 denotes a class 3 pipe operating at 0.6 MPa pressure. The pipe's outer diameter is D, its thickness is t, and the specimen's height is h. Axial compression testing was done. An axial compression test was performed on a column specimen, and the impact of the test on the concrete core was assessed.

2.2 *Material Characteristic*

2.2.1 Concrete

Using Portland Cement, both coarse and fine aggregates, and an admixture, the mix designs M15, M20, and M25 were developed as given in Table 1. A superplasticizer made of cement was added to concrete to improve workability and decrease water content in order to boost strength. Every design is completed in accordance with IS 10262. (2019).

2.2.2 Unplasticized Polyvinyl Chloride (UPVC) Tubes

Unplasticized polyvinyl chloride, a strong material, can be found in nature. Although PVC and UPVC are frequently used interchangeably in the literature, their mechanical and industrial meanings are different. Numerous uses for UPVC exist, including the creation of pipes, doors, and windows. PVC is a material that is used to make siding and fences. UPVC doesn't harm the environment because it doesn't contain phthalates or BPA. It offers strong weather resistance, high corrosion resistance, and

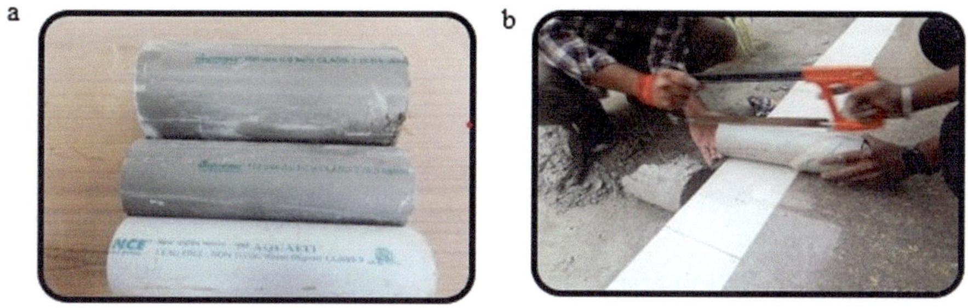

Fig. 3 **a** Different size and class of pipes **b** failure of pipe

Table 2 Concrete mix and property

Grade	M 15	M 20	M 25
Ratio of W/C	0.597	0.545	0.512
Cement (Kg)	3.75	4.60	6.40
Water (L)	2.045	2.27	2.90
Aggregate (Fine) (Kg)	7	6.9	6.40
Aggregate (Coarse) (Kg)	14	13.8	12.80
% Of admixture	0.15	0.20	0.25
Strength of cube in MPa	17.22	23.43	28.42

resistance to chemical attack. The characteristics of a column composed of UPVC tubes are examined in this study.

In this study, class 3 UPVC pipes with sizes of 160, 110, and 90 mm were employed, as shown in Fig. 3. The specimen's length is 300 mm, and the pipe's nominal pressure is 0.6. UPVC pipes are put through a number of tests to ascertain their mechanical characteristics. To find out how UPVC behaves under tension, three specimens were put through a tensile test at 20 °C. 52 MPa is the total tensile stress that was measured.

The mechanical characteristics of UPVC pipes, including elastic modulus, density, ultimate tensile strength, and service life, are listed in the table. The confining pressure of UPVC is assessed using an internal hydrostatic pressure test in a lab. Figure 1 depicts the test setup, and Table 2 displays the confining pressure obtained.

2.3 *Test Setup and Apparatus*

A compressive testing device with a 2000 KN loading capability is used for the test. The setup tools and testing are shown in the Figures 4 and 5

When the loading capacity considerably decreases for all column specimens after loading is started at a displacement control rate of 0.4 mm/min, it continues until a

a b
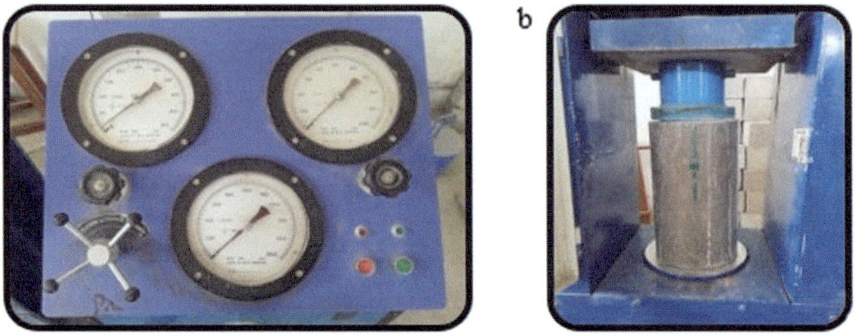

Fig. 4 **a** CTM at loading **b** 160 mm pipe loading condition **c** 110 mm pipe loading condition

a b
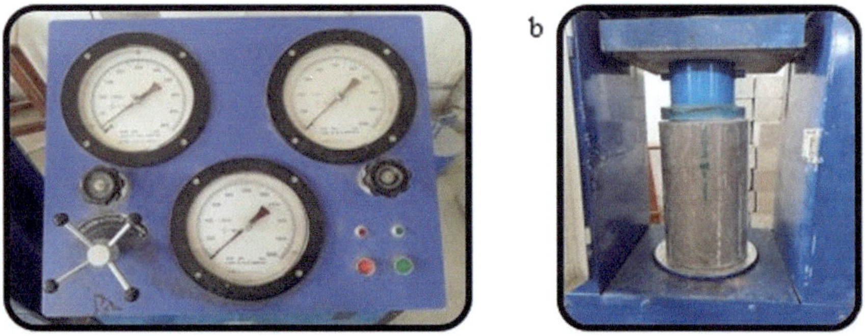

Fig. 5 **a** CTM **b** Specimen setup in CTM

strain of 1.2–2.5% is reached. Concrete was employed for its displacement properties, while UPVC tubes were exclusively used for confinement. By measuring deformation using LVDTs on the specimen's opposing sides, the length changes of the sample are determined. For the analysis, as shown in Table 3, a load–displacement curve will be produced.

Table 3 UPVC's physical properties provided by the company

UPVC's properties	Values
Density of UPVC pipes	1.2 gm/cm^2
Elastic modulus of UPVC pipes	860 MPa
UTS	23.5 MPa
Poisson's ratio	0.32
Life	> 50 years

3 Test Result and Discussion

An experiment was conducted to examine the behavior of a concrete column encased in a UPVC pipe. An axial loading test and an internal hydrostatic pressure measurement were performed to determine the closed pressure. Confinement is the key element in the investigation of the behavior of concrete-filled tubes, as shown in Fig. 6. In the specimen with lower radial strength than the concrete core, failure occurs primarily in the UPVC tube, with minor fractures in the form of shallow hairlines in the concrete. Failure manifests as deformation, such as buckling, drumming, and local bulging, rather than rupture in other cases where the UPVC tube contributes more to column strength than the concrete (see Fig. 7).

The experiment's results are summarized in the table. The deformed forms and associated curves of a few specimens are shown. The 10–14 mm range corresponds to the highest failure deviation. The measured angle between the shear failure and the horizontal surface was between 63 and 65°. In specimens exhibiting ductile behavior, displacement after the peak cannot be disregarded. There is a movement of approximately 5–10 mm following the application of the maximum load, as shown in Table 4. Loads were obtained using test accuracy analysis. In Fig. 7, the post-peak slope re-mains constant, while the initial slope of the curve increases as the diameter of the tubes increases.

Specimen's con figuration	U. Load (kN)	Mode of failure	Load of failure (kN)	Max. strain at experiment failure
160P3PC15	512	Shear cracks with slight bulge	223.45	0.0169
110P3PC15	378		156.89	0.0171
90P3PC15	217		98.73	0.0178
160P3PC20	583		238.34	0.0154

(continued)

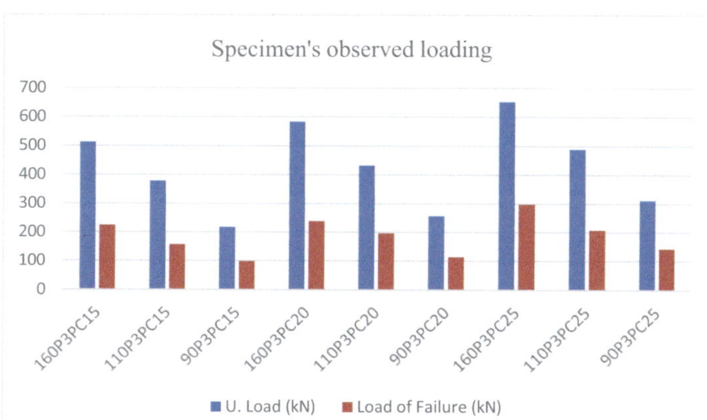

Fig. 6 Graph for load observed on tested specimens

(continued)

Specimen's con figuration	U. Load (kN)	Mode of failure	Load of failure (kN)	Max. strain at experiment failure
110P3PC20	432		196.45	0.0156
90P3PC20	256		113.56	0.0162
160P3PC25	653		298.45	0.0145
110P3PC25	489		207.45	0.0149
90P3PC25	312		143.56	0.0152

4 Conclusion

In this investigation, several columns enclosing UPVC tubes underwent a variety of tests. Tests are conducted on the UPVC composite column. The following conclusion can be drawn from the findings:

- If the tube is stronger than the concrete, a deep fracture will occur at the concrete, and the tube will stay stable without failing or cracking.
- If the tube is weaker than the concrete, a core fracture will happen. All specimens fail due to the formation of macro and shear-cracks, as well as bulging. With increasing strength and decreasing pressure, the absolute slope of the curve rises.
- The strain at break for small-height specimens ranged from 1.25% to 1.70%. This study makes predictions about the ultimate strength of UPVC-coated concrete columns.
- Concrete columns made of UPVC have enhanced durability and corrosion resistance, and they are also more affordable.
- Among all the tested specimens, the specimen with a 160 mm outer diameter and M25 concrete mix (160P3PC25) exhibited the greatest strength, load capacity, and overall performance. This specimen showed the highest ultimate load of 653 kN, the highest load at failure of 298.45 kN, and a maximum strain at failure of 0.0145. This indicates that both a larger diameter of the UPVC tube and a higher concrete mix grade contribute significantly to the structural strength and load-bearing capacity of the composite columns.

The use of UPVC encasement for concrete columns provides several advantages over traditional reinforced concrete columns in terms of buckling resistance. The enhanced confinement effect, uniform stress distribution, lightweight and flexible nature, corrosion resistance, improved post-peak ductility, and economic benefits make UPVC a superior choice for preventing buckling and ensuring long-term structural integrity. These factors collectively contribute to the better performance of UPVC-encased concrete columns under axial loads and buckling conditions.

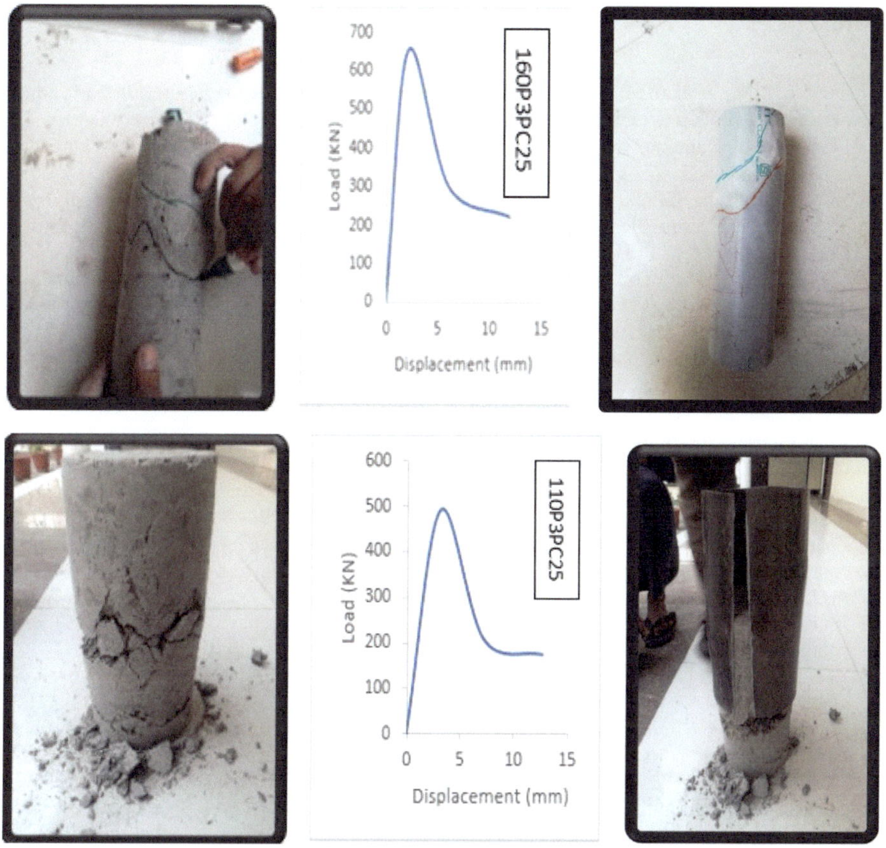

Fig. 7 Load-displacement behavior and failure patterns of UPVC-encased circular concrete columns under axial loading

Table 4 UPVC 's physical properties provided by the company

Grade of Concrete	Dimensions (mm)			Confining pressure (MPa)
	Nominal Outer diameter (D)	Internal diameter (d)	Thickness (t)	
M 15	160.00	148.00	6.0	1.16
	110.00	102.00	4.0	0.98
	90.00	82.00	4.0	0.76
M 20	160.00	148.00	6.0	1.16
	110.00	102.00	4.0	0.98
	90.00	82.00	4.0	0.76
M 25	160.00	148.00	6.0	1.16
	110.00	102.00	4.0	0.98
	90.00	82.00	4.0	0.76

References

1. Ding Y, Zhou Z, Tian H, Peng Z (2022) Compressive behavior of concrete-filled ultra-high performance concrete tube with FRP stirrups. Structures 46:611–624
2. Fleming PH, Ramage MH (2020) Full-scale construction and testing of stress laminated columns made with low-grade wood. Constr Build Mater 230:116952
3. Taraka MR, Balla S, Prakash S, Rajagopal A (2023) Role of size on the compression behaviour of hybrid FRP strengthened square RC columns–experimental and finite element studies. Compos Struct 303:116314
4. Liao JJ, Zeng JJ, Long YL, Cai J, Yi OY (2022) Behavior of square and rectangular concrete-filled steel tube (CFST) columns with horizontal reinforcing bars under eccentric compression. Eng Struct 271:114899
5. Nguyen DL, Thai DK, Tran NT, Ngo TT, Le HV (2022) Confined compressive behaviors of high-performance fiber-reinforced concrete and conventional concrete with size effect. Constr Build Mater 336:127382
6. Chen L, Wu H, Fang Q, Li RW (2021) Full-scale experimental study of a reinforced concrete bridge pier under truck collision. J. Bridge Eng. 26:05021008
7. Ye JB, Wang YT, Cai J, Chen QJ, He A (2022) Evaluation of residual lateral capacities of impact-damaged reinforced concrete members. Buildings 12:669
8. Zhang C, Tan X, Tian H, Chen W (2022) Lateral compression and energy absorption of foamed concrete-filled polyethylene circular pipe as yielding layer for high geo-stress soft rock tunnels. Int J Mining Sci Technol 32(5):1087–1096
9. Jin L, Liang J, Chen F, Du X (2022) Size effect on the seismic behavior of square CFST short columns. J Constr Steel Res 1(199):107589
10. Yi S, Zou J, Wei L (2022) A macroscopic shear angle model for ultimate bearing capacity of glued laminated bamboo hollow columns under axial compression. Structures 45:560–571
11. Jamal-Omidi M, Choopanian BA (2021) A numerical study on energy absorption capability of lateral corrugated composite tube under axial crushing. Int J Crashworthiness 26(2):147–158
12. Wang Z, Li Z, Shi C et al (2019) Mechanical performance of vertex-based hierarchical vs square thin-walled multi-cell structure. Thin-Walled Struct 134:102–110
13. Sharma M, Islam MI (2022) Study of concrete filled Unplasticized poly-vinyl chloride tubes as columns under axial loading. J Mech Mater Mech Res. 5(01) https://doi.org/10.30564/jmmmr.v5i1.4494
14. Abdulla NA (2020) Concrete with an outer plastic protective shell. SN Appl Sci 2(6):1093
15. Saadoon AS, Abbas AM, Khalaf AA (2006) A survey of studies on plastic tubes confined plain concrete as compression members. J Eng Appl Sci 14:2468–2477
16. Abdulrazzaq Azeez A (2018) Behavior of concrete-filled PVC tube columns under axial load (Doctoral dissertation, Universiti Tun Hussein Onn Malaysia)
17. Ahmed AF, Elmannaey AS, Fouad HE (2024) Performance of axially loaded polyvinyl-chloride fibrous concrete filled tube short columns. Ain Shams Eng J, 102709
18. Sharma M, Sharma R, Jaiswal V (2024) Experimental characterization of concrete properties with partial replacement of scrap ceramic tiles. In: Kolathayar S, Sreekeshava KS, Vinod Chandra Menon N (eds) Recent Advances in Building Materials and Technologies. IACESD 2023. Lecture Notes in Civil Engineering, vol 456. Springer, Singapore. https://doi.org/10.1007/978-981-99-9458-8_5
19. Wang Z, Liu J, Yao S (2018) On folding mechanics of multi-cell thin-walled square tubes. Compos B Eng 132:17–27

An Overview on Use of Bio-Bitumen in Bituminous Pavements

K. S. Vinay Kumar, K. Ganesh, and Maitreyi R. Hanmantgad

Abstract The global increase in the construction of flexible pavements and rehabilitation of existing pavements require a huge amount of bitumen produced by crude oil refining. Bitumen industry is always looking for alternative materials that have the ability to improve the properties of asphalt whereas reducing their carbon footprint. Bio-bitumen utilization in the field of pavement engineering has been receiving serious attention due to its eco-friendliness, in comparison with petroleum-based asphalt binders. Bio-bitumen is produced by waste materials or from natural materials, significantly decreases carbon dioxide emission. It can improve low-temperature performance and is economical. The basic tests results indicate better performance of bio-bitumen compared to conventional bitumen due to improvement in the rheological properties of bio-bitumen. The partial replacement of conventional bitumen with bio-waste resulting in bio-bitumen shows improved Marshall properties in bituminous pavements. The Marshall stability value is increased by 10.68%. From the results obtained through Marshall stability indicates bio-bitumen performance is better than conventional bitumen.

Keywords Bio-bitumen · Bio-waste · Bituminous concrete mix · Marshall test

K. S. Vinay Kumar
Department of Civil Engineering, B M S College of Engineering, Bengaluru, India

K. S. Vinay Kumar (✉) · K. Ganesh · M. R. Hanmantgad
Department of Civil Engineering, DSATM, Badamanavarathekaval, India
e-mail: vinaykumarks@bmsce.ac.in

K. Ganesh
e-mail: gkiyer68.civ@bmsce.ac.in

M. R. Hanmantgad
e-mail: maitreyih04@gmail.com

Y. K. Guruprasad et al. (eds.), *Recent Advances in Infrastructure Development and Management—Volume 2*, Lecture Notes in Civil Engineering 566,
https://doi.org/10.1007/978-981-96-1558-2_4

1 Introduction

Bitumen, traditionally used as a binder in flexible pavement construction, poses several challenges including environmental impact, depletion of petroleum reserves, and susceptibility to price fluctuations. To address these issues, bio-bitumen has emerged as a promising alternative. Bio-bitumen involves partially replacing and modifying bitumen with bio-based renewable materials such as bio-char [1].

The depletion of petroleum reserves and rising environmental concerns have significantly impacted the cost and sustainability of petroleum-based binders used in asphalt pavements. In response, road authorities and researchers are exploring alternative binders derived from renewable sources. These alternatives not only aim to reduce dependence on non-renewable resources but also mitigate the environmental impact associated with conventional binders [2].

Bituminous pavement, commonly known as asphalt pavement, is a versatile and widely used type of road surface. It is composed of a combination of bitumen—a sticky, black, and highly viscous liquid derived from crude oil—and aggregates such as gravel, sand, or crushed stone. Bituminous pavements are prevalent in various transportation infrastructures, including highways, roads, airport runways, and parking lots, due to their excellent performance characteristics.

The use of bio-bitumen as a partial substitute for traditional bitumen in pavement construction is an interesting and innovative approach. This method addresses several concerns associated with conventional bitumen use, including environmental impact, petroleum depletion, and price volatility. Here are some key points about the utilization of bio-bitumen in bitumen mixtures:

Environmental Benefits: Bio-bitumen offers environmental advantages compared to conventional bitumen. Additionally, the use of waste cooking oil, plastic, tire-rubber, and other materials as additives further promotes recycling and reduces the environmental impact of these waste materials.

Renewable Energy: The incorporation of bio-bitumen into bitumen allows for the development of a bio-based renewable technology. This aligns with sustainability goals and reduces dependence on finite petroleum reserves. Bio-bitumen provides an alternative solution that is more environmentally friendly and contributes to the development of a circular economy.

Modification of Bituminous Properties: The addition of bio-bitumen modifies certain properties of bitumen. The carbon composition and form of bio-char can alter the rheological and mechanical properties of the binder mixture. This modification may enhance the performance of the pavement in terms of durability, fatigue resistance, and temperature susceptibility.

Reduced Bitumen Quantity: The use of bio-bitumen allows for the reduction of the quantity of bitumen required in the binder mixture. This not only addresses concerns related to petroleum depletion but also helps in mitigating the price spikes associated with traditional bitumen.

Diverse Feedstock Sources: The versatility of bio-bitumen production allows for the utilization of various feedstock sources, including bio-char, bio-oil, waste cooking oil, plastic, tire-rubber, and polymers. This flexibility provides an opportunity to explore and optimize different combinations based on availability and regional considerations.

Research and Development: Ongoing research and development in the field of bio-bitumen are crucial for further understanding its properties and optimizing its use in pavement construction. Testing and evaluation of the performance of bio-bitumen mixtures under various conditions will contribute to the wider adoption of this sustainable technology.

2 Literature Review

Rasman et al. [3]. In this study, incorporation of waste cooking oil (WCO) into pavement material has been explored as an alternative to petroleum-based bitumen for asphalt pavement. WCO can be blended with virgin bitumen or poured directly into heated aggregate or recycled asphalt pavement (RAP). WCO is mainly composed of oleic acid, palmitic acid, and linoleic acid. Adding WCO to bitumen can soften the bitumen and increase temperature susceptibility.

Kalampokis et al. [4]. This study examines the physical and rheological properties of bitumen modified with lignin, as well as the physicochemical and thermal properties of Kraft lignin powder. Compared to solid biomass modifiers, liquid-phase biomass modifiers, or “bio-oils,” are more compatible with bitumen. By adding biomass as a modifier, bitumen’s stiffness, resilience to deformation, and ability to rut at high temperatures can all be increased.

Pandit et al. [1]. The study clearly indicates that penetration decreases with increasing ductility values, softening increases, and flash and fire point rise with increasing bio-char concentration. This indicates that temperature susceptibility rose as the softening point rose, and that stiffness increased as the penetration value dropped. Further reducing the creation of depressions is the stiffness of the mixture that resists the rutting action.

Penki et al. [5], It has been noticed that the bio-oil up gradation is not required when oil is extracted from agricultural and herbaceous biomass, but the performance is poor compared to upgraded bio-oil. The treated or upgraded oil performance is better than untreated biomass. In this study, the results clearly reveal that bitumen’s biomass content increases penetration, ductility, phase angle, fatigue resistance, and low-temperature thermal resistance while decreasing viscosity, complex shear modulus, softening point, and creep stiffness.

Ren et al. [6]. The study indicates addition of Lignin to the bitumen reduced the softening point. The blended viscosity was reduced when compared to the conventional one. The density of the blended bitumen was increased over the original.

From the above literature review, we can observe that only basic tests were conducted for bio-bitumen. The Marshall stability test can be conducted for bio-bitumen to analyze the performance of bio-bitumen in bituminous pavements.

3 Objectives of the Study

To study basic properties of bitumen and aggregates.
To evaluate the effect of Lignin powder addition on the basic properties of bitumen.
To study the Marshall properties of conventional bitumen and bio-bitumen in Bituminous Concrete mix (Grade I)
To compare the Marshall test results obtained using conventional bitumen and bio-bitumen in Bituminous Concrete mix.

4 Methodology

The major aim of the current investigation is to compare basic and Marshal properties of conventional VG 10 bitumen with bio-bitumen properties at various percentages. The methodology for the current study entails evaluating fundamental and Marshall properties of conventional and bio-bitumen in Bituminous Concrete mix as per IS codes and MoRTH specifications.

4.1 Material Collection

Natural aggregates were procured from Jyothi Crushers, a granite quarry in Bidadi, Bengaluru. VG10 Grade bitumen was supplied from a local supplier in Bengaluru named Zabi Constructions. Bio-bitumen was procured from Praj industries, Pune. Tests were conducted on natural aggregates, VG 10 Grade bitumen, and bio-bitumen and are shown in Tables 1, 2, and 3, respectively.

4.2 Marshall Test

The Marshall specimen was prepared according to the gradation obtained. The proportion of different sizes of aggregates and filler as mentioned below: 20 mm–20%, 12 mm–30%, 6 mm–35% and dust–15%. 1200 g of aggregates was taken according to the proportionate. Aggregates and bitumen were heated separately and then blended together to form a bituminous mix. Specimen was casted by giving 75 blows on each side by using automatic compaction machine. After 24 h, the specimen

Table 1 Tests on natural aggregates

Property	Test	Specification (%)	Results (%)	Standards
Particle shape	Combined flakiness and elongation index	Max 35	17.26	IS:2386 Part 1
Toughness	Aggregate impact value	Max 30	18.82	IS:2386 Part 1
Strength	Crushing value	Max 30	17.98	IS:2386 Part 4
Water absorption	Water absorption	Max 2	0.48	IS:2386 Part 3
Specific gravity	Specific gravity test	Min 2.6	20 Down = 2.65 12 Down = 2.67 6 Down = 2.77 Dust = 2.78	IS:2386 Part 3

Table 2 Tests on VG 10 grade bitumen

Sl. No	Tests on conventional bitumen	Results	Specifications
1	Softening point	48.20 °C	40 °C
2	Penetration test	92.00 mm	80–100 mm
3	Ductility (cm)	96.80 cm	Min 75 cm
4	Specific gravity	1.02	0.97–1.02
5	Flash and fire point	Flash point: 290 °C Fire point: 300 °C	Min. 220 °C for all grade of bitumen

Table 3 Basic tests on bio-bitumen for varying percentage of lignin powder

Tests	Lignin powder					
	0 (%)	5 (%)	10 (%)	15 (%)	20 (%)	25 (%)
Softening point (°C)	48.20	45.60	44.30	42.20	41.20	39.70
Penetration (mm)	92.00	84.00	75.00	68.00	61.00	58.00
Ductility (cm)	96.80	91.20	86.60	82.70	76.20	70.30
Flash point (°C)	290	305	320	335	340	350
Fire point (°C)	300	325	340	355	360	370

was demolded and Marshall Test was conducted according to ASTM D 1559–65 [6]. The Marshall test results are shown in Table 4.

Table 4 Marshall test results on bituminous concrete mix using VG 10 grade bitumen

Bitumen content (%)	Stability (kN)	Flow value (mm)	Bulk density Gb(g/cc)	Voids in the mix Vv (%)	VFB (%)
5.2	10.31	2.48	2.35	6.86	63.68
5.4	10.95	2.89	2.36	6	67.64
5.6	10.24	3.71	2.37	5.04	72.09
5.8	9.42	4.11	2.38	3.94	77.47

5 Results and Discussions

The results obtained showed decrease in penetration values and ductility as bio-bitumen content increased. There was an increase in softening point and flash and fire points as the content of bio-bitumen increased. This indicates that temperature susceptibility has increased as softening point increased along with increase in stiffness as penetration value decreased (Fig. 1).

From the above result the optimum (Table 4), binder is 5.4% and stability value for 5.4% bitumen content is 10.95%. Bulk density and VFB value are slightly increasing as bitumen content increasing (Fig. 2). Tables 4 and 5 give the Marshall Test results for the Conventional BC mix and BC mix using Bio-bitumen respectively.

For the optimum binder content, the bio-bitumen was added at various percentages from 5 to 25%. The stability value has been increased to 12.12% from10.95% for 15% bio-bitumen content. Similarly bulk density and VFB value are slightly increasing as bio-bitumen content increasing.

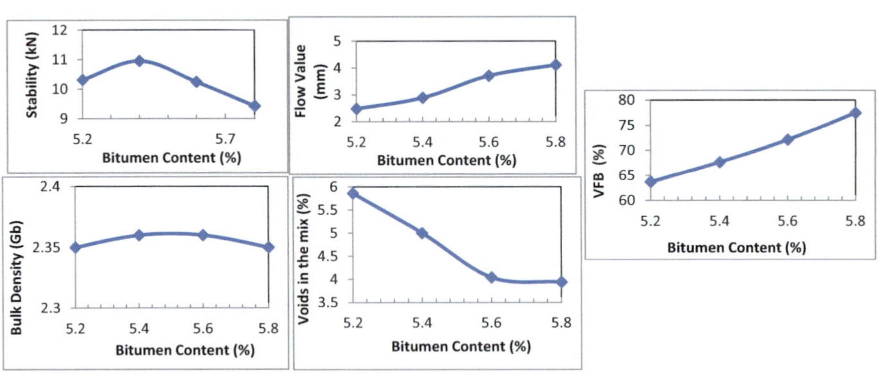

Fig. 1 Graphical representation of Marshall parameters in Bituminous Concrete mix using VG10 Grade Bitumen

Table 5 Marshall test results of bituminous concrete mix for varying percentage of lignin powder in bio-bitumen

Bio-bitumen content (%)	Stability (kN)	Flow value (mm)	Bulk density Gb, (g/cc0	Voids in the mix Vv (%)	VFB (%)
0	10.95	2.9	2.36	5	67.64
5	11.12	2.95	2.37	4.88	67.69
10	11.48	3.08	2.39	4.62	67.79
15	12.12	3.16	2.4	4.41	68.23
20	11.14	3.28	2.42	4.3	68.43
25	10.32	3.44	2.41	4.04	68.93

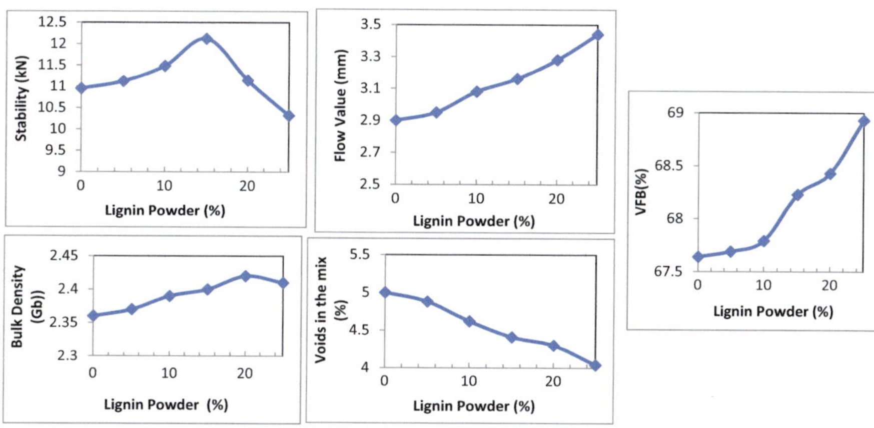

Fig. 2 Graphical representations of Marshall Parameters in Bituminous Concrete mix using bio-bitumen for varying Percentage of Lignin Powder

6 Conclusion

The basic test results of bio-bitumen indicate that temperature susceptibility has increased as softening point increased along with increase in stiffness and decrease in penetration value.

The addition of bio-bitumen at optimized value of 15% compared to conventional bitumen in Bituminous Concrete mix has yielded better Stability value satisfying the MoRT&H requirements.

The stability value of bituminous Concrete mix using bio-bitumen has increased by 9.6%.

The increase in stability value indicates that bio-bitumen yields better result than conventional bitumen in Bituminous Concrete mix.

References

1. Pandit SU, Marathe A, Patil R, Karanje A (2022) Bio-bitumen, Int Res J Eng Technol (IRJET)
2. Zahoora M, Nizamuddina S, Madapusib S, Giustozzia F (2021) Recycling asphalt using waste bio-oil: a review of the productionprocesses, properties and future perspectives, Process Saf Environ Prot
3. Rasmanl M, Hassan NA, Hainin MR, Jaya RP, Haryati Y, Shukry NAM, Abdullah ME, Kamaruddin NHM (2018) Engineering properties of bitumen modified with bio-oil, MATEC web of conferences
4. Kalampokis S, Papamoschou M, Kalama DM, Pappa CP, Manthos E, Triantafyllidis KS (2022) Investigation of the characteristic properties of lignin-modified bitumen. Civil Eng 3(3): 734–747
5. Penki R, Rout SK (2021) Next-generation bitumen: a review on challenges and recent developments in bio-bitumen preparation and usage, Biomass Convers. Bio refinery
6. Marshall Test Procedure (1995) Standard marshall test for bituminous mixture, (ASTM D 1559–65), Am Soc Test Mater
7. IS 2386 Part I (1997) Methods of test for aggregates for concrete particle size and shape
8. IS 2386 Part III (1997) Methods of test for aggregates for concrete specific gravity, density, voids, absorption and bulking
9. IS 2386 Part IV (1997) Methods of test for aggregates for concrete mechanical properties
10. IS 1201–1220 (1978) Methods for testing tar and bituminous materials

Building Energy Optimization Using Sustainable Materials

D. T. Jayant and Akshayakumar V. Hanagodimath

Abstract With the growing concern for environmental sustainability and energy efficiency, optimizing the energy consumption of buildings has become a critical research and practical focus. This abstract presents a comprehensive review and analysis of various strategies and approaches employed in building energy optimization. The building energy optimization helps in highlighting the significance of energy optimization in the context of increasing global energy demand and the associated environmental impacts. It underscores the role of buildings as significant contributors to energy consumption and explores the potential benefits of optimizing their energy usage. Subsequently, the abstract delves into a systematic overview of key strategies for building energy optimization. This includes discussing architectural and passive design techniques that leverage building orientation, insulation, shading, and natural ventilation to reduce energy demand and also the integration of energy-efficient technologies such as smart lighting, HVAC (Heating, Ventilation, and Air Conditioning) systems, and building automation to actively manage energy consumption.

Keywords Building energy · Optimization · Efficiency · Design efficiency · Water efficiency · Material efficiency

1 Introduction

Energy crisis has evolved into its dynamic form from the mid 1970's due to the extensive use of resources. Buildings, industries, and transportation systems are the major factors that influence energy consumption. Buildings are already responsible for more than 30% of the total energy consumption of mankind, and this figure is

D. T. Jayant (✉) · A. V. Hanagodimath
Department of Civil Engineering, M S Ramaiah University of Applied Sciences, Bangalore, Karnataka, India
e-mail: jayantdt20091999@gmail.com

A. V. Hanagodimath
e-mail: akshay.g9807@gmail.com

Y. K. Guruprasad et al. (eds.), *Recent Advances in Infrastructure Development and Management—Volume 2*, Lecture Notes in Civil Engineering 566,
https://doi.org/10.1007/978-981-96-1558-2_5

expected to grow. With the increasing demand for more energy-efficient buildings, the construction industry is facing the challenge to ensure that the energy performance predicted during the design is achieved once the building is in use. It is expected that 72% of the electricity consumed in this planet is generated by fossil fuels which are on the verge of extinction. In addition, buildings are among the most important elements of our daily life. Normally, a person spends over 70% office live span inside buildings. A comfortable indoor environment can not only improve occupants working efficiency but also preserved their health. To achieve a comfortable indoor environment in an energy-efficient building is a goal that draws the attention of the professionals in the architecture and in civil, mechanical, and electrical engineering.

Building energy optimization refers to the process of systematically and strategically reducing the energy consumption of a building while maintaining or improving its performance and comfort for occupants. The goal of building energy optimization is to minimize energy usage, reduce greenhouse gas emissions, and lower operating costs while ensuring that the building remains functional and comfortable for its occupants.

There are various approaches and strategies involved in building energy optimization, which can be categorized into several key areas:

1. **Architectural and passive design**: Designing buildings with careful consideration of factors such as site orientation, building shape, insulation, window placement, and shading. Passive design techniques aim to naturally regulate indoor temperatures and lighting, reducing the need for active heating, cooling, and lighting systems.
2. **Energy-efficient technologies**: Incorporating energy-efficient technologies and systems into the building, such as LED lighting, high-efficiency HVAC systems, energy-efficient appliances, and smart controls. These technologies use less energy to achieve the same or even improved performance.
3. **Building automation and controls**: Implementing advanced building automation systems that monitor and control various building systems, including lighting, HVAC, and ventilation, based on real-time data and occupancy patterns. This ensures that energy is used only when and where it's needed.
4. **Renewable energy integration**: Integrating renewable energy sources like solar panels, wind turbines, and geothermal systems to generate on-site clean energy, which can offset the building's energy consumption from the grid.
5. **Advanced energy modeling and simulation**: Using sophisticated computer simulations to model the building's energy performance under various scenarios. This helps identify the most effective energy-saving measures and predict potential outcomes.
6. **Occupant behavior and engagement**: Educating building occupants about energy-saving practices and encouraging responsible energy use. Behavioral changes, such as adjusting thermostats or turning off lights when not needed, can contribute significantly to energy savings.
7. **Data analytics and machine learning**: Employing data analytics and machine learning algorithms to analyze energy consumption patterns and predict future

energy usage. This can lead to more informed decision-making and the implementation of proactive energy-saving measures.

8. **Continuous monitoring and maintenance**: Regularly monitoring and maintaining building systems to ensure they operate efficiently. Preventive maintenance can help identify and address issues before they lead to energy waste.
9. **Regulations and certification programs**: Adhering to energy codes, standards, and certification programs (such as LEED or BREEAM) that promote energy-efficient building design and operation.

Building energy optimization is a multidisciplinary effort that involves architects, engineers, building managers, occupants, and other stakeholders. By implementing a combination of these strategies and approaches, buildings can significantly reduce their energy footprint, contribute to environmental sustainability, and create more comfortable and cost-effective spaces for their occupants.

2 Materials and Methodology

2.1 Building Selection

The Building that is selected for this particular study is located at 10th Main Rd, near Sainik Public School, Vidhana Soudha Layout, Laggere, Bengaluru, Karnataka, 560058.The building is a multi-story building that is spread over an area of 1200 sqft. The building is oriented toward north direction, and the building is used for commercial purpose. The building is constructed using concrete blocks, OPC 53 cement, and steel bars (fe 500). The cost estimation and quantity estimation of the carried out (Figs. 1 and 2).

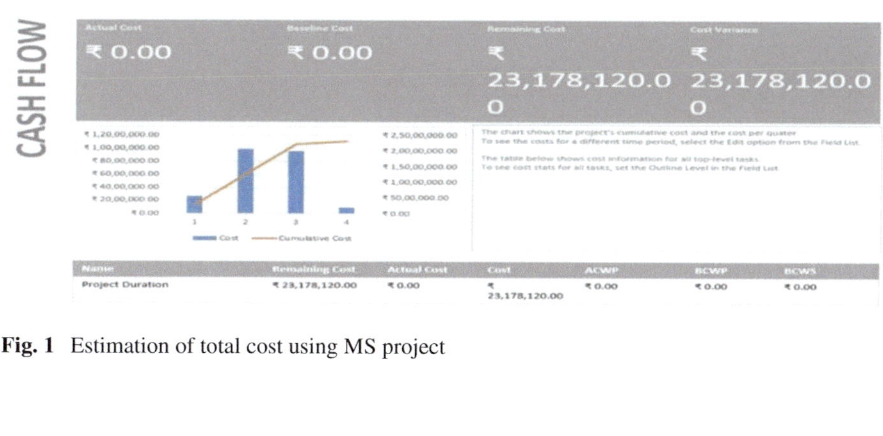

Fig. 1 Estimation of total cost using MS project

Sl no	Type of work	
1	Excavation	2794.1 cft
2	Column	10.045 cum
3	Wall	200.25 cum
4	Slab	65.74 cum
5	Internal plastering	18.95 cum
6	Ceiling plastering	3995 Sqft

Fig. 2 Quantity estimation

2.2 Methodology

The optimization was carried out using Edge software. The software was used to calculate design efficiency, water efficiency, material efficiency, and CO_2 emissions. First, design efficiency, water efficiency, and material efficiency were carried out for present condition, and the results were found. Next, design efficiency, water efficiency, and material efficiency were carried out for changed materials, and the results were found and compared (Figs. 3 and 4).

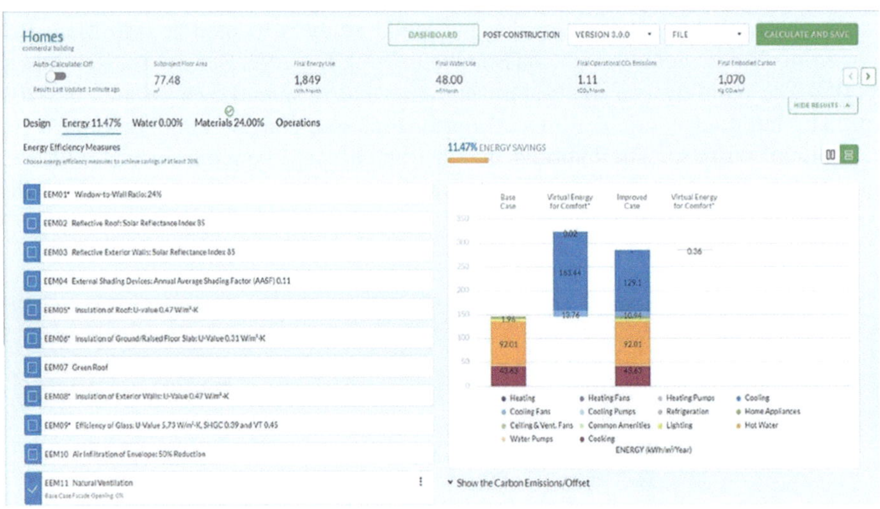

Fig. 3 User interface of edge software

SL No	Parameters	Results
1	Floor area	77.48 m^2
2	Final energy use	1849 kWh/month
3	Final water use	48m^3/month
4	Final operational CO2 emissions	1.11tCO_2/month
5	Final embodied carbon	1070kgCO_2e/m^2

Fig. 4 Results obtained for present condition

3 Test Results and Discussion

3.1 Test Results of Improved Efficiency due to Change of materials

Changes made to improve energy efficiency

- Window to wall ratio increased to 15.1%.
- Transforming the normal roof to green roof.
- Providing adequate natural ventilation.
- Usage of energy-efficient ceiling fans.

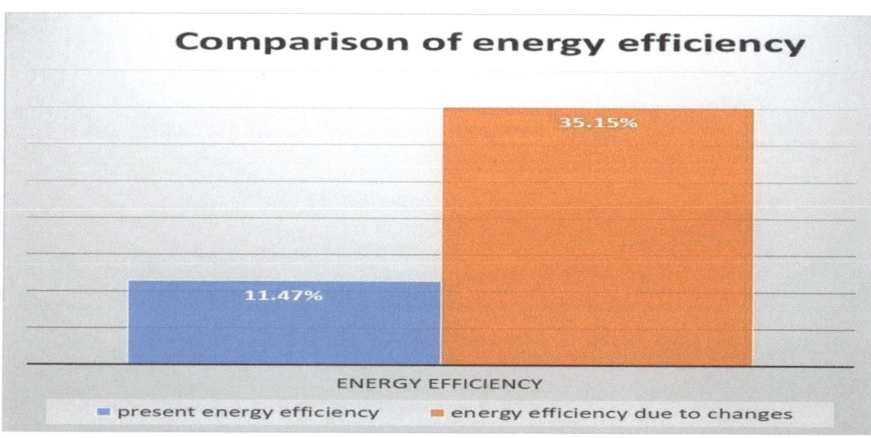

Changes made to improve water efficiency

- Usage of water-efficient shower heads with a flow rate of 9L/minute.
- Usage of water-efficient faucets for all bathroom with a flow rate of 8L/minute.
- Efficient closets for all bathroom with 5L capacity for high volume flush and 2.5L capacity for low volume flush.
- Water-efficient faucets for kitchen sink with a flow rate of 8L/minute.

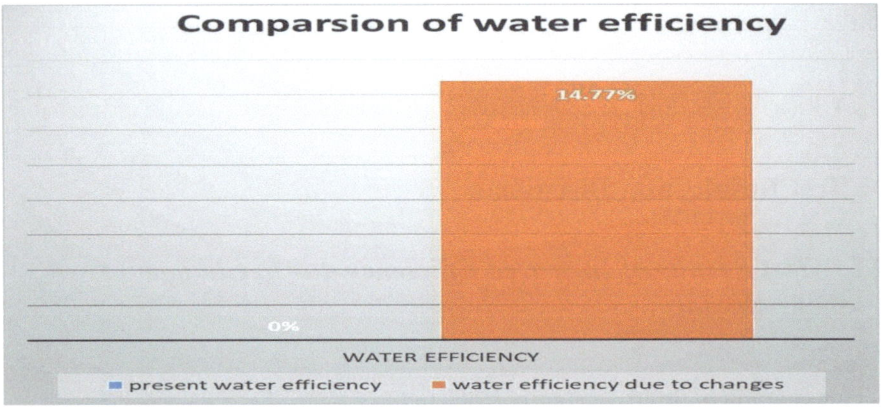

Changes made to improve material efficiency

- Replacing the conventional slab with filler slab.
- Replacing the concrete blocks with autoclaved aerated concrete blocks.

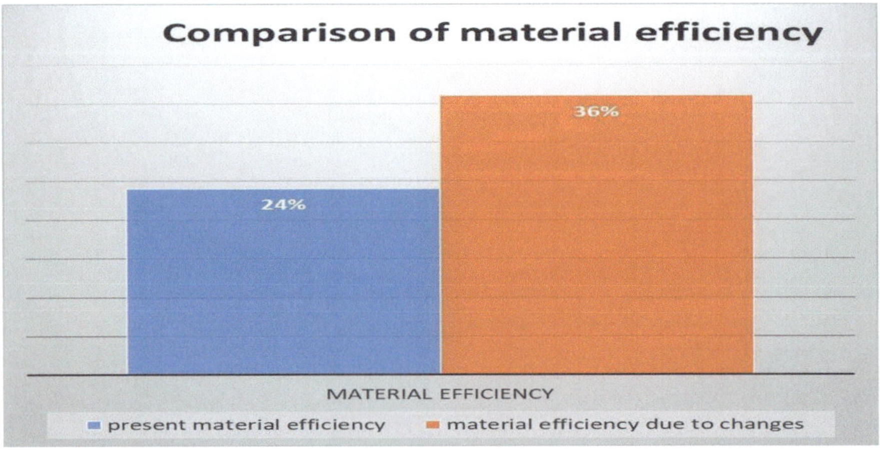

Results Obtained after Change of Materials

See Figs. 5, 6, 7, 8 and 9.

SL No	Parameters	Results
1	Floor area	77.48 m^2
2	Final energy use	1245 kWh/month
3	Final water use	36m^3/month
4	Final operational CO2 emissions	0.75tCO_2/month
5	Final embodied carbon	902kgCO_2e/m^2

Fig. 5 Results obtained after change of materials

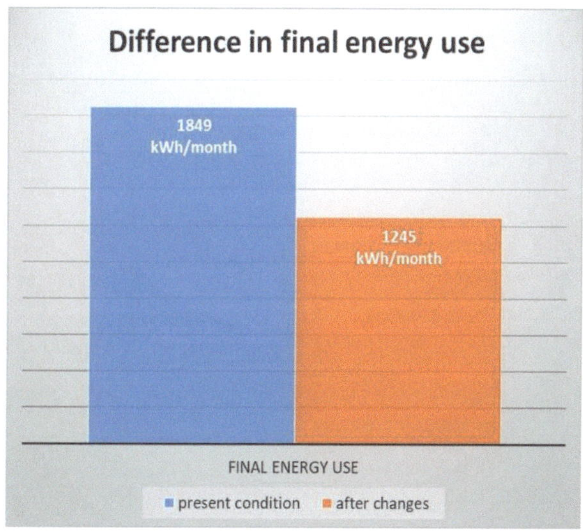

Fig. 6 Difference in water use

4 Conclusion

This experimental investigation studies about building energy optimization techniques and the influence of materials on energy efficiencies. After changing the original materials with the sustainable materials, the following results were obtained. The final energy use was reduced from 1849 kWh/month to 1245 kWh/month. The final water use was reduced from 48 m3/month to 36 m3/month. The final operational CO_2 emissions was reduced from 1.11 tCO_2/month to 0.75 tCO_2/month. The final embodied carbon was reduced from 1070 kgCO_2e/sqm to 902 kgCO_2/sqm.

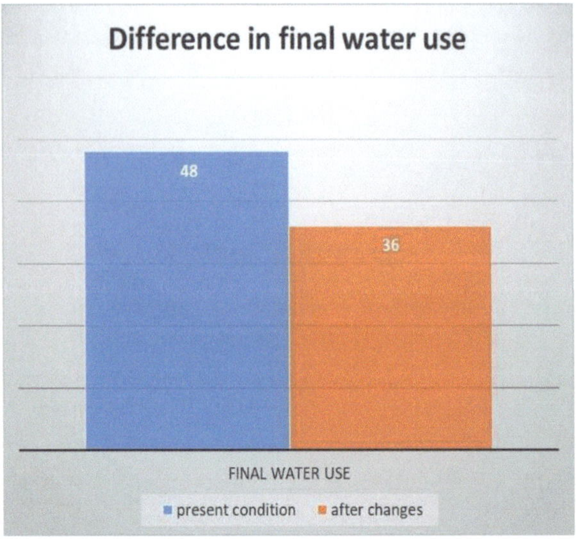

Fig. 7 Difference in final energy use

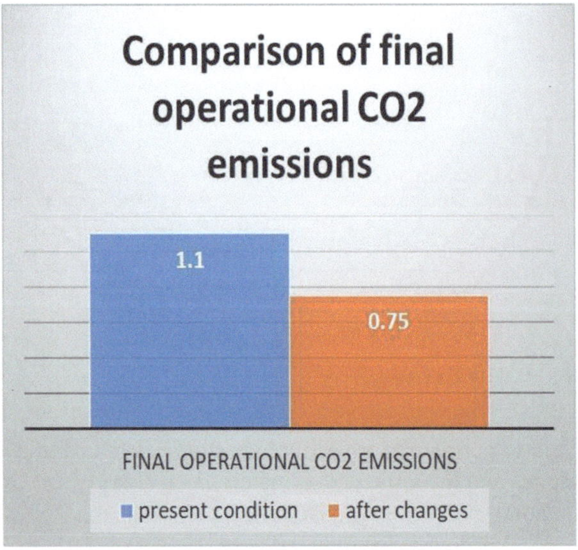

Fig. 8 Comparison of final operation CO_2 emissions

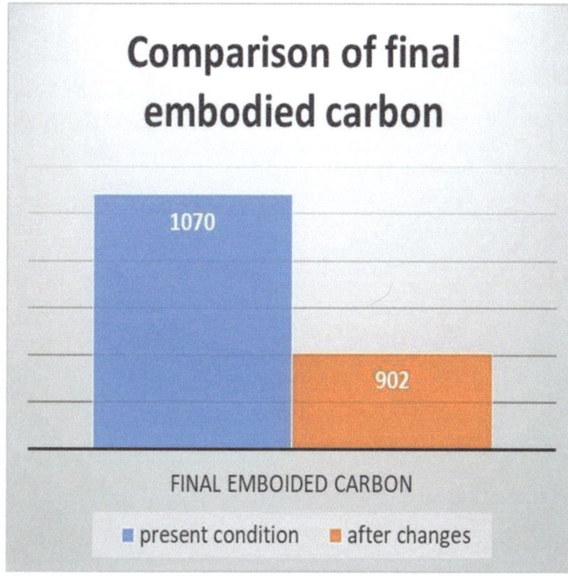

Fig. 9 Comparison of final embodied carbon

References

1. Yuan X, Karbasforoushha MA, Syah RB, Khajehzadeh M, Keawsawasvong S, Nehdi ML (2023) An effective metaheuristic approach for building energy optimization problems. Buildings 13(1):80
2. Tian ZC, Chen WQ, Tang P, Wang JG, Shi X (2015) Building energy optimization tools and their applicability in architectural conceptual design stage. Energy Proced 78:2572–2577
3. Ghalambaz M, Jalilzadeh YR, Davami AH (2022) Building energy optimization using butterfly optimization algorithm. Therm Sci 26(5):3975–3986
4. Chen J, Dai X, Cai X (2022) Optimization of zero-energy buildings based on multicriteria optimization method. Adsorpt Sci Technol
5. Kumar R, Aggarwal RK, Sharma JD (2013) Energy analysis of a building using artificial neural network: a review. Energy Build 65:352–358
6. Zhang H, Feng H, Hewage K, Arashpour M (2022) Artificial neural network for predicting building energy performance: a surrogate energy retrofits decision support framework. Buildings 12(6):829
7. Bojić M, Nikolić N, Nikolić D, Skerlić J, Miletić I (2010) Simulation of solar energy use for building heating by a vapor compression heat pump. In: 3rd Lasted African international conference on modelling and simulation (Africa MS 2010) pp 179–186
8. Mishra S, Usmani JA, Varshney S (2012) Energy saving analysis in building walls through thermal insulation system. Int J Eng Res. Witte A, Henninger MJ, Glazer RHJ, Crawley DB (2001) Testing and validation of a new building energy simulation program. Proceedings, building simulation, international building performance simulation association (IBSPA), Rio de Janeiro, Brazil Applications, 2(5), pp 128-135

Structural Response and Design of RC Frame with Flexible Base Under Gravity Loading

K. S. Vivek, B. Sravani, and C. Ravi Kumar Reddy

Abstract The soil-foundation interaction (SFI) plays a vital role in the force response of the RC frame elements, which duly needs to be accounted during modeling to obtain accurate internal force distribution for which the design should be carried out. In this study, a typical G + 3 building symmetric in plan is considered. The modeling and analysis for gravity loading were carried out using STAAD software. The analyses were performed for frames (without plinth beams) with both fixed and flexible base conditions. Three different soil strata, viz., soft clay, loose sand and dense sand, were considered for the flexible base condition. The spring stiffnesses of the isolated footings resting on soil were calculated as per the available expressions in ASCE / SEI 41-17 and incorporated in the models accordingly for accounting SFI. The analysis results are presented for an intermediate frame. The influence of SFI on the force response and design of beams, columns and footings was explained in great detail. On comparing the results, it was found that the forces on the footing decreased significantly in the case of flexible bases. A maximum reduction in the required area of steel reinforcement by 24. 65% and 24.36% of footings of interior and exterior columns, respectively, was noticed in the case of a frame resting on loose sand. A maximum increase in hogging moment and corresponding required area of steel reinforcement by 33.47 % and 44.29 % was observed in the beams. A maximum increase in hogging moment and corresponding required area of steel reinforcement by 33.47 % and 44.29 % was observed in the end span beams. The maximum increase and decrease in the required area of main steel reinforcement in the columns was noted as 195.82% and 37.79%.

Keywords Soil-foundation interaction · Structural response · Design · Footing · Soft clay · Loose sand · Dense sand

K. S. Vivek (✉) · B. Sravani
Department of Civil Engineering, VVIT, Nambur, AP, India
e-mail: saivivek@vvit.net

C. R. K. Reddy
Department of Civil Engineering, CBIT, Gandipet, Telangana, India

Y. K. Guruprasad et al. (eds.), *Recent Advances in Infrastructure Development and Management—Volume 2*, Lecture Notes in Civil Engineering 566,
https://doi.org/10.1007/978-981-96-1558-2_6

1 Introduction

In practice, the analysis of Reinforced Concrete (RC) frames is carried out by assuming fixed support for the columns at base, neglecting the influence of soil settlement. This is a common conventional practice adopted by structural designers over the world. However, the settlement of the soil beneath the footing of the columns alter the force distribution in the frame elements, i.e., columns and beams. In general, for any symmetric plan of the building, the interior columns are subjected to greater loads in comparison with the exterior columns, and hence all the columns do not settle by same amount leading to uneven settlement. This uneven settlement alters the internal force distribution of the columns and beams, which is ignored in the conventional practice of modeling and analysis of the building frames. Due to the flexibility imparted by soil, the foundation can undergo vertical translation (settlement under gravity loading), horizontal translation and rotation [1] as depicted in Fig. 1. Extensive research was available ranging from shallow foundations to deep foundations in the literature.

Allam et al. [2] carried out numerical study to investigate the effect of soil settlement on the force response of building frames considering Winkler and half-space models. It was reported that the hogging moments in the beams were very sensitive to relative stiffness of footing and soil beneath, though, the sagging moments were marginally varied. The variation in axial force of columns is directly depended upon the settlement of the foundations. Roy and Dutta [3] reported that for a plane frame without tie beams, the maximum increase in axial force in central and corner columns was found to be 14% and 25%, respectively, for the case of footing resting on loose sand in comparison to the fixed base condition. The maximum increase in bending moment was found to be 55% and 75% in columns of first story and second story, respectively, for the case of footing resting on loose sand. Several analyses were carried out considering frames with different stories and soil types. It was

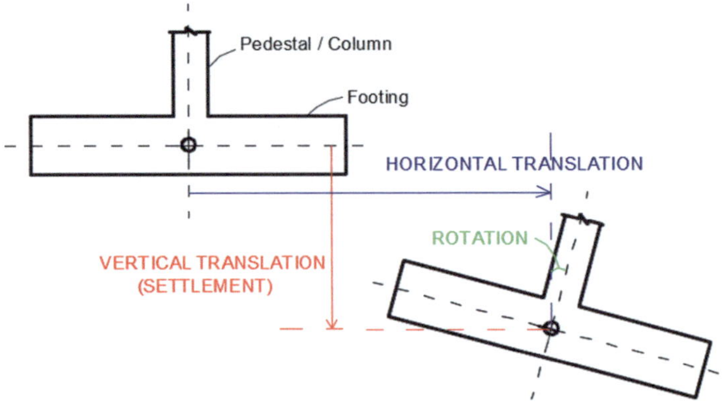

Fig. 1 Displacement of isolated footing

also reported that providing tie/plinth beams reduced the effect of soil-foundation interaction.

2 Aim and Scope of Present Study

The aim of the study is to investigate and present in detail the influence of soil-foundation interaction on the structural response of reinforced concrete (RC) plane frame (without plinth/tie beams) subjected to gravity loading from the structural analysis and design point of view. The scope of the study is limited to linear-elastic behavior (both RC frame and soil) and static gravity loading.

3 Problem Statement and Analysis Procedure

3.1 Problem Statement

A typical G + 3 residential building plan (symmetric) was considered as shown in Fig. 2a. The elevation of an intermediate frame was depicted in Fig. 2b. The story height of FL 1 (Floor Level—1) was assumed as 3.2 m and for the rest of floors as 3 m. The bay spacing is 4 m. The initial size of the columns and beams was assumed as 300 × 450 mm and 230 mm × 400 mm. The thickness of slabs (at all floor levels) was 150 mm. The dead load due to slab and walls was calculated in accordance with IS 875: Part I [4]. The thickness of external and internal (partition) walls was assumed as 230 mm and 150 mm, respectively. The live load at every floor level was considered as 4.5 kN/m^2 except at roof level (FL 4) for which 1.5 kN/m^2 was assumed as per IS 875: Part II [5]. No live load reduction was considered in the analysis.

3.2 Frame with Fixed Base

The modeling, analysis and design as per IS 456 [6] of the frame was carried out with the help of STAAD. Pro software for the load combination of 1.5 * (Dead Load + Live Load). The assigned beam and column numbering for the considered plane frame was shown in Fig. 2b. The grade of concrete and main steel reinforcement was assumed M25 (F_{ck} = 25 MPa) and Fe 500 (F_y = 500 MPa), respectively. Fe 415 (F_y = 415 MPa) was assumed for secondary reinforcement. After the analysis and design, the assumed column and beam sizes were assumed to be unchanged. Based on the obtained support reactions, the design of isolated rectangular footings was done manually by assuming safe bearing capacity of soil as 210 kN/m^2. The

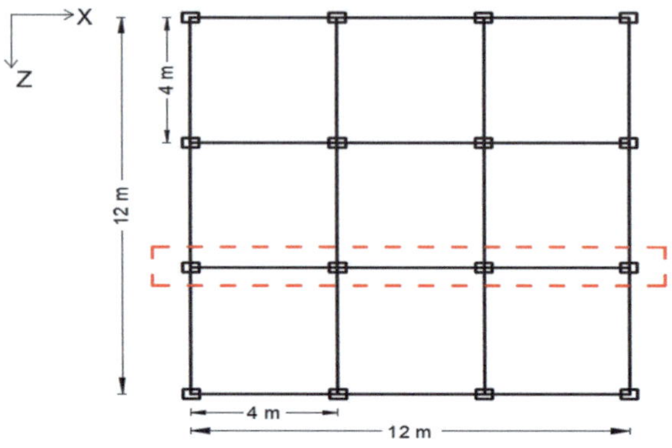

(a) Plan

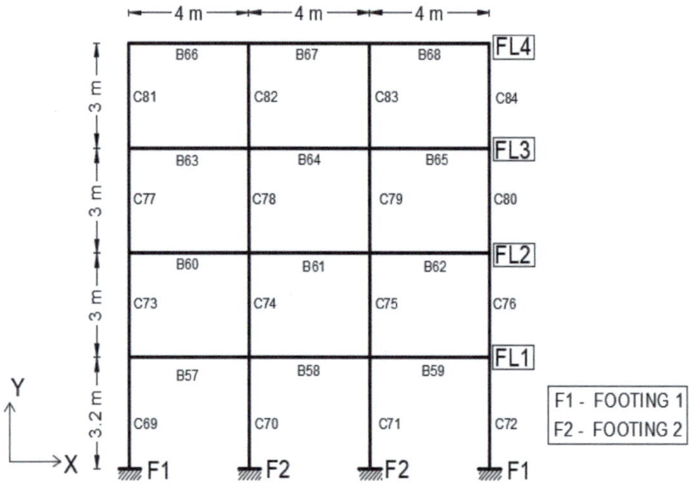

(b) Elevation of intermediate frame

Fig. 2 G + 3 building

support reactions for the both exterior columns and the both interior columns were same, respectively, and hence identical footings were obtained, i.e., the footing size for exterior columns (designated as F1 in Fig. 2b) was obtained as 2.1 m (L_f) × 2 m (B_f) × 0.5 m (t_f) and 3 m × 2 m × 0.65 m for the interior columns (designated as F2 in Fig. 2b). The diagrammatic representation of the footings was depicted in Fig. 3.

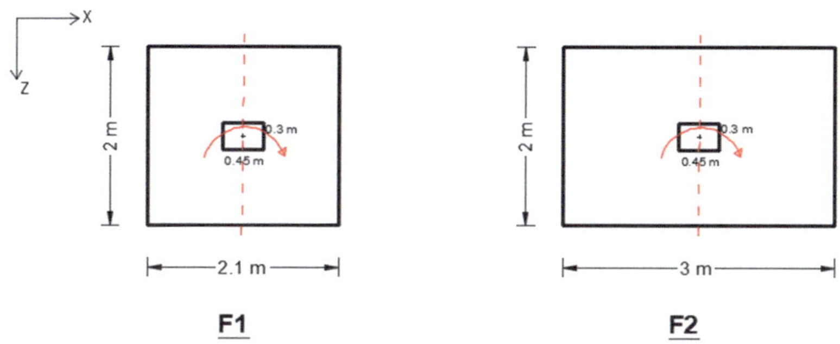

Fig. 3 Footings F1 and F2

3.3 *Frame with Flexible Base*

For the case of flexible base, three different types of soil viz. soft clay, loose sand and dense sand were considered. The soil-foundation interaction can simply be incorporated by replacing the fixed supports at base of columns with 'elastic springs.' For a plane frame, a total of 3 springs corresponding to vertical translation (settlement), horizontal translation and rotational degrees of freedom (with corresponding stiffnesses K_y, K_x and K_θ, respectively) should be provided at each column base [1] as depicted in Fig. 4. However, as gravity loading alone was considered in the present study, the footing was assumed to be rigid against horizontal translation (along X-axis), i.e., $K_x = \infty$, and hence horizontal springs were not included in the model.

The spring stiffnesses depend upon the elastic properties of the soil and footing pad dimensions. The spring stiffnesses K_y and K_θ were calculated based on the expressions from ASCE / SEI 41-17 [7] as stated below.

$$K_y = \frac{G * B}{(1 - \mu_s)}[1.55(\frac{L}{B})0.75 + 0.8] \tag{1}$$

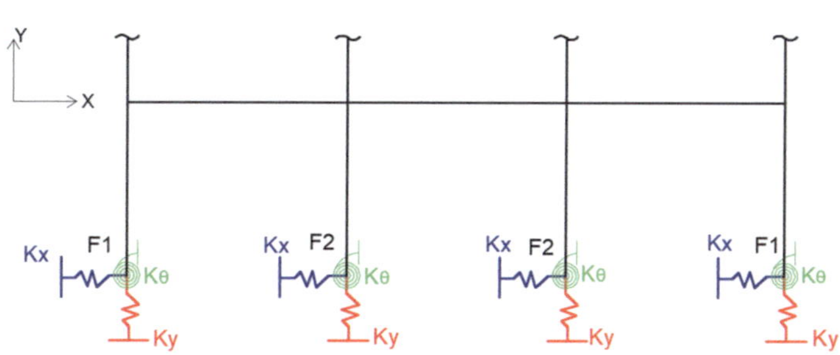

Fig. 4 Flexible base

Table 1 Elastic properties of soil and spring stiffnesses

Type of soil	E_s (kN/m^2)	μ_s	K_y (kN / m)		K_θ (kN.m / deg)	
			F1	F2	F1	F2
Soft clay	25000	0.45	75478.66	90936.51	1230.694	2796.049
Loose sand	18000	0.20	45145.67	54391.40	736.1087	1672.387
Dense sand	66000	0.40	189181.86	227925.90	3084.646	7008.098

$$K_\theta = \frac{G * B^3}{(1 - \mu_s)}[0.47(\frac{L}{B})2.4 + 0.034] \tag{2}$$

Where E_s = Modulus of elasticity of soil (kN/m^2)

G_s Shear modulus = E / 2(1 + μs)
μs Poisson's ratio of soil
L Length of footing (m)
B Breadth of footing (m)

The elastic properties [1] and the calculated spring stiffnesses for the considered soils were presented in Table 1.

4 Results and Discussion

The footing settlements were presented in Table 2. The forces and required long span reinforcement of footings were tabulated in Table 3. Similarly, the internal forces and required area of longitudinal steel reinforcement ($A_{st,req}$) for critical beams and columns were presented in Tables 4 and 5, respectively. The bending moment diagram for the members at FL2 and axial force diagram for columns were depicted in Figs. 5 and 6, respectively. The support settlement diagram was shown in Fig. 7.

1. Significant differential settlement of the footings was observed in the case of flexible base which would not be accounted in the case of fixed base.
2. The footings of the interior columns were prone to larger settlement due to the action of greater load in comparison with the exterior columns. The maximum settlement was observed for the case of loose sand due to which maximum

Table 2 Settlement of foundations

Type of soil	Vertical settlement (mm)	
	F1	F2
Fixed	0	0
Soft clay	– 10.79	– 13.26
Loose sand	– 15.41	– 18.93
Dense sand	– 5.08	– 6.25

Table 3 Forces and long span steel reinforcement area required in footings

Type of base	F_y (kN)		M_z (kNm)		d_{req}		$A_{st,req}$	
	F1	F2	F1	F2	F1	F2	F1	F2
Fixed soft clay loose sand dense sand	1090.70	1617.42	14.89	0.15	353	546	1765 1475	2730 2290
	814.26	1205.52	10.74	2.62	295	458	(-16.43%) 1330	(-16.12%) 2065
	695.69	1029.46	9.08	3.15	266	413	(-24.65%)	(-24.36%)
	960.64	1423.34	12.88	1.53	327	507	(-7.37%)	(-7.14%)

() = Change in partner with respect to fixed base condition. '+/−' indicate increase decrease (F_y = Axial force; M_z = Moment about Z-axis; d_{req} = effective depth required)

Table 4 Bending moments and required flexural reinforcement in beams

Beam No.	Type of base	M^-max (kNm)	$A_{st,req}^-$ (mm^2)	M^+ max (kNm)	$A_{st,req}^+$ (mm^2)
57, 59	Fixed	79.09	585.64	48.11	330.5
	Soft clay	88.92	691.83 (18.13%)	48.37	332.53 (0.61%)
	Loose sand	96.22	765.26 (30.67%)	48.46	333.28 (0.84%)
	Dense sand	79.84	598.12 (2.13%)	48.24	331.51 (0.31%)
60, 62	Fixed	77.15	568.28	47.41	325.04
	Soft clay	95.27	754.99 (32.86%)	47.42	325.12 (0.02%)
	Loose sand	102.97	819.96 (44.29%)	47.44	325.18 (0.04%)
	Dense sand	85.72	651.81 (14.70%)	47.42	325.07 (0.01%)
63, 65	Fixed	80.14	600.56	47.1	322.62
	Soft clay	98.03	780.22 (29.92%)	47.1	322.93 (0.10%)
	Loose sand	105.63	839.53 (39.79%)	47.51	323.06 (0.14%)
	Dense sand	88.6	688.43 (14.63%)	47.09	322.77 (0.05%)
66, 68	Fixed	61.37	433.1	40.75	276.04
	Soft clay	69.96	502.89 (16.11%)	40.96	277.43 (0.50%)
	Loose sand	75.82	556.66 (28.53%)	41.05	278.03 (0.72%)
	Dense sand	62.67	443.36 (2.37%)	40.85	276.7 (0.24%)

'+ / −' indicate 'Sagging / Hogging'

Table 5 Internal forces and required reinforcement in columns

Column No.	Type of base	F (kN)	M (kNm)		Ast, req (mm^2)
			Top	Bottom	
70	Fixed	1617.42	– 0.07 (0.07)	– 0.15 (0.15)	1188
	Soft clay	1205.52	6.35 (– 6.35)	– 2.62 (2.62)	865 (– 27.19%)
	Loose sand	1029.46	8.91 (– 8.91)	– 3.15 (3.15)	739 (– 37.79%)
	Dense sand	1423.34	3.03 (– 3.03)	– 1.53 (1.53)	1021 (– 14.06%)
74	Fixed	1191.02	0.66 (– 0.66)	0.25 (– 0.25)	855
	Soft clay	886.96	9.2 (– 9.2)	– 9.06 (9.06)	636 (– 25.61%)
	Loose sand	756.98	12.85 (– 12.85)	– 13.11 (13.11)	543 (– 36.49%)
	Dense sand	1047.75	4.69 (– 4.69)	– 4.11 (4.11)	752 (– 12.05%)
	Dense sand	465.38	40.71 (– 40.71)	– 41.89 (41.89)	334 (– 11.64%)
78	Fixed	770.54	3.32 (– 3.32)	– 2.69 (2.69)	553
	Soft clay	574.59	10.96 (– 10.96)	– 10.72 (10.72)	412 (– 25.50%)
	Loose sand	490.83	14.29 (– 14.29)	– 14.11 (14.11)	352 (– 36.35%)
	Dense sand	678.21	6.93 (– 6.93)	– 6.49 (6.49)	487 (– 11.93%)
81	Fixed	231.56	55.83 (– 55.83)	– 43.00 (43.00)	287
	Soft clay	171.38	68.81 (– 68.81)	– 52.46 (52.46)	648 (125.78%)
	Loose sand	145.58	74.33 (– 74.33)	– 56.48 (56.48)	849 (195.82%)
	Dense sand	203.25	61.97 (– 61.97)	– 47.47 (47.47)	426 (48.43%)

Clockwise moment is '+' and anti-clockwise moment is '–'

reduction in required effective depth and area of longitudinal steel by 24. 65% and 24.36% corresponding to interior column and exterior column footings, respectively.

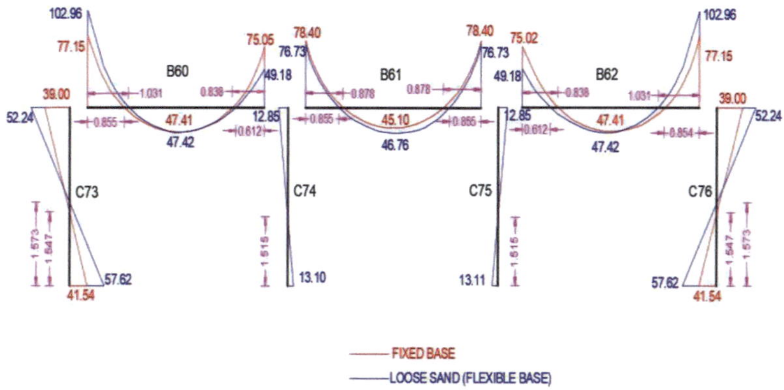

Fig. 5 Bending moment diagram and points of contraflexure *(not to scale). (Moments in 'kN.m.' and lengths in 'm')*

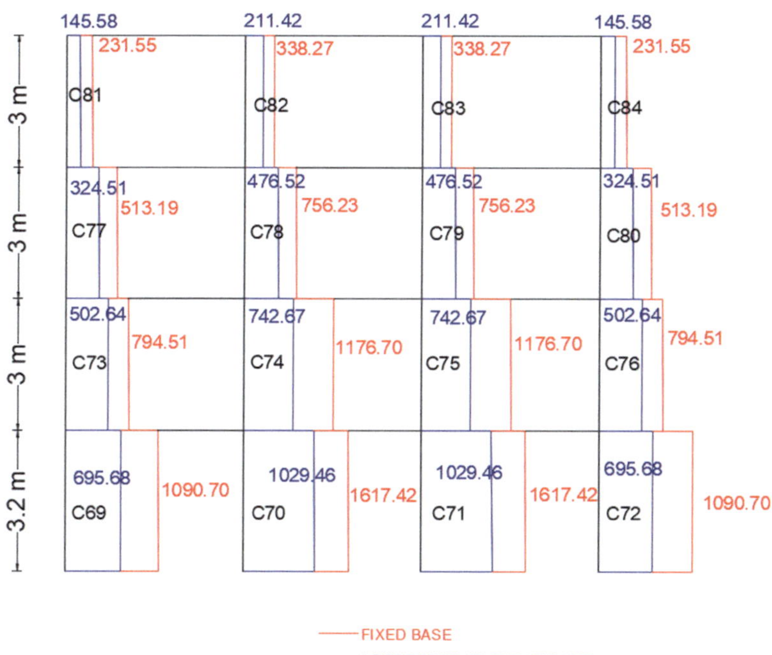

Fig. 6 Axial force (kN) diagram for columns *(not to scale)*

3. Due to the greater settlement of footings of interior columns, the hogging moments increased significantly in the end span beams, in comparison with the intermediate span beams at any floor level in the case of flexible base. A maximum increase in hogging moment and corresponding area of steel reinforcement required by 33.47 % and 44.29 % was observed in the end span beams

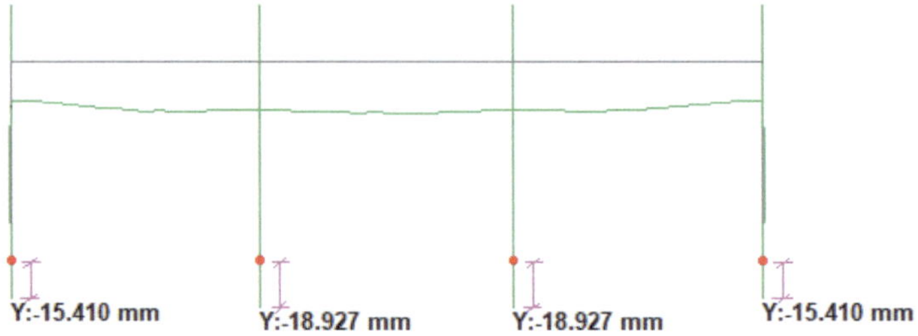

Fig. 7 Support settlement diagram

at floor level 1 for the case of frame resting on loose sand. Not much variation in sagging moments was noticed.

4. Significant variation in the profile of bending moment diagram and location of points of contraflexure was expected (see Figs. 5 and 6).
5. The axial compressive force in the columns was reduced for the case of flexible base in comparison with fixed base frame. A maximum reduction of 36.44% was noticed for the case of frame resting on loose sand.
6. However, the bending moments in the majority of the columns were increased depending upon the type of soil. The maximum increase and decrease in the required area of main steel reinforcement in the columns was noted as 195.82% (in the exterior columns of topmost floor) and 37.79% (interior columns beneath floor level 1).

5 Conclusion

A simplified modeling of soil-foundation interaction (SFI) was carried out for a typical G + 3 building symmetric in plan. The results (force response and required steel reinforcement area) were presented for an intermediate frame. From the obtained results and presented discussion in Sect. 4, it may be concluded that the conventional assumption of fixed base for columns of frame resting on soil leads to inaccurate forces and design especially in the case of loose sand for which the modulus of elasticity is very low, and thereby resulting in improper design and detailing.

References

1. Bowles JE (1996) Foundation analysis and design, 5th edn. McGraw-Hill, Singapore
2. Allam MM, Subba Rao KS, Subramanya BVV (1991) Frame—soil interaction and Winkler model. Proceed Inst Civil Eng Part 2(91):477–494
3. Roy R, Dutta SC (2001) Differential settlement among isolated footings of building frames: the problem, its estimation and possible measures. Int J Appl Mech Eng 6(1):165–186

4. IS 875 (part 1) (1987) Indian standard code of practice for design loads for buildings and structures (Dead Loads). BIS, New Delhi
5. IS 875 (part 2) (1987) Indian standard code of practice for design loads for buildings and structures (Imposed Loads). BIS, New Delhi
6. IS 456 (2000) Plain and reinforced concrete—code of practice. BIS, New Delhi
7. ASCE / SEI 41-17 (2017) Seismic evaluation and retrofitting of existing structures. USA

Pavement Evaluation—A Case Study at Peenya Industrial Area–Phase I, Bengaluru

K. S. Meenakshi and K. S. Jayadeep

Abstract The pavements constructed in Industrial areas should be structurally stable to sustain heavy wheel loads imposed by commercial vehicles. Unstable pavements should be subjected to appropriate and timely maintenance intervention for their long-term sustainability. Hence, pavement evaluation studies are necessary to assess both the functional and structural adequacy of existing pavements against traffic loading and review them to improve the transportation infrastructure in industrial areas. This paper presents detailed structural and functional evaluation studies conducted for selected industrial roads in the Peenya Industrial Area, Bangalore. The structural evaluation of pavement is performed by Benkelman Beam Deflection (BBD) studies. Unevenness of road surface was measured using MERLIN, and pavement surface friction was determined by Simple Portable Pendulum Skid Resistance Tester. Pavement distress surveys were conducted through a visual approach as per ASTM-D6433-11 specifications, and the Pavement Condition Index (PCI) is rated for roads according to the standard PCI rating scale.

Keywords Pavement surface assessment · Benkelman beam deflection analysis · Overlay thickness · Functional evaluation · PCI model · PSI

1 Introduction

Roadways provide infrastructure facilities and enhance communication between communities all over the country. Since there is demand for the use of roadways, it is necessary to check the routine condition of pavement and to carry out necessary maintenance actions to ensure the long-term performance of pavements.

Every structure loses its stability and strength over a time period. Hence, the pavement should be checked and assessed every two years in terms of functional

K. S. Meenakshi (✉) · K. S. Jayadeep
Department of Civil Engineering, Raja Reddy Institute of Technology, Chikkabanvara, Bangalore, India
e-mail: meenakshiks2312@gmail.com

Y. K. Guruprasad et al. (eds.), *Recent Advances in Infrastructure Development and Management—Volume 2*, Lecture Notes in Civil Engineering 566,
https://doi.org/10.1007/978-981-96-1558-2_7

and structural parameters. The importance of regular maintenance and rehabilitation is necessary due to the deterioration of pavement conditions due to traffic operation and environmental characteristics.

The pavement is said to be satisfactory when it is free from unevenness, cracking, depression, and other distress. Initially, the distress is found along the wheel path and its magnitude increases over the duration of time due to poor adoption of quality control during construction, traffic volume, environmental condition, and lack of maintenance during design life.

Pavement indicators are key measures to overview the present condition serviceability and performance. Hence, the assessment of the functional and structural circumstances of the pavement is important to contemplate the distressed elements.

The objectives and scope of the present study are.

- To conduct the structural and functional evaluation of selected pavement stretches to determine the causes and magnitude of distress.
- To determine the causes and magnitude of distress and recommend suitable methods of maintenance measures.
- The pavement evaluation predicts the structural adequacy of existing pavement and reviews its maintenance.
- To improve the infrastructure and transportation in industrial areas.

2 I Methodology

2.1 A Methods

1. Functional Evaluation of Pavement

The pavement surface features the intensity of the comfort, and safety of drivers and road services is known as functional evaluation. The primary attributes examined in the functional assessment are surface roughness, road unevenness, and skid resistance.

i. Pavement Distress Survey

The distress on pavement surface of the current road is identified and classified into several categories depending on the type, location, and intensity of the distress. The ASTM-D6433-11 guidelines are followed in identifying the distress and measuring its extent.

ii. Unevenness of Pavement Surface Using MERLIN

The unevenness of the pavement indicates undulations on the pavement surface. It affects vehicle operating cost, safety, and comfort of ride. International Roughness Index (IRI) is a standard term for indicating lateral and longitudinal unevenness of pavement surface. According to IRC: SP-16-2004, guidelines are described for surface unevenness of flexible pavements to establish maintenance at the earliest levels. The above specification also recommends the use of a Machine for evaluating Roughness using Low-cost Instrumentation (MERLIN).

The IRI scale and the Merlin scale are related by the following equation,

IRI = 0.593 + 0.0471D

42 > D > 312 (2.4> IRI > 15.9)

UI = 630*(IRI) 1.12

IRI = International Roughness Index in m/km

D = roughness in Merlin scale measured in mm

UI = Unevenness index in mm/km

iii. Pavement Condition Index (PCI):

Pavement distress surveys were conducted through visual approach as per ASTM-D6433-11 specifications

iv. Present Serviceability Index (PSI)

PSI is the common technique to relate the objective measures of surface condition of existing pavement to the public's assessment of serviceability. The functional evaluation of the existing pavement is mainly associated with the pavement performance from the road user's point of view. The valuation of riding eminence and safety is counted to be the most important constituent of pavement surface condition evaluation. The evaluation of pavement riding quality may be carried by physical measurement of distress and pavement surface undulations and through the measurement of riding eminence or pavement undulations by computerized instruments available.

i. **Co-efficient of friction using simple portable pendulum skid resistance tester** for dry and wet pavement conditions.

Frictional co-efficient or skid resistance (f) = (9.4081 sin Θ)/N

N = (self-weight of sliding shoe + centrifugal force induced by measuring equipment) kg

Θ = angular values obtained by using the tester.

2. Structural Evaluation of Pavement

Structural evaluation is the condition of pavement service analysed through structural capacity to cumulative traffic load.

i. **Benkelman Beam Deflection Studies**: The rebound deflection of the flexible pavement was measured by Benkelman Beam deflection studies (BBD). The temperature of the pavement was checked for everyone-hour interval during the study. Traffic volume count was done to assess the number of commercial vehicles plying on these roads. The physical characteristics of subgrade soil were tested in the laboratory to assess the physical properties of soil required for evaluating the overlay requirement for selected stretches as per IRC 81–1997 (Table 1).

Table 1 Details of selected pavement stretches in Peenya industrial area for case study

Road no	No. of lanes	Width (m)	Width of carriage way (m)	Length (m)
1	2 lanes divided	21	7(L) 11.8(R)	700
2	2 lanes divided	15.5	6.7(L) 8(R)	1300
3	2-way traffic lanes with undivided	14	13	1000
4	2-way traffic lanes with undivided	9.8	9.2	1000

Table 2 Overall PCI for selected roads

Road no	Road description		PCI	Standard PCI
1	Volvo road	Left	79	Satisfactory
		Right	71	Satisfactory
2	Pipeline road	Left	79	Satisfactory
		Right	74	Satisfactory
3	Kirloskar road		58	Fair
4	BFW road		12	Serious

2.2 B Experimental Studies

1. Pavement Condition Index

In this study, the PCI is assessed by visual inspection method as per ASTM-D-6433-11 Standards (Fig. 1) (Table 1).

2. Unevenness Index of Pavement

The roughness of selected stretches obtained from MERLIN satisfies IRC: SP-16-2004 provisions mentioned in Table 4.12.

1. The UI for roads 1 and 2, varies from 1758 mm/km to 2686 mm/km.
2. The UI of road 3 is 4000 mm/km from 0 to 500 m chainage and 2318 mm/km 500–1000 m in "average" state (Tables 3, 4 and 5).
3. The UI for road 4 is above 4000 mm/km (Figs. 2, 3, 4 and 5).

3. Skid Resistance

Co-efficient of friction

4. Present Serviceability Index (PSI)

5. Structural Evaluation of Pavement

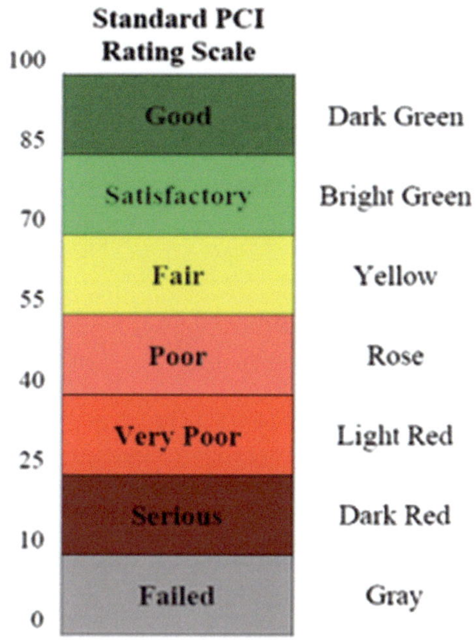

Fig. 1 PCI rating scale mentioned in the ASTM-D-6433-11 code book

Table 3 Co-efficient of friction obtained for dry and wet condition of pavement surface by Portable Pendulum skid resistance tester

Road No	Co-efficient of friction (f)	
	Dry	Wet
Road 1 (L)	1.956	1.58
Road 1 (R)	1.993	1.701
Road 2 (L)	2.632	2.383
Road 2 (R)	2.044	1.818
Road 3	2.003	1.596
Road 4	2.151	1.872

2.3 *D. Results and Discussion*

PCI Results

The pavement condition was found to be 'satisfactory' for roads 1 and 2 with a PCI value in the range of 71–79. It is found to be 'fair' in road 3 with a PCI value of 58, and the pavement condition is found to be 'failed' for road 4 with the PCI value 12.

Unevenness indices of pavement.

The unevenness for roads 1 and 2 was found to be in 'Good' perceptions. The UI for both dividing lanes of roads 1 and 2 varies from 1758 mm/km to 2686 mm/

Table 4 PSI for road

Road no	Road description		PSI Equations by multiple linear regression analysis
Road no 1	Volvo road	Left	PSI = 10.728—0.0 (IRI)–0.473 (Ravelling)—1.155 (Potholes)—1.232 (Depressions) + 0.205 (Bumps & Sags)—0.483 (Patching)
		Right	PSI = 6.705—0.0 (IRI) + 1.99 (Ravelling)—0.374 (Potholes)—1.66 (Depressions) – 2.596 (Bumps & Sags) + 1.702 (Patching)
Road no 2	Pipeline road	Left	PSI = 8.952–119 (IRI) + 0.218 (Cracking)-1.041 (Ravelling) – 0.265 (Potholes) – 1.169 (Rutting) + 1.541 (Depressions) + 1.476 (Bumps & Sags)—1.780 (Patching)
		Right	PSI = 9.874–0.253 (IRI)—1.471 (Cracking)-0.180 (Ravelling) – 0.1026 (Potholes)—0.141 (Rutting) – 22.508 (Depressions)—1.158 (Bumps & Sags) + 3.228 (Patching)
Road no 3	Kirloskar road		PSI = 8.23—0.248 (IRI) – 0.123 (Cracking)—0.0047 (Ravelling) + 0.0689 (Potholes) – 0.3883 (Rutting) + 0.545 (Depressions) – 0.001 (Bumps & Sags)—0.122 (Patching)
Road no 4	BFW Road		PSI = 5.11–0.198(IRI) + 0.194 (Cracking)—0.0008 (Ravelling)—0.3786 (Potholes) – 0.268 (Rutting) + 0.515 (Depressions) + 0.0579 (Bumps & Sags)

Table 5 Overlay thickness for existing pavement

Road No	Traffic volume in MSA	Characteristic deflection (D_C)	Overlay thickness in terms of BC (mm)
Volvo road (L)	16.7	0.58	–
Volvo road (R)	20	0.556	–
Pipeline road (L)	16.7	0.973	25
Pipeline road (R)	20	0.867	14
Kirloskar road	20	1.11	50
BFW road	20	1.306	90

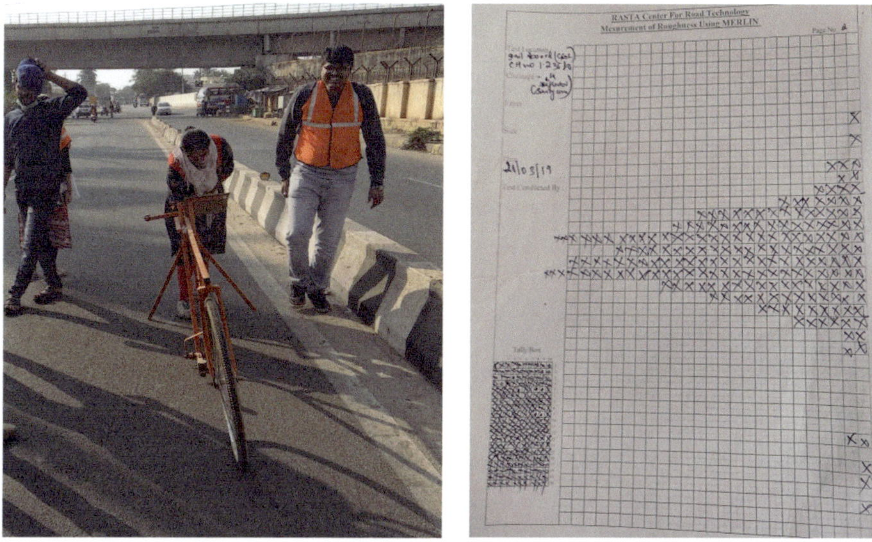

Fig. 2 Histogram obtained by MERLIN studies

Fig. 3 Portable simple pendulum skid resistance tester used during the case study

km. It displays slight flaws in the road surface. The UI for road 3 indicates 'poor' surface condition from 0 to 500 m chainage and 500–1000 m in an 'average' state. The UI for road 4 is above 4000 mm/km; hence, the surface is categorized as poor in condition.

PSR and PSI Model

PSR values are obtained by visual rating given by five highway professionals, and the ratings are used to determine mean PSR

Fig. 4 Potholes found in the road no. 3 during case study

Fig. 5 Conducting BBD test on road no.2

The PSI model for the case study is represented in the form of a multiple linear regression equation with the distress variables, roughness index, and mean PSR values.

The PSI equation represents correlation between the roughness and serviceability of selected stretches.

Co-efficient of friction

The co-efficient of friction on all roads is said to be satisfactory in both the dry and wet conditions.

Design of overlay thickness

The BBD study concludes that road 1 is structurally adequate and does not require an overlay. Road 2 is structurally inadequate and requires 25 mm and 15mm of BC overlay for left and right side of pipeline road, respectively. For road 3, the stretch requires BC overlay of thickness 50 mm, and for road 4, the total overlay thickness obtained to be 90 mm for corresponding traffic 20msa.

3 Conclusions

1. The pavement inventory survey recommends precise periodic maintenance for distresses found in roads 1, 2, and 3. The immediate counteractive measures are to be considered for road 4 as it is failed condition. The periodic maintenance such as filling potholes, sealing the minor cracks, and repairing the other distresses such as rutting, undulations, and severe cracks.
2. From BBD studies, the road 1 found to be structurally adequate as the study indicates minimum deflection, hence zero thickness of bituminous overlay.
3. Road 2 is found to be structurally inadequate as pavement layers showing noticeable deflections and overlay thickness obtained from study is minimum and impractical. However, to enhance the stability of pavement structure, the preventive measure is recommended to provide minimum thickness 30 mm of BC for both lanes.
4. The road no 3 is determined to be in 'fair' condition by PCI studies. But the study concludes that the pavement layers are structurally inadequate. Hence, the BC overlay of 50 mm is required to improve the bearing adequacy.
5. The road no 4 is found to be in failed condition in all aspects of geometric specifications. Hence, the road is recommended for overlay construction from binding course that is DBM of thickness 90 mm and BC of thickness 40 mm with confined shoulders and drainage facilities.
6. The guidelines of IRC recommend a skid resistance value of 0.35 is assumed for designing stopping sight distance value for speed 60kmph and higher. In this study, the obtained skid resistance value for all existing pavement is greater than 0.35 in both wet and dry surface conditions. Hence, it is concluded that all the selected pavement stretches have adequate skid resistance.
7. The obtained PSI model from the study is exercised for future pavement management system (PMS) and also suitable economic analysis can be carried out for the selected stretch of pavement under distress condition.

References

1. The Handbook of Highway Engineering, by T. F. FWA, Taylor & Francis groups. Modern Pavement Management, by Ralph Haas—W. Ronald Hudson—John Zaniewski, Krieger Publishing Company
2. Cundill MA The MERLIN road roughness machine: user guide, TRL Report 229, Overseas unit, TRRL, Crowthrone, Brekshire, UK

3. Gunashekar TK (1996) Skid resistance characteristics of pavements—a thesis
4. Khanna SK, Justo CEG (2001) Highway engineering, 8th edition. Nem Chand & Bros., Civil Lines, Roorkee
5. Kadyali LR, Lal NB (2006) Highway engineering, 4th edition, Khanna Publications
6. Indian Roads Congress, Report containing recommendations of the regional workshops in highway safety, IRC Special Publication 27
7. Nanyang Technological University, School of civil and structural engineering article on skid resistance
8. ASTM—D6433–11 Standard practice for roads and parking lots pavement condition index surveys
9. IRC—81–1997 Guidelines for strengthening of flexible pavements using BBD technique
10. IRC—SP 16–2004 Guidelines for surface evenness of highway pavements
11. IRC-37–2018 Guidelines for design of flexible pavements

Gold Mineralization and Their Lithological Controls at Nagavi Area, Gadag Schist Belt, in Karnataka, India

Chandrashekharappa Agasnalli, M. S. Deepak, Anupkumar G. Ekbote, and T. K. Lakkundi

Abstract Both parts of the Dharwar Craton, the Western (WDC) and the Eastern (EDC), house a number of schist belts. Many of these schist belts, such as the Kolar schist belt, Hutti schist belt, Chitradurga schist belt and the Gadag schist belt, are auriferous. In such schist belts, gold occurs in association with sulphides such as pyrite, pyrrhotite, arsenopyrite, and chalcopyrite. Gold mineralization in these belts is found to occur in different lithologies. The Gadag schist belt is composed dominantly of metabasalt in its western half and metasediments in the eastern half. Quartz-carbonate veins are observed to cut across these lithologies as well as the banded iron formations (BIFs). This paper aimed to study the gold mineralization in different lithologies of the Gadag schist belt and found that metabasalts and BIFs are the dominant rocks hosting gold. It is also observed that the mineralization is strata bound. Gold concentration varies from < 21 to < 26 ppb (parts per billion).

Keywords Gold mineralization · Lithological control · Geological Structures · Nagavi area · Gadag schist belt

1 Introduction

In India, Dharwar Craton hosts many of the country's mineral resources, including the famous Hutti, Kolar, Gadag goldfields, etc. The Dharwar Craton is divided into two blocks, viz., Western Dharwar Craton (WDC) and Eastern Dharwar Craton (EDC), which are based on the abundance and nature of the greenstone belts and also the age of the gneissic basement rocks. These two Cratons are considered to have different

C. Agasnalli · M. S. Deepak · A. G. Ekbote
Faculty of Civil Engineering, BMS Institute of Technology and Management, Bengaluru, India

T. K. Lakkundi (✉)
Department of Geology, School of Earth Sciences, Central University of Karnataka, Kalaburagi, Karnataka, India
e-mail: tejaswikl@gmail.com

Y. K. Guruprasad et al. (eds.), *Recent Advances in Infrastructure Development and Management—Volume 2*, Lecture Notes in Civil Engineering 566,
https://doi.org/10.1007/978-981-96-1558-2_8

evolutionary histories and metallogenic characteristics [1]. Gadag schist belt is well-known in India and the rest of the world for ancient British gold mining activities. In Nagavi, lithological controls of gold mineralization have gained more prominence [2], Napoleon et al (2006). Geological Survey of India has done exploration work for Gold and associated mineral-bearing tracts and sulphide mineralization. The detailed study has been carried out in Nagavi area especially on petrology and mineralogy, and it is showing good results as it concerned with gold values of rock samples.

2 Location and Accessibility

Nagavi study is located in the northern part of the Gadag schist belt (GSB) and about 6 km south of Gadag taluk and district. Nagavi area is bound by latitudes from 75° 35′ 21.15" to 75° 38′ 13.20" E and latitudes from 15° 20′ 25.14" to 15° 24′ 9.65" N in top sheet 48 M/11 of 1:50,000 scale. The total study area is about 13 sq km. The study area is about 350 km by road towards northwest Bangalore and well connected by railway and road. The study area Nagavi is well served by all-weather roads including the National Highway southeast of Mulugunda (Fig. 1).

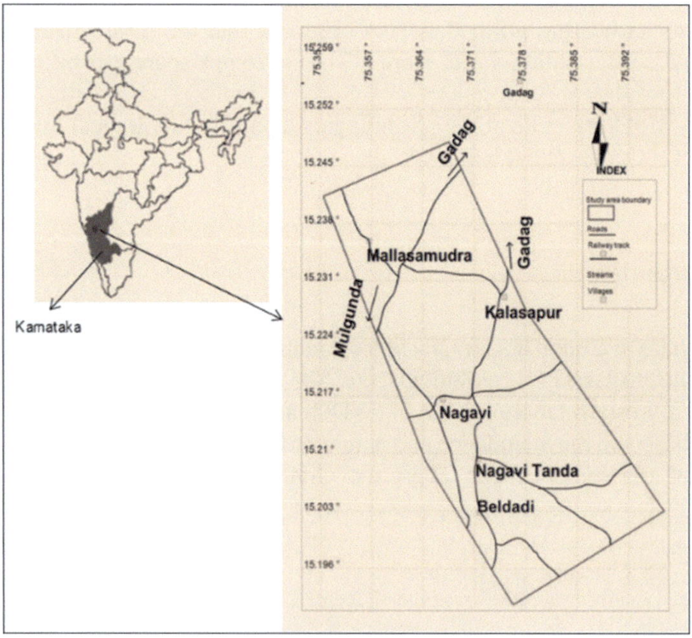

Fig. 1 Location map of Nagavi area

3 Objectives

The following are the main objectives:

(1) Detailed geological mapping of the study area,
(2) To study the various lithounits through geochemistry to understand the gold mineralization,
(3) To study the different geological structures to identify the auriferous shear zones, and
(4) To study rockchip samples through petrological microscope to understand gold mineralization

3.1 Scope

There is limited work carried out particularly on gold mineralization in Nagavi area. The geochemistry and petrology plays a vital role to identify the auriferous sheared zones. Therefore, there is an urgent need for detailed research in Nagavi area.

4 Geology of Nagavi Area

The gold and other mineralisation are localized in the reef at the tope of the hill of Nagavi area. The gold mineralization is localized in the reef at the top of the hill of Nagavi area, which is well defined by the BIFs prominently forming the ridges [3] (Fig. 2). The alternating silica and iron-rich layers in Banded Iron Formations [2, 4, 5]. These different layers are observed in the field from millimetre scale to metre scale bands (Fig. 16). The Gadag schist belt consists of a 2000-m-thick pile of meta-volcanic and meta-sediments (Chakrabarti et al. 1993 and 1994) and a Banded Iron Formation [6]. The structural disposition of the belt is the result of an overall E-W compressional regime with uplift and diaperism of the sialic basement within which the N-S trending Archaean shear systems have caused buckling and refolding of earlier fold belts, making all the linear elements parallel to the direction of shear (Fig. 3) [1].

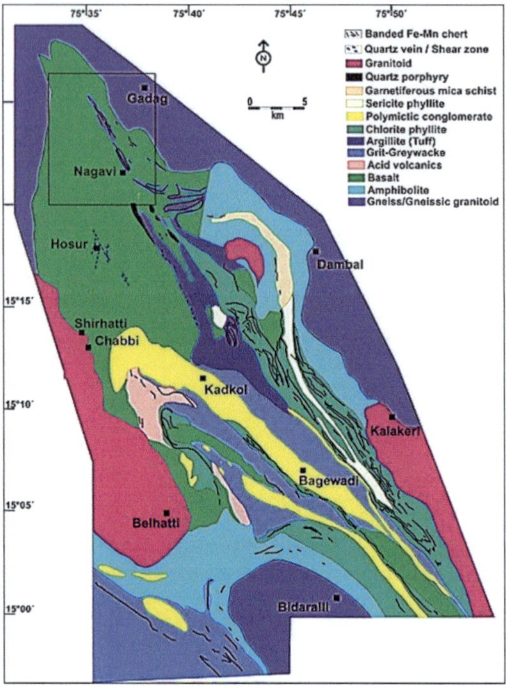

Fig. 2 Geological map of Gadag Schist Belt. *Source* GSI, Gold ores in India (1994)

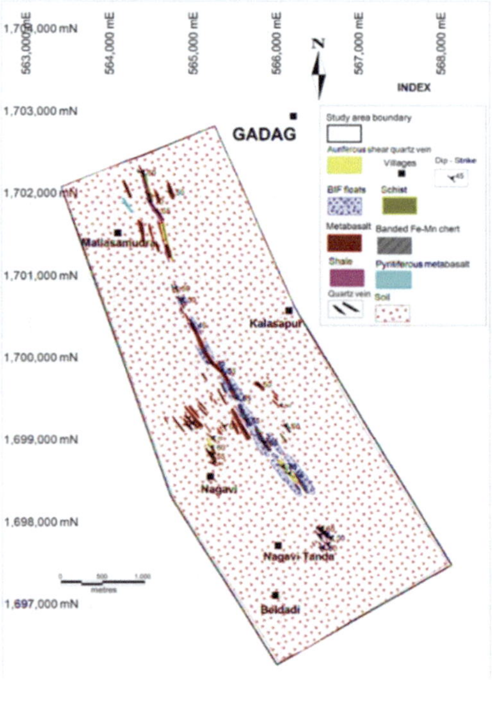

Fig. 3 Geological map of Nagavi study area [5]

5 Methodology

5.1 Field Techniques

Geological mapping of the area has been carried out by traversing (200 m interval) in field across regional strike of the area along with a Global Positioning System (GPS) (Garmin made Map 62S model with 3–4 m accuracy) and the compass. Noted down all the lithounits and rock exposure, the litho-contacts, rock exposure, sampling rock exposure trends (Dip and Strike), sheared quartz zones, and also the sulphide zones and descriptions are recorded during the geological traverses.

5.2 Rockchip Sampling

During field visit, has been we have collected four rockchip samples. The sherred quartz vein and outcrop samples were also collected from the Nagavi area at various locations and packed properly with sealed cover, and submitted to GSI Lab, Bangalore.

5.3 Quality Control During Sampling and in Laboratory

In the field, we collected samples at fresh face of rocky area to avoid contamination for sampling. During geological traversing in the field, we carefully observed the litho-contacts and noted down it systematically. Systematic sample preparation has been done in the laboratory to avoid contamination.

5.4 Rock-Chip Sample Analysis for Gold Mineralization

The rock samples tests are performed in GSI Lab Bangalore under Carl Zeiss / Nikon binocular petrological microscope for the mineralogy, texture, alteration, and photogrammetry. The rock chip samples have been collected from shear zones, trial pits, and sheared quartz veins with BIF contacts, and these samples were analysed by Graphite Tube Atomizer (GTA) and Atomic absorption spectroscopy (AAS) method.

6 Petrography and Mineralogy

6.1 *Banded Magnetite/Hematite Chert/Quartzite*

The study area consists of Banded Iron Formation (BIF), metabasalt, schist, sheared quartz veins, shale, and carbonated sheared anorthosite. The boundary between the ferruginous and siliceous layers is very thin and even on a microscopic scale. The alternate bands of hematite, magnetite, and chert are identified. The banded magnetite/hematite chert/quartzite exposure (Sample No. NNB 1) (Fig. 4). Under megascopic examination, it is compact and greyish brown colour. The quartz-rich bands are highly recrystallized, equigranular (~ 100 μm) with triple-point junctions (Fig. 4). This quartz vein is intruded within BIF formation and appears grey, indicating the Au mineralization [7] as shown in Fig. 5.

Fig. 4 Photomicrograph showing feldspar porphyroblasts and recrystallized feldspar and quartz (under plane polarized)

Fig. 5 Photomicrograph showing feldspar porphyroclasts, chlorite and pyroxene (under plane polarized)

6.2 Metabasalt

These metabasalt exposures are trending NW–SE direction N 500–550 W–S 500-550E and dipping towards NE direction (N45^0–50^0 E). At some places, fine-grained quartz veins are cutt across the metabasalt. The appearance of chlorite in the metabasalts and sericite in the greywackes/argillites implies the regional grade of metamorphism or the green-schist facies metamorphism (Fig. 9). The sheared carbonated metabasalt rock exposure (Sample No. NNB 13) near Mallasamudra has been examined under megascopic study, and it is compact, weathered, greenish-grey coloured, fine to medium-grained, and equigranular. Under the microscope, it consists of medium-sized subhedral grains of twinned to un-twinned plagioclase (~ 40%), anhedral carbonate (~ 40%), and actinolite mainly altered to chlorite (~ 17%) with minor opaque (~ 3–5%). Carbonate appears to be after pyroxene. Carbonate also occupies fracture in plagioclase and rarely forms bands parallel to tectonic foliation. The rock is sheared with the development of feldspar porphyroblasts and recrystallized feldspar and quartz as shown in Figs. 6 and 7.

6.3 Carbonated Sheared Anorthosite

Microscopic study of the polished thin section of sulphides bearing carbonated sheared anorthite samples shows sulphide assembles of pyrite and arsenopyrite (Figs. 6 and 7). Thin section shows mutual boundary.

The carbonated sheared anorthosite rock exposure near Mallasamudra and the same rock sample (sample number NNB-10) has been collected and properties are studied. The rock is very fine to medium-grained and in equigranular. It consists of simple to polysynthetically twinned subhedral (100–200 by 400–600 μm) laths (~40–45%) set in a groundmass of irregular anhedral grains of secondary carbonate

Fig. 6 Photomicrograph exhibiting plagioclase with pyrites (under plane polarized)

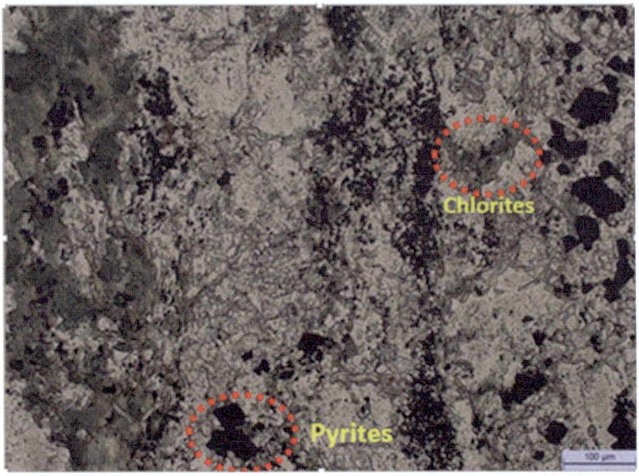

Fig. 7 Photomicrograph showing chlorite and opaque pyrite crystals (under plane polarized)

Fig. 8 Photograph showing fine-grained metabasalt with disseminated pyrites

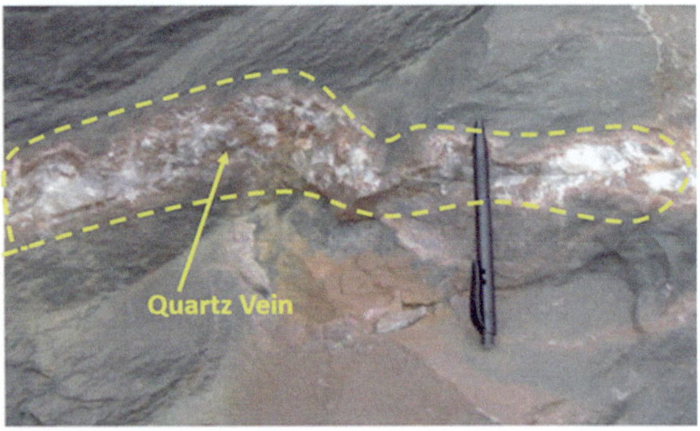

Fig. 9 Photograph showing quartz vein within metabasalt (photo facing East–West direction)

(~35–40%), chlorite (~8–10%), opaque (~5–7%), and accessory epidote and quartz. Opaque mainly occurs as euhedral to subhedral, square to rhombus shaped 100 by 100–150 by 250 μm randomly distributed discrete grains. Tectonic foliation is defined by the presence of recrystallized feldspar, bent plagioclase lamellae, and tapering deformation twins in plagioclase as shown in Fig. 8, 9, 10 and 11. Pyrite occurs as characteristic cubic crystals and as clusters of small grains. Coarse grains are also reported by Rao et al. [8].

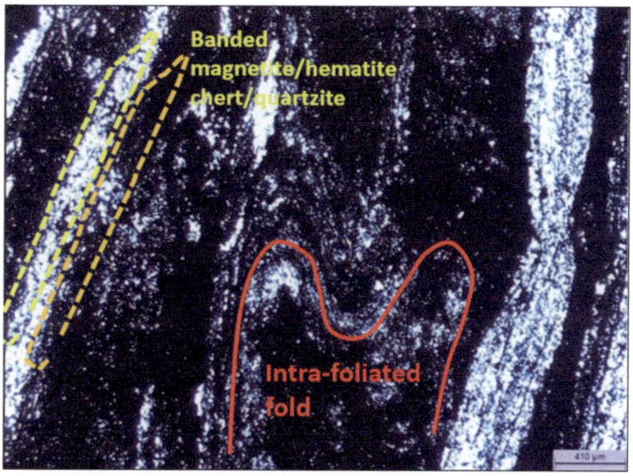

Fig. 10 Photomicrograph exhibiting banded magnetite/hematite chert/quartzite exhibiting intrafolial fold (under plane polarized)

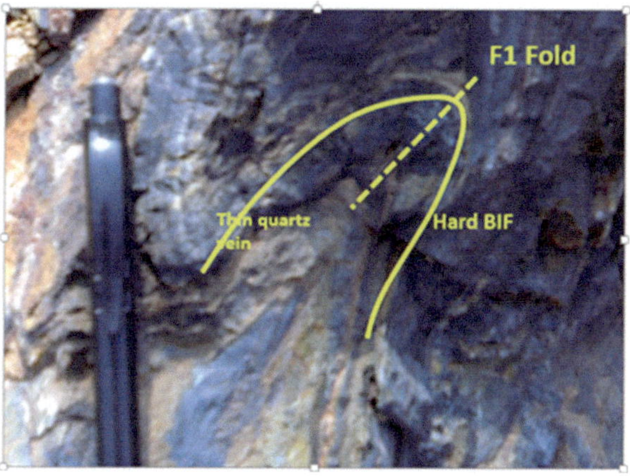

Fig. 11 Photograph showing outcrop of folded banded magnetite/hematite chert/quartzite

Fig. 12 Sheared/fractured milky quartz veins intruded within BIF (photo facing NW direction)

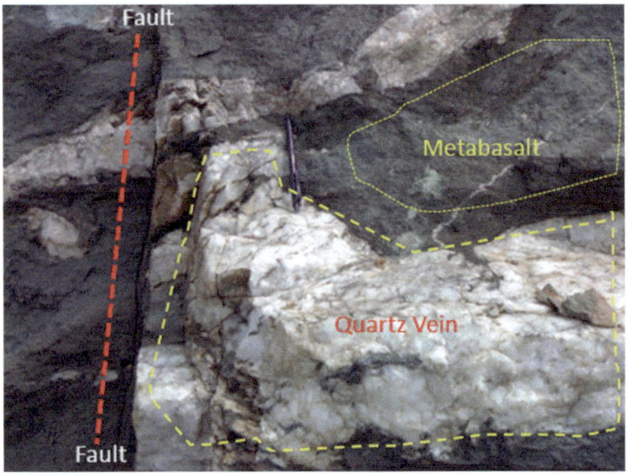

Fig. 13 Minor faults across and along the strike in metabasalt (photo facing NW direction)

7 Rock Sample Analysis Results

The various styles of gold mineralisation in BIF have led to the proposition of three different genetic models, namely, syngenetic, epigenetic, and a hybrid of both the epigenetic and syngenetic (multistage) models for its deposition [9–11] Groves et al. (1987); Kishida (1987).

Four samples were collected analysed from different locations in the study area for Au by GTA-AAS method (Table 1). NNB 1sample was collected from top of the hill where quartz vein intruded into the BIF and gold analytical results showing < 22 ppb. Sample NNB 10 was collected near Mallasamudra village Carbonated Sheared Anorthosite (Pyrite grains) and gold analytical results showing < 26 ppb. BIF-hosted gold mineralization forms an integral part of many Archaean greenstone belts in countries such as Australia [12], Canada [13], South Africa [14], and Zimbabwe [10]. These results show positive signature for further exploration in the Nagavi area.

Table 1 Details of rockchip sample analysis

S. No	Sample No	*Au by GTA—AAS (in ppb)	Rock type	Sample location
1	NNB—1	< 22	Banded magnetite / hematite quartzite	Near Nagavi village on top of the hill and about 4 km from Gadag city towards SW)
2	NNB—10	< 26	Carbonated Sheared Anorthosite (Pyrite grains)	Near Mallasamudra village and 3 km from Gadag city
3	NNB—11	< 21	Banded magnetite/ hematite /quartzite	Near Nagavi village, about 4.5 km from Gadag city, towards SW on top of the hill
4	NNB—13	< 20	Sheared carbonated Metabasalt	Near Mallasamudra village and about 2.3 km from Gadag city toward SW on top of the hill

If pure gold particles are present, they show bright yellow colour and typical greenish tint under crossed nicols. Aurich arsenopyrite crystallizes at 200–2500 C, and Au-poor arsenopyrite crystallizes between 2500 and 5000 C [15]. The earliest ore mineral to form is pyrrhotite, a simple sulphide to iron (FeS) (Fig. 10 and 11).

8 Nagavi Structures

Geological structures play a vital role for gold mineralization in Nagavi area. The fold pattern can be observed in the photo-micrograph and field outcrop. The contacts between chert bands and quartz veins and chilled thin quartz veins across chert bands are observed. The banded magnetite/hematite chert/quartzite exposure is observed between Gadag and Haveri main road. The rock formation consists of compact, fine-grained, greyish brown coloured, and equigranular. These bands are mostly bounding, and one band shows intra-folial folds and also sheared / fractured milky quartz veins intruded within BIF Minor faults across and along the strike in met abased in Fig. 8 and Fig. 11.

The greyish white quartz vein with an associated wall-rock of predominantly BIF and sericitic nature may indicate a somewhat more meso-thermal type of mineralization than a deep-seated, high temperature, hypo-thermal type such as those of Kolar and Hutti [11, 16] Browning et al. (1987) versus Syngenetic origin [9].

9 Ore Localization in Nagavi Area

The auriferous shear zones especially ferrogounours quratz with interpreted as Banded Iron formats. The Au-bearing veins crosscut ferruginous quartzite which is interpreted as banded iron formations (BIFs). BIFs contain a major portion of the world's iron ores and especially in Archaean terrines, host gold mineralization [14, 17, 18].

10 Conclusions

In the Nagavi area, the general trend of the massive banded hematite quartzite with fractured quartz veins' intrusion indicates the solution entering into BIF with mineralization. The BIF sample has been examined under the microscope, and it is composed of alternate bands of quartz-rich and opaque-rich bands. The rock is compact and greyish brown in colour. In the microscopic examination, the quartz-rich bands consist of highly recrystallized equigranular (~ 40–50 μm) quartz grains and subordinates amount of opaque.

Acknowledgements Authors are thankful to Mr. Mahesh M. Karekoppa, Sr. Geologist, PPOD Division, Geological Survey of India, Southern Region, Bangalore, for providing the Petrography and Mineralogical studies facility. Authors are thankful to Dr. M. Basavanna, Professor (Retired), Department of Studies in Geology, Karnatak University, Dharwad. Authors are also thankful to Chairman and all faculties, Department of Studies in Geology, Karnatak University, Dharwad who has extended their required facilities to carry out this research work in the department.

References

1. Puranik SC (2011) Primary sedimentary structures in banded iron formations from Gadag schist belt, Karnataka, India. Int J Earth Sci Eng 4(2):111–121
2. Ugarkar AG, Panaskar DB, Ranganath Gowda G (2000) Geochemistry, petrogenesis and tectonic setting of metavolcanics and their implications for gold mineralisation in Gadag gold field Southern India. Gondwana Res 3(3):371–384
3. Agasnalli C, Chandrappa KD, Lakkundi TK (2021) A study of primary sedimentary structures in Banded iron formations (BIFs) of Nagavi area, Gadag schist belt, Karnataka India. J Mines Metals Fuels 69:11
4. Agasnalli C, Basavanna M, Lakkundi TK (2015) Geochemistry of Banded iron formations and associated gold mineralisation of Nagavi, Gadag Schist Belt, Karnataka. Int J Earth Sci Eng 8(4):1880–1886
5. Agasnalli C, Basavanna M, Lakkundi TK (2014) Textural and fluid inclusions studies on banded iron formations (BIFs) hosted gold mineralisation in the Nagavi area Gadag schist belt, Dharwar craton. Int J Earth Sci Eng 07(04):1593–1600
6. Radhakrishna BP, Vaidyanadhan R (1997) Geology of Karnataka. Geol Soc India Bangalore, p 354

7. Agasnalli C, Deepak MS, Lakkundi TK, Ajey Kumar VG (2022) An integrated study of landsat ETM and cartosat DEM data in identification of banded iron formations (BIFs) associated with sulphide mineralization, Acta Geodyn Geomater, 19;1 (205). https://doi.org/10.13168/AGG.2021.0041
8. Bhaskar Rao YJ, Sivaraman TV, Pantulu GVC, Gopalan K, Naqvi SM (1992) Rb-Sr ages of late Archaean metavolcanics and granites, Dharwar craton, South India and evidence for early Proterozoic thermotectonic events. Precambr Res 59:145–170
9. Fripp REP (1976) Strata-bound gold deposits in archaean banded iron fomation, Rhodesia. Econ Geol 71:58–75
10. Saager R, Oberthür T, Tomschi H (1987) Geochemistry and mineralogy of Banded iron formation—hosted gold mineralization in the Gwanda Greenstone Belt, Zimbabwe. Econ Geol 82:2017–2032
11. Phillips GN, Groves DI, Martyn JE (1984) An epigenetic origin for archean banded iron formation-hosted gold deposits. Econ Geol 79:162–171
12. Vielreicher RM, Groves DI, Ridley NJ, McNaughton NJ (1994) A replacement origin for the BIF-hosted gold deposit at Mt. Morgans, Yilgarn block, W.A. Ore Geol Rev 9:325–347
13. Lhotka PG, Nesbitt BE (1989) Geology of unmineralized and gold bearing iron formation, Contwoyto lake point lake region, Northwest territories, Canada. Can J Earth Sci 26:46–64
14. Pretorius AI, van Reenen DD, Barton JM Jr (1988) BIF-hosted gold mineralization at the fumani mine, Sutherland greenstone belt South Africa. S Afr J Geol 91:429–438
15. Cathelineau M, Boiron MC, Hollinger P, Marion P, Denis M (1988) Gold in arsenopyrites, crystal chemistry, location and state, physical and chemical conditions of deposition. Econ Geol Monograph 6:328–341
16. Beeraiah MB, Sengupta S (2001) Venkateswamy; Ramachandran T. V Exploration for gold in sulphidic banded magnetite chert of Nagavi-Mallasamudra area, Gadag schist belt. Spec Publ Series Geol Surv India 58:339–343
17. Gilligang JM, Foster RP (1987) Gold mineralisation in iron-formation: the importance of contrasting modes of deformation at the Lennox Mine, Zimbabwe. In: African mining, inst. of mining and metallurgy, London, pp 127–138
18. Groves DI (1993) The crustal continuum model for late-archaean lode-gold deposits of the Yilgarn Block, Western Australia. Miner Depos 28:366–374
19. Hammond NQ, Moore JM (2006) Archaean lode gold mineralisation in Banded Iron formation at the Kalahari Goldridge Deposit. Kraaipan Greenstone Belt South Afr Miner Deposita 41:483–503. https://doi.org/10.1007/s00126-006-0074-6
20. Chadwick B, Vasudev VN (1996) Nazeer Ahmed: The Sandur schist belt and its adjacent plutonic rocks implications for late Archaean crustal evolution in Karnataka. J Geol Soc India 47:37–57
21. Naqvi SM (1997) The Sandur schist belt and its adjacent plutonic rocks: implications of late Archean crustal evolution in Karnataka. J Geol Soc India 49(4):459–460
22. Ram Mohan M, Sarma DS, Charan SN, Balaram V, Rajasekhar VB, Ahmad T (2008) Geochemistry and petrogenesis of amphibolites from the Southern part of Gadag Greenstone Belt, Karnataka. J Geol Soc India 72:484–494
23. Chakabarti S, Reddy US, Natarajan WK (1993) Sedimentary structures in the Archaean sediments of Gadag schist belt Karnataka. J Geol Soc India 41:523–528
24. Chakabarti S (1994) Ugarkar, Devraju, Sedimentary structures in the Archaean sed iments of Gadag schist belt Karnataka. J Geol Soc India 41:523–528
25. Brown PE, Lamb WM (1988) P-V-T properties of fluids in the system H2O-CO2-NaCl: new graphical presentation and implications for fluid inclusion studies. Geochim Cosmochim Acta 53:1209–1221

Influence of Metakaolin on Self-Compacting Geopolymer Concrete with GGBFS at 8 Molarity Concentration

B. R. Arun, K. Kavyashree, Naveen Revanna, and V. Hemanth Kumar

Abstract Ordinary Portland cement (OPC) production raises environmental concerns due to high energy consumption and CO_2 emissions. Geopolymer concrete (GC) emerges as a sustainable alternative utilizing industrial by-products like Ground Granulated Blast Furnace Slag (GGBFS) and Metakaolin (MK) activated by alkaline solutions. This study investigates Self-Compacting Geopolymer Concrete (SCGC) development using GGBFS/MK blends and superplasticizers for improved workability. Fresh properties were evaluated through Slump flow, V-Funnel, T-50 slump flow, and L-Box tests. Compressive strength served as the primary indicator of mechanical properties. Metakaolin content influenced workability, with a notable decrease in slump flow at a 30% replacement rate. Water absorption tests revealed variations in concrete quality based on GGBFS/MK ratios. At 90 days, higher replacement of metakaolin resulted in decreased compressive strength of around 12.63% for mix 4, whereas water absorption decreased with lower replacement levels of metakaolin, i.e., around 1.8% for mix 1. The study highlights the potential for enhanced strength through denser microstructure and reduced absorption. It contributes to the development of sustainable construction materials while promoting improved concrete performance for various applications.

Keywords Self-compacting geopolymer concrete (SCGC) · Super plasticizer (S.P) · Ordinary portland cement (OPC) · Manufactured sand (M-Sand) · Fluid to binder ratio (F/B) · Ground granulated-blast furnace slag (GGBFS) · Molarity (M) · Metakaolin (MK) · Water absorption

---

B. R. Arun · N. Revanna · V. Hemanth Kumar
Department of Civil Engineering, Dr. AIT, Bangalore, Karnataka, India

K. Kavyashree (✉)
Department of Civil Engineering, GPT, Mulabagilu, Karnataka, India
e-mail: kavyagptk@gmail.com

Y. K. Guruprasad et al. (eds.), *Recent Advances in Infrastructure Development and Management—Volume 2*, Lecture Notes in Civil Engineering 566,
https://doi.org/10.1007/978-981-96-1558-2_9

1 Introduction

1.1 Environmental Concerns and the Need for Sustainable Alternatives

The construction industry plays a vital role in global development, but its reliance on ordinary Portland cement (OPC) raises significant environmental concerns. OPC production is highly energy-intensive, contributing a staggering 8% of global CO_2 emissions [1]. This has fueled a pressing need for eco-friendly alternatives in construction materials, prompting researchers to explore innovative solutions with a lower environmental footprint [1].

1.2 Geopolymer Concrete: A Promising Path Forward

Geopolymer concrete (GC) emerges as a promising alternative to OPC, offering superior mechanical strength and durability while boasting a more sustainable profile [2]. GC harnesses industrially derived products rich in silicate materials, such as Ground Granulated Blast Furnace Slag (GGBFS) and Metakaolin (MK), as its primary binding agents. This significantly reduces the environmental impact associated with traditional cement production, aligning with the growing demand for sustainable construction practices [2].

1.3 Research Gap and Objectives: Enhancing Workability in SCGC

While GC offers a compelling solution, workability remains a challenge compared to OPC. Self-compacting concrete (SCC), a revolutionary advancement pioneered in Japan, addresses workability and durability concerns in traditional concrete while maintaining high strength [3]. However, a crucial gap exists in the comprehensive understanding of developing Self-Compacting Geopolymer Concrete (SCGC) utilizing GGBFS/MK blends and superplasticizers for enhanced workability.

This study aims to bridge this knowledge gap by investigating the influence of various GGBFS/MK combinations and percentages on the fresh and hardened properties of SCGC. We employ established testing methods such as slump flow, V Funnel, T-50 slump flow, and L-Box to comprehensively evaluate the workability and flow characteristics of the developed concrete [4, 5]. Additionally, compressive strength testing, a fundamental indicator of concrete performance, is conducted to assess the hardened properties [6].

1.4 Significance: Paving the Way for a Sustainable Future

This research on 8 M SCGC development aspires to contribute significantly to the field of sustainable construction materials. Through rigorous testing and analysis, we aim to demonstrate how this innovative concrete formulation, combining GGBFS/MK blends and superplasticizers, can address the critical issues of CO_2 emissions, durability, and environmental impact associated with traditional cement production. The findings of this study have the potential to influence how we perceive and utilize construction materials, paving the way for a more eco-friendly and sustainable future in the building industry.

2 Literature Review

2.1 Environmental Concerns and the Rise of Geopolymer Concrete

The construction industry's reliance on ordinary Portland cement (OPC) poses a significant environmental challenge. OPC production is highly energy-intensive and a major contributor to global warming, accounting for a staggering 8% of global CO_2 emissions [1]. This has fueled a pressing need for eco-friendly alternatives in construction materials, prompting researchers to explore innovative solutions with a lower environmental footprint [1].

2.2 Geopolymer Concrete: A Sustainable Path Forward

Geopolymer concrete (GC) emerges as a promising alternative to OPC, offering superior mechanical strength and durability while boasting a more sustainable profile [2]. GC harnesses industrially derived products rich in silicate materials, such as Ground Granulated Blast Furnace Slag (GGBFS) and Metakaolin (MK), as its primary binding agents. This significantly reduces the environmental impact associated with traditional cement production, aligning with the growing demand for sustainable construction practices [2].

2.3 Shifting Focus: Enhancing Workability in SCGC

While GC offers a compelling solution, workability remains a challenge compared to OPC. Self-compacting concrete (SCC), a revolutionary advancement pioneered in Japan, addresses workability and durability concerns in traditional concrete

Table 1 Summary of existing research on SCGC

Study	Key findings	Focus area
Davidovits [7] Global warming impact on the cement and aggregates industries	OPC production is a major contributor to global warming	Strengthens the need for sustainable alternatives (not SCGC specific)
Provis and Van Deventer [8] Geopolymers: structure, processing, properties and industrial applications	Comprehensive exploration of geopolymers	Provides a foundation for understanding geopolymer applications in SCGC (not specific research)
Bernal [9] Production of geopolymer foams from metakaolin	Metakaolin-based geopolymers have potential in sound insulation	Focuses on a specific application (geopolymer foams) and not SCGC properties
Duxson et al. [10] Recent trends in geopolymer technology	Provides an overview of state-of-the-art geopolymer technology	Informs the approach to SCGC development (background information)
Khan et al. [11] Properties of fly ash-based geopolymer mortar	Fly ash-based geopolymer mortars show good strength and durability	Explores a different geopolymer material (fly ash) and not GGBFS/MK blends
Other research	Explored impact of metakaolin on various geopolymer concrete aspects	Provides a broader understanding of metakaolin's influence, often limited to specific properties

while maintaining high strength [3]. Combining these two innovations—geopolymer concrete and self-compacting properties—is a promising path forward for sustainable construction.

2.4 Critical Analysis of Existing Research on SCGC

Several studies have explored the development of Self-Compacting Geopolymer Concrete (SCGC) utilizing GGBFS/MK blends and superplasticizers [4–6]. However, a critical analysis reveals a gap in comprehensively understanding the influence of these combinations and percentages on both fresh and hardened properties. Existing research tends to focus on specific aspects of SCGC, as summarized in Table 1.

2.5 Gaps in Knowledge and the Current Study's Focus

As observed in Table 1, existing research primarily focuses on specific aspects of SCGC, such as optimizing strength development or flow characteristics. There is a lack of comprehensive studies investigating the combined influence of GGBFS/MK ratios and percentages on both the fresh and hardened properties of SCGC.

This study aims to bridge this knowledge gap by investigating the influence of various GGBFS/MK combinations and percentages on a comprehensive range of fresh and hardened properties of SCGC. We employ established testing methods like slump flow, V Funnel, T-50 slump flow, L-Box, and compressive strength testing to comprehensively evaluate the performance of the developed concrete [4–6].

Through rigorous testing and analysis, we aim to establish a more holistic understanding of how these GGBFS/MK blends and superplasticizers influence the workability, strength, and overall performance of SCGC. The findings of this study will contribute to the development of optimized and sustainable SCGC formulations for the construction industry.

3 Experimental Studies

3.1 Materials

Ground Granulated Blast Furnace Slag (GGBFS). The GGBFS employed in this study originates from RMC Ready Mix (India), located in the Kumbalgod Industrial Area, Bangalore, Karnataka. This material exhibits a particle size distribution that closely aligns with that of cement and conforms to the specifications outlined in IS: 12089-1987 [13].

Metakaolin. The metakaolin utilized in this research was sourced from Shree Ram Minerals in Gujarat, India. It is a highly reactive, amorphous aluminosilicate pozzolana characterized by its white color and ability to form stable hydrates with lime and water, thereby imparting hydraulic properties to the mortar. The metakaolin possesses a specific gravity of 2.01, a surface area of 1200 m^2/kg, and a bulk density of 390 g/cm^3. Its chemical composition primarily consists of 52% SiO_2, 42.2% Al_2O_3, 0.7% Fe_2O_3, and a loss on ignition (LOI) of 0.3% by mass.

Preparation of Alkaline Activator. The alkaline activator employed in this study is an 8 M solution prepared by combining sodium hydroxide (NaOH) and sodium silicate solution (Na_2SiO_3). The NaOH component consists of 99% purity, industrial-grade sodium hydroxide pellets with a specific gravity of 2.13. A total of 320 g of NaOH solids are dissolved per liter of water to achieve the desired 8 M concentration. The sodium silicate solution exhibits a 2.06 SiO_2/Na_2O ratio and a pH of 12. The specific properties of the sodium silicate solution are as follows: specific gravity of 1.55, viscosity of 74 s, Na_2O content of 14.34% by weight, SiO_2 content of 29.55% by weight, and water content of 56.11% by weight. To facilitate the heat-generating polymerization process, the mixed alkaline solution is allowed to stand for 20 min prior to use.

Superplasticizer (SP). This study employs Master Glenium SKY 823 as the superplasticizer. It is a light brown, liquid admixture with the following properties:

Relative density	1.08 ± 0.01 at 25 °C.
pH	> 6.
Chloride ion content	< 0.2%

Fine Aggregate. The fine aggregate utilized in this research is M-SAND sourced from Tavara Mines and Minerals located in the Jigani Industrial Area, Anekal Taluk, Bangalore, Karnataka. Physical testing confirms that the M-SAND complies with the requirements for "Zone 2" grade as specified in IS: 383-1970 (Reaffirmed 2011) [12].

Coarse Aggregate. The coarse aggregate employed in this study was sourced from a local supplier located on Magadi Road, Bangalore, Karnataka. Its physical properties comply with the relevant Indian standards.

3.2 Specimen Preparation, Casting, and Curing

The preparation of the fresh SCGC involved dry mixing of GGBFS, metakaolin, and manufactured sand in predetermined proportions. Subsequently, coarse aggregates in a saturated surface dry (SSD) condition were added and blended for approximately 2.5 min to achieve a uniform distribution. A well-shaken liquid mixture (referred to as F/B) was then introduced. This F/B solution comprised the alkaline activator, superplasticizer, and additional water. The mixture was further blended for at least 3 min to ensure homogeneity.

To evaluate the workability of the fresh SCGC, various tests were conducted following the guidelines established by EFNARC [13]. These tests included slump flow, V-funnel, L-box, and T50 slump flow time. For the compressive strength and water absorption tests, cubic specimens measuring 100 mm × 100 mm × 100 mm were cast. To facilitate testing, the fresh concrete was briefly hand-mixed again and poured into molds without any compaction effort. Following a 24-h oven curing process at 70 °C alongside the wooden molds (as depicted in Fig. 1a), the samples were transferred to a curing room maintained at ambient temperature conditions testing as depicted in Fig. 1b.

3.3 Mix Proportions

This study employed four concrete mixtures designated as M_1, M_2, M_3, and M_4. As detailed in Table 2, the proportion of Ground Granulated Blast Furnace Slag (GGBFS) was systematically varied by replacing it with metakaolin at 0, 10, 20, and 30% by weight, while maintaining a constant 8 Molar alkaline activator solution.

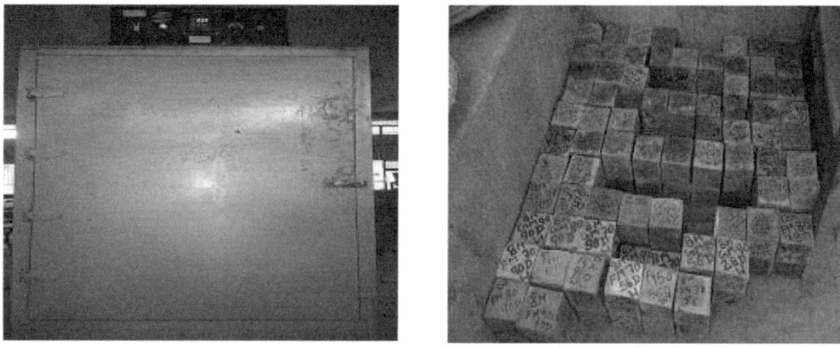

(a) Oven curing at 70 ⁰C [14-18] (b) Ambient Curing [14-18]

Fig. 1 **a** Oven curing at 70 °C [14–18] **b** Ambient curing [14–18]

Table 2 Details of different mixes with respective molarities

Mix	Binder components	GGBFS (Kg/m^3)	MK (Kg/m^3)	M-Sand (Kg/m^3)	Coarse aggregate (Kg/m^3)	(FB) Ratio	Molarity (M)	SP (%)	Extra water (%)
M_1	GGBFS 100% and MK 0%	512	0	872	706	0.37	8	12	6
M_2	GGBFS 90% and MK 10%	460.8	38.2	872	706	0.37	8	12	8
M_3	GGBFS 80% and MK 20%	409.6	76.4	872	706	0.37	8	12	10.5
M_4	GGBFS 70% and MK 30%	358.4	114.6	872	706	0.37	8	12	12.5

4 Test Procedures

4.1 Workability Test

The fresh properties of the SCGC mixtures were evaluated based on the three key characteristics of SCC defined by the EFNARC guidelines [13] filling ability, passing ability, and segregation resistance. In this study, these characteristics were assessed using four standardized tests: slump flow, T50 slump flow time, V-funnel, and L-box.

4.2 Compressive Strength Test

Compressive strength testing was conducted on 100 mm × 100 mm × 100 mm cubic specimens in accordance with IS 516-1959 [19]. The testing employed a digital 2000 kN compression testing machine (CTM). Specimens were tested at ages of 3, 7, 28, 56, and 90 days. The compressive strength value for each mixture was determined as the average of three specimens tested.

4.3 Water Absorption Test

Water absorption testing of the SCGC mixtures was conducted on 100 mm × 100 mm × 100 mm cubic specimens, following the guidelines of ASTM C642-82 [20]. The fresh concrete was cast into molds without any compaction effort and oven-cured at 70 °C for 24 h. Subsequently, the specimens were transferred to a room temperature environment for further ambient curing at designated periods: 28, 56, and 90 days. The percentage water absorption of each mixture was evaluated at two specific time points: 30 min and 24 h.

5 Results and Discussion

5.1 SEM Analysis of Eco-friendly Materials

Scanning Electron Microscope (SEM) and XRD of GGBFS. SEM micrographs of the GGBFS (see Fig. 2a) reveal the presence of particles with angular, flaky, and elongated morphologies. Figure 2b illustrates the significant amorphous phase containing calcium, silica, and alumina as identified by X-ray diffraction (XRD) analysis of the GGBFS.

Scanning Electron Microscope (SEM) and XRD of Metakaolin. SEM micrographs of the powdered metakaolin (see Fig. 3a) show an amorphous morphology. The corresponding XRD pattern (see Fig. 3b) exhibits broad humps characteristic of amorphous phases in the 20° to 35° 2θ range, along with sharper peaks indicative of crystalline phases such as microcline, quartz, and illite.

5.2 Fresh Properties and Test Results of SCGC

Slump Flow Test. Analysis of the GGBFS and metakaolin particles revealed contrasting morphologies: GGBFS exhibited a flaky and elongated shape with a

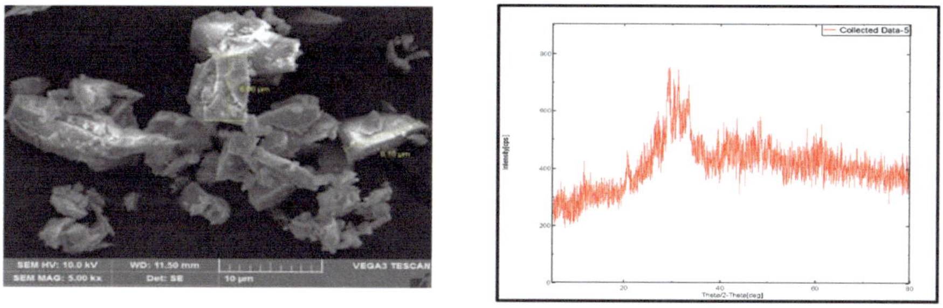

(a) SEM image of GGBFS (b) XRD Diffract gram patterns of GGBFS

Fig. 2 **a** SEM image of GGBFS **b** XRD Diffract gram patterns of GGBFS

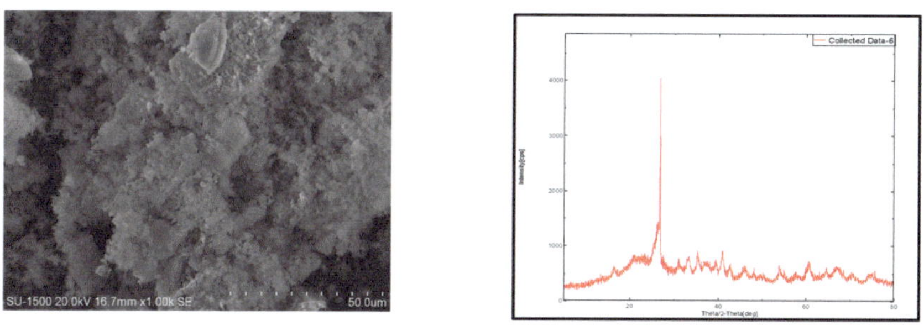

(a) SEM image of Metakaolin (b) XRD Diffractogram patterns of Metakaolin

Fig. 3 **a** SEM image of metakaolin **b** XRD Diffractogram patterns of metakaolin

large surface area (see Fig. 2), while metakaolin appeared amorphous with a similarly significant surface area (see Fig. 3). To achieve the desired flow characteristics in the SCGC mixtures, a constant superplasticizer dosage of 12% was employed. The remaining water content was then adjusted through an iterative trial-and-error process to optimize workability. As depicted in Fig. 4, the slump flow results exhibited the lowest values for the mixture with 30% metakaolin replacement (M_4). Additionally, compared to the control mix (GGBFS100, M_1), a decrease in slump flow of 3.6%–4.3% was observed for mixtures with 10% and 30% metakaolin replacement (M_2 and M_3), respectively, across the 8 Molar alkaline activator solution. These findings highlight the influence of the GGBFS/metakaolin ratio on the workability of the SCGC, as evidenced by the variations in slump flow observed for reference mixes M_1, M_2, M_3, and M_4.

T50 Slump Flow. A consistent trend was observed between slump flow and T50 slump flow time (see Fig. 5). As slump flow decreased, the T50 slump flow time correspondingly increased. Notably, the T50 values for the 30% metakaolin replacement mix (M_4) exceeded the measurable range. Compared to the control mix (GGBFS100, M_1), the T50 slump flow time increased by 0% to 2% and 8% for the 10% and 30%

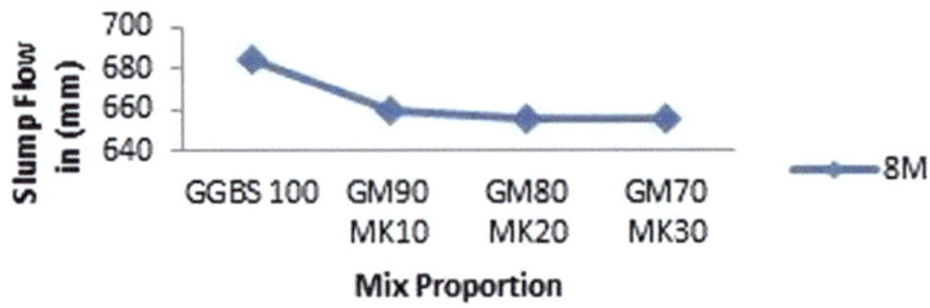

Fig. 4 Variation of slump flow w.r.t 8 molarity

metakaolin replacement mixes (M_2 and M_3), respectively. These findings further support the influence of the GGBFS/metakaolin ratio on the workability of SCGC.

V-Funnel Test. The V-funnel test assesses the ability of the SCGC to fill and flow through a continuously decreasing section without segregation or blockage. As observed in Fig. 6, the V-funnel flow time increased with increasing metakaolin content. Compared to the control mix (GGBFS100, M_1), the V-funnel flow time exhibited increases of 5.2% and 16.15% for the 10% and 30% metakaolin replacement mixes (M_2 and M_3), respectively.

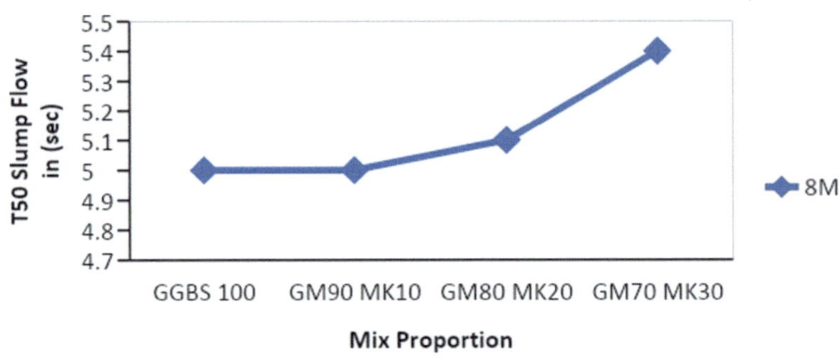

Fig. 5 Variation of T50 slump flow w.r.t 8 molarity

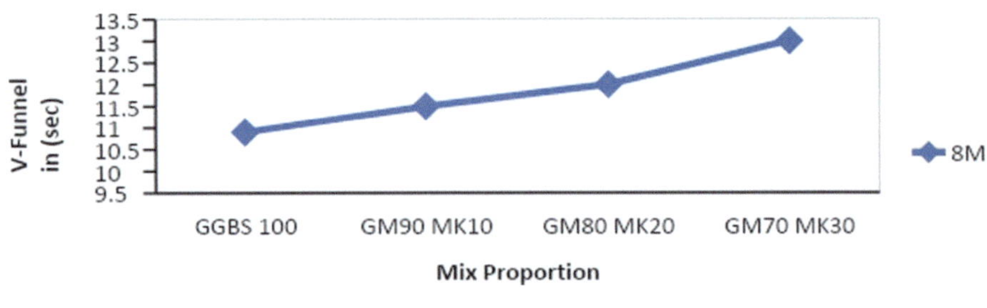

Fig. 6 Variation of V-Funnel test w.r.t different Molarities

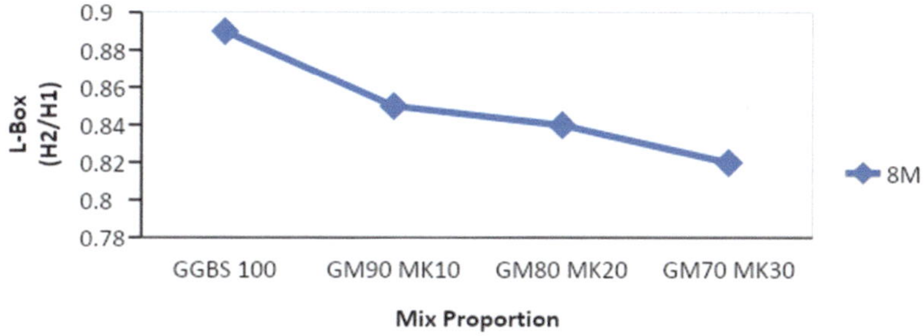

Fig. 7 Variation of L-Box test w.r.t 8 molarity

L-Box. The L-box test results (see Fig. 7) demonstrate a decreasing trend in the blocking ratio (H_2/H_1) with increasing metakaolin and sodium hydroxide concentrations. This trend aligns with the observations from the slump flow and V-funnel tests, suggesting similar underlying mechanisms influencing workability. Compared to the control mix (GGBFS100, M_1), the L-box blocking ratio decreased by 4.50% and 7.0% for the 10% and 30% metakaolin replacement mixes (M_2 and M_3), respectively.

5.3 *Strength Performance Studies*

Compressive Strength Test. The average compressive strength of the SCGC mixtures is presented in Fig. 8. As observed, an increase in the metakaolin replacement percentage led to a decrease in compressive strength. The 30% metakaolin replacement mix (M_4) exhibited the lowest compressive strength at all curing ages. Compared to the control mix (GGBFS100, M_1), the percentage decrease in compressive strength after 90 days curing for the 10%, 20%, and 30% metakaolin replacement mixes (M_2, M_3, and M_4) was approximately 2.10%, 4.21%, and 12.63%, respectively.

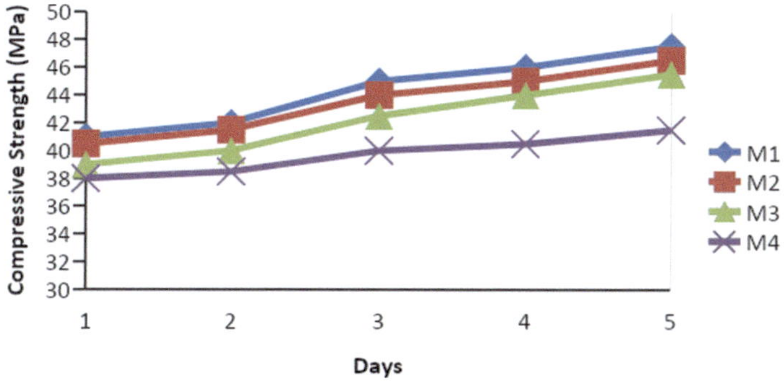

Fig. 8 Variation of compressive strength with binder composition

5.4 Durability Performance Studies

Water Absorption Test. Table 3 shows concrete society initial absorption criterion and the average findings of GGBFS-metakaolin SCGC specimens.

Table 3 presents the water absorption data for the SCGC mixtures. Analyzing the initial absorption results, it is evident that all mixtures except M_4 (30% metakaolin replacement at 8 Molarity after 28 days) exhibited values below 3%. This indicates good quality concrete based on initial absorption.

For the 24-h water absorption results, all mixtures except the control mix (GGBFS100, M_1) displayed values within the range of 3–5%. The 30% metakaolin replacement mix (M_4) exceeded 5% at both 28 and 56 days. However, during the 90-day observation period, the water absorption for M_4 also decreased to below 5%, placing it within the category of average quality concrete.

As illustrated in Fig. 9, a trend is observed where decreasing GGBFS content (increasing metakaolin replacement) corresponds with a decrease in compressive strength and an increase in water absorption. This trend is particularly evident in the 30% metakaolin replacement mix (M_4), which exhibits the highest water absorption values.

Table 3 Water absorption values

Mx	28 days		56 days		90 days	
	Initial absorption @ 30 min (%)	24 h Water absorption (%)	Initial absorption @ 30 min (%)	24 h Water absorption (%)	Initial absorption @ 30 min (%)	24 h Water absorption (%)
M_1	1.72	2.96	1.6	2.58	1.25	1.8
M_2	3.04	3.47	2.57	3.46	2.55	3.4
M_3	3.12	4.58	2.75	3.57	2.65	3.53
M_4	5.02	5.77	4.46	5.47	2.76	4.91

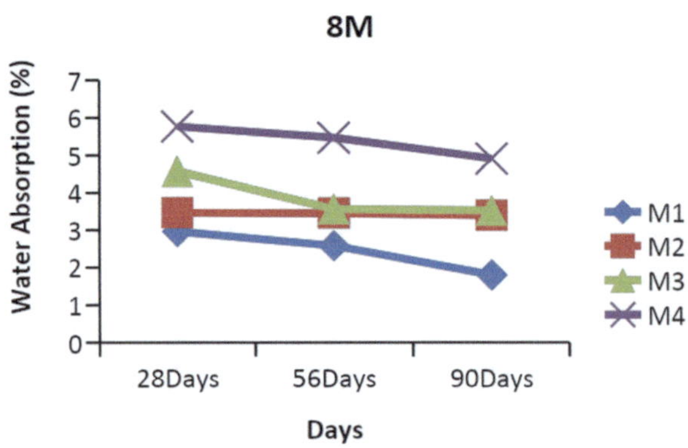

Fig. 9 Different SCGC mix proportions for 8 molarity-variation in water absorption

For the 30% metakaolin replacement mix (M4), a decrease in water absorption is observed over time. Specifically, compared to the 28-day results, water absorption decreased by 5.19% and 10.23% at 56 days and 90 days, respectively.

6 Conclusion

This study investigated the influence of GGBFS/metakaolin ratios on the properties of Self-Compacting Geopolymer Concrete (SCGC). The findings contribute to the field of SCC by demonstrating the potential of SCGC as a sustainable alternative to traditional concrete while offering valuable insights into optimizing its workability and mechanical performance.

6.1 Key Findings and Contributions

- The study established a clear correlation between GGBFS content and SCGC workability. Lower GGBFS levels, achieved through replacement with metakaolin, resulted in decreased workability as measured by slump flow, V-funnel, T50 slump flow, and L-box tests.
- While compressive strength decreased with reduced GGBFS content, the mixtures maintained acceptable values. This highlights the need for a balance between workability and strength when formulating SCGC mixes.
- Water absorption analysis revealed varying degrees of water uptake depending on the metakaolin content within the 8 M SCGC mixes. Further investigation is warranted to fully understand the long-term durability implications.

6.2 Future Research and Practical Considerations

Future research should explore methods to enhance the workability of SCGC with lower GGBFS content. This could involve the incorporation of supplementary cementitious materials or workability-enhancing admixtures. Additionally, studies on the long-term durability and life cycle assessment of SCGC compared to conventional concrete are crucial for wider adoption in construction projects.

6.3 Call to Action

The promising results obtained in this study encourage further research and development of SCGC formulations. By addressing workability challenges and comprehensively evaluating long-term performance, SCGC can become a viable and sustainable construction material, mitigating the environmental impact of traditional cement production.

References

1. Hendriks C, Crijns-Graus W, Bergen F (2004) Global carbon dioxide storage potential and costs
2. Naik A, Patel Y, Baheti R, Vasava A, Desai Y (2020) Study on implementation of metakaolin & foundry sand on self-compacting concrete properties. Int J Pure Appl Math 07(08)
3. Okamura H, Ouchi M (2003) Self-compacting concrete. J Adv Concr Technol 1(1)
4. Nematollahi B, Sanjayan J (2014) Effect of different superplasticizers and activator combinations on workability and strength of fly ash based geopolymer. Mater Des 57:667–672. https://doi.org/10.1016/j.matdes.2014.01.064
5. Nandanam K, Biswal US, Dinakar P (2021) Effect of fly ash, GGBFS, and metakaolin on mechanical and durability properties of self-compacting concrete made with 100% coarse recycled aggregate. J Hazard Toxic Radioact Waste 25(2):04021002. https://doi.org/10.1061/(ASCE)HZ.2153-5515.0000595
6. Ahmed MF, Nuruddin MF, Shafiq N (2011) Compressive strength and workability characteristics of low- calcium fly ash-based self-compacting geopolymer concrete. Int J Civil Environ Eng
7. Davidovits J (1994) Global warming impact on the cement and aggregates industries. In: Presented at the World Resource Review
8. Provis J, Van Deventer J (2009) Geopolymers: structures, processing, properties and industrial applications
9. Bernal SA (2015) Effect of the activator dose on the compressive strength and accelerated carbonation resistance of alkali silicate-activated slag/metakaolin blended materials. Constr Build Mater 98:217–226. https://doi.org/10.1016/j.conbuildmat.2015.08.013
10. Duxson P, Fernández-Jiménez A, Provis JL, Lukey GC, Palomo A, van Deventer JSJ (2007) Geopolymer technology: the current state of the art. J Mater Sci 42(9):2917–2933. https://doi.org/10.1007/s10853-006-0637-z
11. Khan MdNN, Kuri JC, Sarker PK (2021) Effect of waste glass powder as a partial precursor in ambient cured alkali-activated fly ash and fly ash-GGBFS mortars. J Build Eng 34:101934. https://doi.org/10.1016/j.jobe.2020.101934
12. IS 383 (1970) Specification for coarse and fine aggregates from natural sources for concrete
13. Efnarc.pdf. (2023) Accessed. (Online). Available: https://wwwp.feb.unesp.br/pbastos/c.especiais/Efnarc.pdf
14. Arun BR, Nagaraja PS, Srishaila JM (2019) An effect of NaOH molarity on fly Ash—metakaolin-based self-compacting geopolymer concrete. In: Das B, Neithalath N (eds) Sustainable construction and building materials. Lecture notes in civil engineering, vol 25. Springer, Singapore. https://doi.org/10.1007/978-981-13-3317-0_21
15. Arun BR, Nagaraja PS, Mahalingasharma S (2018) Combined effect of flyash & GGBS on workability and mechanical properties of self compacting geopolymer concrete. Int J Pure Appl Math 119(15):1369–1380

16. Mahalingasharma SJ, Parasivamurthy P, Das VR, Arun BR (2021) The combined effect of GGBS and low volumefibres on the low-velocity impact and mechanical properties of self-compacting concrete. Mater Today Proc 49(2022), Part 5. https://doi.org/10.1016/j.matpr.2021.09.339
17. Kavyashree K, Kumar DN, Arun BR, Kumar H (2024) enhancing strength and durability of self-compacting geopolymer concrete by incorporating metakaolin and GGBS with varied molarity concentration, Feb 23, 2024. Acta Sci J, Acta Sci. (Canoas), 26(1):102–117, Q2, Scopus Indexed. Available at SSRN: https://ssrn.com/abstract=4767677
18. Arun BR. A feasibility study on self compacting geopolymer concrete. http://hdl.handle.net/10603/346943
19. IS 516 (1959) Method of tests for strength of concrete
20. Astm C 642|PDF (2023) Scribd. Accessed: (Online). Available: https://www.scribd.com/document/502849922/ASTM-C-642

The Effects of Agave Fibres on the Compressive Strength of Concrete

Kanchan R. Abhang, Aditi A. Pansare, Aysha N. Shaikh, Bhakti S. Jagdale, and Atteshamuddin S. Sayyad

Abstract Concrete is widely used in construction due to its strength, durability, and cost-effectiveness. However, traditional concrete has notable drawbacks, such as brittle behaviour, low tensile strength, and limited ductility. To address these issues, various reinforcing materials like steel, glass, and synthetic fibres have been employed. However, these materials are challenging to work with and harmful to the environment. Concrete has been researched as an alternative reinforcement material using natural fibres including hemp, sisal, and jute. This study examines the viability of employing agave fibres as a low-cost, environmentally friendly way to improve the strength and durability of concrete used in civil construction. Agave fibres were chosen because of their superior strength, low density, and global availability. The report provides a thorough description of the characteristics of agave fibres and analyses earlier studies on the use of natural fibres in concrete. A variety of concrete mixtures including various amounts of agave fibres are tested as part of the experimental programme to see how they affect the material's compressive strength, workability, and durability. In areas where agave plants are common, agave fibres may be an effective substitute for steel in the reinforcement of concrete. According to the results, agave fibres can greatly increase the strength and durability of concrete, offering a practical answer for the building sector. The results of this study shed important light on the potential of agave fibres as a long-lasting and environmentally friendly replacement for conventional reinforcement materials in concrete.

Keywords Agave fibres concrete · Compressive strength · Natural fibres

K. R. Abhang (✉) · A. A. Pansare · A. N. Shaikh · B. S. Jagdale · A. S. Sayyad
Department of Structural Engineering, Sanjivani College of Engineering, Savitribai Phule Pune University, Kopargaon, Maharashtra, India
e-mail: kanchan.abhang.shirdi@gmail.com

A. S. Sayyad
e-mail: sayyadst@sanjivani.org.in

Y. K. Guruprasad et al. (eds.), *Recent Advances in Infrastructure Development and Management—Volume 2*, Lecture Notes in Civil Engineering 566,
https://doi.org/10.1007/978-981-96-1558-2_10

1 Introduction

The performance and durability of concrete structures have been improved by the addition of fibres, revolutionizing the construction sector. Although commonly used in construction, concrete has inherent shortcomings in terms of tensile strength, flexibility, and crack resistance. Concrete is known for its compressive strength. Different kinds of fibres have been added to concrete mixtures as reinforcements to get around these restrictions. A composite material that combines the advantages of fibres with concrete is referred to as fibre-reinforced concrete (FRC). Fibres are a group of materials that resemble hair and come in single elongated segments or continuous threads, much like a thread. They could be included in composite materials. In recent years, agave fibres have drawn a lot of attention as a viable and environmentally friendly alternative to conventional cementitious materials for reinforcing concrete. They are the perfect choice for improving the structural performance and sustainability of concrete because of these qualities. Agave fibres can replace or reduce the requirement for traditional reinforcement materials like steel or synthetic fibres, which has a positive impact on the environment and lowers construction costs. The ability to be recycled is one of the main benefits of Agave fibre. Sonmali and Autade [1] have performed a research study on the impact of agave fibres on the compressive strength of concrete. The impacts of Agave Vera-Cruz Mill fibre on the mechanical characteristics of the concrete matrix in wet and hardened state tests have been discussed by the authors. The M30 grade of concrete was tested for mechanical strengths with fibre-cement ratios of 0.05, 0.1, and 0.15%. With the addition of fibre, workability was shown to decrease as measured by the Vee-Bee instrument. The Vee-Bee time was recorded as 7, 10, 13, and 17 s, respectively, with various fibre-to-cement ratios of 0, 0.05, 0.1, and 0.15%. Before compounding, Langhorst et al. [2] studied the impact of heat treatment on the mechanical performance and morphology of blue agave fibre. The authors observed that when fibres were heated to 180 °C, their secondary modulus increased by around 150% while their strain at break decreased by about 80%. Heat-treated fibres failed in a more brittle way than untreated fibres, according to a morphological study of fibre fracture surfaces, which is consistent with an increase in fibre crystallinity and stiffness. Tameem et al. [3] studied the mechanical strengths of M25 grade concrete using agave fibres, varying the dosage of the fibres from 1 to 4% by volume of cement with an ideal length of 40 mm. The mechanical characteristics of M25-grade conventional concrete, including its compressive strength and split tensile, were compared. Contrary to conventional concrete, it has been found that adding fibre to concrete increases its compressive and split tensile strengths. The fibre extraction, tensile, and flexural characteristics of agave fibre-reinforced epoxy hybrid composites were studied by Sathiamurthi et al. [4]. The relative fibre weight proportion can be 10, 15, 20, or 25%, and the fibre lengths can range from 1 to 4 cm. The hybrid composites with randomly mixed fibre reinforcement were created using the simple layup technique and compression moulding. The specimen's micro-structure was examined using scanning electron microscopy. According to an experimental study, fibres with a length of 3 cm and a weight percentage of 20% produce the greatest tensile and flexural strength values. A sustainable composite material made of Portland cement and reinforced with *Agave lecheguilla* fibres with a high tensile strength was developed by Juarez et al. [5]. The findings showed that the fibres, also known as "*Agave lechugilla*" or simply "fibres," exhibits a high tensile capacity but can suffer significant degradation in the alkaline environment of the composite. The authors conclude that the composite performs adequately in harsh environments and fluctuating humidity and temperature conditions when the fibre is protected with paraffin and the composite matrix is modified with a pozzolan admixture, such as fly ash. Mehrez et al. [6] have optimized the preparation method to achieve the best physicomechanical properties for composite materials. By using an experimental design method, they enhanced the mechanical properties and water uptake of the mortar/unidirectional Agave Americana fibre composite. The ideal preparation conditions for the mortar Americana composite, ensuring maximum mechanical qualities and minimal moisture impact, were identified using a central composite design. Li et al. [7] presented a review article on natural fibres, critically examining and discussing the science behind the treatment methods of natural fibres and suggesting intriguing areas for further investigation. Shadheer Ahamed et al. [8] focused on the similarities and differences among various natural fibres, aiming to analyse the factors influencing overall performance and reliability. Their study involved comparing proportions and drawing conclusions. Selvaraj et al. [9] conducted an experimental study on the morphology characterization and thermal-physical-mechanical properties of Ficus Carica bark fibre. Valadez-Gonzalez et al. [10] explored improving the effective mechanical properties of a thermoplastic matrix reinforced with natural fibres by enhancing the physicochemical interactions at the fibre-matrix interface. The primary objective is to study the effect of sisal fibre reinforcement in concrete and determine the optimal fibre mix percentage. Fibres are mixed in varying proportions and cut into small pieces of 25 mm. The mechanical and transport properties of the concrete are studied through compressive tests on concrete cubes (150 × 150 × 150 mm).

2 Material and Properties

Cement, fine aggregates, coarse aggregates, water, and agave fibres are used to prepare the M20 concrete in the present study. The concrete is made with ordinary Portland cement of 53 grades. Table 1 shows the cement characteristics. Small, inert particles of material with a particle size typically ranging from 0.075 to 4.75 mm are known as fine aggregates. Usually, natural sand or gravel is crushed or ground to produce them. Fine aggregates should comprise a variety of particle sizes and should have a well-graded particle size distribution. Concrete relies on coarse particles to provide strength, durability, and volume stability. Larger inert material particles, usually larger than 4.75 mm, are referred to as coarse aggregates. The density of the aggregate as contrasted to the density of water is measured by its specific gravity. It helps in determining the volume of voids in the aggregate, which can affect the yield and strength of the concrete mix. To ensure properly graded aggregates in the concrete mix, sieve analyses of both the conventional and recycled aggregates were performed using accepted practices. Tables 2 and 3 show the sieve analysis results of aggregates. A sample of 500 g is taken for the sieve analysis of fine aggregates, whereas a sample of 6000 g is taken for the sieve analysis of coarse aggregates. The sieve analysis results shown in Tables 2 and 3 show that both aggregates confirm the standard requirements of well-graded aggregates. Figure 1 shows the sieve analysis results of aggregates.

Concrete should only be mixed with water that is pure, uncontaminated, and safe to drink. To promote the best possible chemical reactions while hydrating, the pH of the water should be within the prescribed range, which is often between 6 and 8. Because of their comparatively high tensile strength, agave fibres are a good choice for concrete reinforcement. Agave fibres are an environmentally friendly substitute for synthetic reinforcing materials since they are renewable and biodegradable. The quantity of cellulose and the orientation of the microfibres affect the failure strength and elastic modulus. The properties of agave fibres are shown in Table 4.

Table 1 Properties of cement

Properties of cement	Properties of fine aggregate	Properties of coarse aggregate
Grade-OPC 53	Sieve analysis result-Zone I	Sieve analysis result-20 mm
Specific gravity-3.15	Specific gravity-3.12	Specific gravity-3.07
Fineness modulus-95%	Water absorption-2.06%	Water absorption-1.53%

Table 2 Sieve analysis results of fine aggregate (500 gm Sample)

Sr. No.	IS sieve	Weight retained	Individual % retained	Cumulative % retained	% Passing
1	4.75	150	30	30	70
2	2.36	75	15	45	55
3	1.18	140	28	73	27
4	600	95	19	92	8
5	300	25	5	97	3
6	150	10	2	99	1
7	75	5	1	100	0
8	pan	–	–	–	–

Table 3 Sieve analysis results of coarse aggregate (6000 gm Sample)

Sr. No.	IS sieve	Weight retained (gm)	Individual % retained	Cumulative % retained	% Passing
1	80	–	–	–	–
2	63	70	1.16	1.16	98.8
3	50	325	5.41	6.57	93.4
4	40	220	3.66	10.23	89.7
5	31.5	135	2.25	12.48	87.5
6	25	190	3.16	15.64	84.3
7	20	560	9.33	24.97	75.0
8	16	360	6.0	30.97	69.0
9	12.5	1290	21.5	52.47	47.5
10	10	1360	22.66	75.13	24.9
11	8	1270	21.16	96,029	3.71
12	4.75	185	3.08	99.37	0.63
13	pan	35	0.58	99.95	0.05

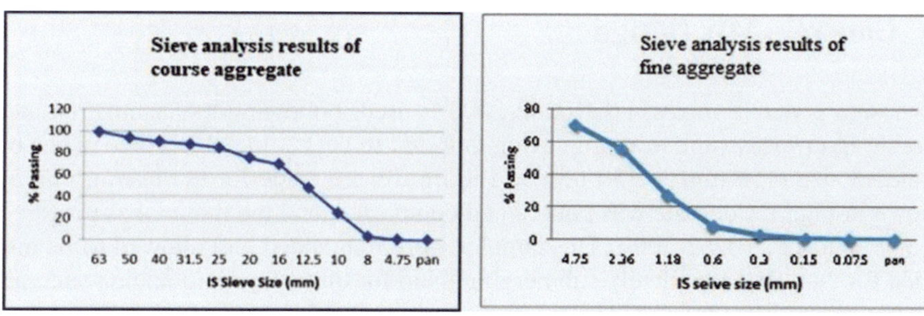

Fig. 1 Sieve analysis of fine and coarse aggregates

Table 4 Properties of agave fibres

Sr. No.	Quantities	Results
1	Diameter	0.2 mm
2	Elongation	4%
3	Water absorption	1.02%
4	Cellulose	70%

3 Experimental Methodology

Dried agave fibres used for the preparation of the present M20 concrete are extracted from the plant using the following process: (1) At the very initial step, we just soak the freshly cut leaves in water at normal room temperature for 7–8 days. (2) After

Fig. 2 Extraction process of agave fibres

7 days, we just take out the leaves from the water and scratch the leaves with a wire brush, just to remove the green skin over the leaves as shown in Fig. 2b. (3) Then, we cut the fibres into 25 mm length with the help of scissors. (4) After cutting the fibres to the desired length, we just dried the fibres for 3 days in the shade as shown in Fig. 1c.

4 Concrete Mix Design

To design concrete mixes, I.S. 10262-2009 is used. Following the calculation and mix design process, the mix proportion is found to be 1:2.14:3.17. Three cubes of standard size (150 mm × 150 mm × 150 mm) were made for the testing mix as shown in Fig. 3. Concrete was poured and compacted, and the mould was prepared in line with I.S. 10262-2009. The samples were remoulded and allowed to be air-dried for 24 h. By completely submerging them for three days in potable water, all the specimens were treated. The specimens were air-dried for 24 h after the curing period was complete.

4.1 Compression Test

To ascertain the compressive strength of concrete, a compression test, commonly referred to as a concrete cube test, is carried out. Concrete's capacity to support loads or endure compression is determined by its compressive strength. To evaluate the strength and quality of concrete constructions, this test is crucial. For the compression test on the concrete cube, the steps below must be followed:

(1) Usually, these samples are taken from the batch of concrete (Fig. 4). A cube measuring 150 mm × 150 mm × 150 mm is the most typical sample shape.

Fig. 3 Preparation of concrete cube

Fig. 4 Concrete cube specimen

(2) The samples must be properly cured after being taken (Fig. 5a). This normally entails keeping the samples in a moist setting for a predetermined amount of time, typically 28 days as depicted in Fig. 4.

(3) The samples are evaluated in a compression testing machine after the curing period. A hydraulic piston is used in the device to provide a compressive load to the concrete sample until it breaks. The sample cracks when the load is gradually raised.

Curing of Specimen | Testing on CTM | Concrete cube after failure

Fig. 5 Curing and testing of concrete cube on CTM

(4) Align the concrete sample correctly and set it on the compression testing apparatus (Fig. 5b). Apply a compressive load on the sample that increases progressively until it fails. The testing standard normally calls for a consistent rate of load application.

(5) As demonstrated in Fig. 5c, the failure of concrete samples is typically signalled by a sharp decrease in load or a characteristic cracking sound. Once the sample fails, note the highest load that was put on it.

A compression testing machine (CTM) was used to conduct the compressive strength test (Fig. 5b). Table 5 provides the compressive strength calculated from the failure load. Table 5 shows that after the testing of 3 cubes from the calculated proportion, we do not meet the required target compressive strength of M20 concrete. Therefore, modifications in the mix design proportion are done according to the trial and error process. The final proportion calculated is 1:1.87:3.56. Using this final mix proportion, 15 cube specimens were cast with varying percentages of Agave fibres. The total quantity of materials required for the preparation of 15 cubes is shown in Table 6. The specimens were tested after 7 days of curing. In each category, there should be three specimens to be tested and the average value is reported. All specimens were curved into the potable water for 7 days. After 7 days, the blocks were ready to test.

Table 7 shows the values of the average compressive strength of concrete cubes prepared with the inclusion of different percentage of agave fibres. The effects of inclusion of agave fibres (1, 2, 3, and 4%) in the concrete are examined with normal concrete (0%). Table 7 shows that the compressive strength increases with an increase

Table 5 Cube test results (Compressive strength after 7 days)

Block number	Weight (kg)	Days	Force (kN)	Compressive strength (N/mm^2)
1	8.962	7	350	15.6
2	8.635	7	332	14.8
3	9.001	7	368	16.4

Table 6 Total quantity of material required for different percentage of agave fibres

% of Agave fibre					
Material quantity (%)	0	1	2	3	4
Fibres (gm)	0	52.2	104.4	156.6	208.8
Cement (kg)	5.22	5.16	5.115	5.063	5.011
Fine aggregate (kg)	9.78	9.78	9.78	9.78	9.78
Coarse aggregate	18.56 kg	18.56 kg	18.56 kg	18.56 kg	18.56 g

Table 7 Compressive strength of cubes for different percentages of agave fibres

Sr. No.	% of fibres	Number of blocks	Wt. of block before curing (kg)	Compressive strength (MPa)	Average compressive strength (MPa)
1	0	A	8.900	17.8	17.1
2		B	8.730	18.7	
3		C	8.775	14.9	
1	1	A	8.375	18.9	18.3
2		B	8.605	17.3	
3		C	8.700	19.5	
1	2	A	8.420	18.2	18.9
2		B	8.505	18.0	
3		C	8.355	20.5	
1	3	A	8.305	20.6	19.7
2		B	8.350	19.4	
3		C	8.340	19.0	
1	4	A	8.370	21.8	20.9
2		B	8.485	21.2	
3		C	8.895	19.8	

in the percentage of agave fibres. It is observed from Table 7 that the addition of 4% of agave fibres gives maximum compressive strength.

5 Conclusions

Since agave fibres are a naturally occurring and renewable resource, adding more of them to concrete will make it more environmentally friendly. It is concluded from the present study that the compressive strength increases when Agave fibre content in concrete is increased from 1 to 4%. Even if it becomes more ecologically friendly by lowering the dependency on non-renewable resources, it still has a positive impact on the various engineering properties. The ductility and flexibility of concrete both can be improved using agave fibres. Agave fibres are typically lightweight, so increasing their percentage in concrete results in a reduction in overall weight. This can be advantageous in applications where weight reduction is desired, such as in transportation infrastructure or precast elements.

References

1. Sonmali PN, Autade PB (2015) Agave Vera-Cruz mill fiber reinforced concrete. Int J Eng Res Gen Sci 3(3):578–592
2. Langhorst A, Ravandi M, Mielewski D, Banu A (2021) Technical agave fiber tensile performance: the effects of fiber heat-treatment. Ind Crops Prod 171:113832
3. Tameem MN, Mohiuddin MY, Haleem MA (2016) Experimental investigation for the effects of Agave fiber on the properties of concrete. J Adv Sci Technol 11(22):1–7
4. Sathiamurthi P, Karthi Vinith KS, Sathishkumar TP, Arunkumar S, Anaamalaai AS (2021) Fiber extraction and mechanical properties of Agave Americana/Kenaf fiber reinforced hybrid epoxy composite. Mater Today Proc 46(17):8594–8601
5. Juarez C, Duran A, Valdez P, Fajardo G (2007) Performance of "*Agave lecheguilla*" natural fiber in Portland cement composites exposed to severe environment conditions. Build Environ 42(3):1151–1157
6. Mehrez I, Hachem H, Gheith R, Jemni A (2023) Optimization of mortar/agave Americana fibers composite behavior based on experimental design. J Nat Fib 20(1):2152149
7. Li Q, Ibrahim L, Zhou W, Zhang M, Yuan Z (2021) Treatment methods for plant fibers for use as reinforcement in cement-based materials. Cellulose 28:5257–5268
8. Shadheer Ahamed M, Ravichandran P, Krishnaraja AR (2021) Natural fibers in concrete—a review. IOP Conf Ser Mater Sci Eng 1055:012038
9. Selvaraj M, Pannirselvam N, Ravichandran PT, Bhuvaneshwaran M, Samson S (2023) Extraction and characterization of a new natural Cellulosic fiber from bark of ficus Carica plant as potential reinforcement for polymer composites. J Nat Fib 20(2):2194699
10. Valadez-Gonzalez A, Cervantes-Uc JM, Olayo R, Herrera-Franco PJ (1999) Effect of fiber surface treatment on the fiber–matrix bond strength of natural fiber reinforced composites. Compos Part B Eng 30(3):309–320

Impact of Use of Marble Waste on the Properties of Light Transmitting Concrete

Abhishek Patel, Vikas Babu, Shivam Singh Chauhan, Manish Sharma, and Vineet Jaiswal

Abstract In numerous Indian states, marble mining and block production are both quite popular. However, because of the enormous amount of marble sludge produced by the processing businesses, its broad expansion has had a negative influence on the environment. The effects of using marble debris on the characteristics of light transmitting concrete (LTCM) are the main topic of this experimental investigation. This experimental program's primary objective is to determine whether using marble waste as an aggregate would increase light transmittance in concrete and determine the best ratio of quarry dust and marble powder to replace river sand in concrete. To measure compressive strength, permeability, and flexural strength, six proportions of mix with marble waste substituted for fine aggregate were tested. As a consequence, all these qualities were enhanced to allow for a complete replacement of fine aggregate with marble powder from the experiment. In order to determine if LTCM can improve its light transmittance, 75 and 100% substitutions of fine aggregates were conducted. A 100% marble powder mix had a light transmittance of 44% compared to a 0% marble powder control mix.

Keywords Fiber · Marble power · Waste materials · Light transmitting concrete · Plastic optical fibers (POF)

1 Introduction

Due to the rise of the world's advanced infrastructure, concrete is now the most common and regular construction material utilized worldwide. Due to the increasing need for infrastructure, there is a great deal of pressure on the concrete industry to produce enormous amounts of concrete [1, 2]. The three primary raw materials

A. Patel · V. Babu · S. S. Chauhan
Department of Civil Engineering, Rajkiya Engineering College Kannauj, Kannauj, UP, India

S. S. Chauhan · M. Sharma · V. Jaiswal (✉)
Department of Civil Engineering, Rajkiya Engineering College Mainpuri, Mainpuri, UP, India
e-mail: vineetjaiswal72.jt@gmail.com

Y. K. Guruprasad et al. (eds.), *Recent Advances in Infrastructure Development and Management—Volume 2*, Lecture Notes in Civil Engineering 566,
https://doi.org/10.1007/978-981-96-1558-2_11

required to manufacture concrete—cement, coarse and fine aggregate, and water—determine the price of the finished product. One of the raw materials that contributes to around 35% of the volume of concrete, natural river sand, is essential in determining the cost of concrete [2]. Due to the dwindling availability of natural river sand and stringent environmental rules on mining, the concrete industry has gradually shifted its attention toward finding a suitable fine aggregate replacement for the already-used natural river sand [3, 4]. The necessity to find alternative materials for component materials used as aggregates in the manufacturing of concrete has risen due to the diminishing supplies of natural sand and the requirement to reduce the cost of concrete production. In both developing and industrialized nations, several materials have been studied for this purpose with varying degrees of success. By-product materials, or occasionally produced aggregates, are the materials that are often investigated for this purpose [5, 6]. Utilizing waste resources such as by-products or aggregates can reduce environmental impact and waste management expenses, lower the cost of making concrete, and possibly even improve some concrete properties. One of the materials being examined, particularly as a sand alternative for fine aggregates, is quarry dust, a by-product of the crushing process during quarrying activities [7, 8]. Quarry dust has been utilized for a variety of construction-related tasks, including constructing roads and producing lightweight aggregates, bricks, tiles, and autoclave blocks [9, 10]. The extraction of marble and its subsequent processing into marble blocks are among the most prevalent mining activities in numerous Indian states [11]. But because of the enormous quantity of marble sludge produced by the processing businesses, its phenomenal rise has had a negative influence on the environment. Marble blocks are carved into various sizes and thicknesses at factories or workshops. Water is utilized throughout the cutting operation to minimize dust and overheating of the rock saw [12–15]. By combining dust and water, marble sludge is created as a result of this procedure. When material is processed, waste is created, and either the residual waste or the waste after dewatering in the treatment plant is thrown into the environment [16]. Sludge is the waste product of around 30% of a marble block that has been quarried. In the processing factories, it's difficult to store this waste marble sludge. The industries will thus consider the revaluation of this waste material as advantageous from a financial and environmental standpoint. The possibility of using quarry dust and marble dust as fine aggregate in concrete has emerged as a potential substitute for natural river sand. Studies have shown that incorporating marble waste as a partial replacement for fine aggregates enhances the sustainability of concrete while maintaining its structural integrity [17]. Additionally, the inclusion of optical fibers in concrete enables light transmission, making it a potential material for architectural and energy-efficient applications [18]. Light-transmitting concrete (LTC) is a unique material that combines concrete's strength with optical fibers to allow light to pass through, enhancing architectural aesthetics and reducing light energy demand in buildings. However, the high consumption of natural aggregates in concrete production raises environmental concerns. Incorporating marble waste as a partial replacement for fine aggregates in LTC can provide a sustainable solution by reducing waste and promoting circular economy principles [19]. The use of marble waste in conventional concrete has been studied extensively, demonstrating improvements in workability, strength, and durability. However, its impact on LTC remains relatively unexplored. This study aims to assess the effects of marble waste on LTC's compressive strength, light transmittance, and overall performance.

2 Objective of Study

Considering the increasing growth of infrastructure, concrete is now the most commonly used building material. The sources of natural river sand have reduced its availability. Natural river sand extraction upsets the ecosystem. Industrial garbage disposal takes up space and pollutes the environment. Investigating the mechanical characteristics of concrete and substituting the ideal ratio of these waste materials for river sand to best replace river sand in concrete with a mixture of marble powder and quarry dust to assess if adding marble debris as an aggregate can increase concrete's ability to transmit light.

3 Methodology

The preparation of the mold is the first stage in the process of creating light transmitting concrete. The mold for the prototype might be made out of many materials, such as wood or tin. A cube must be at least 5 × 5 × 5 cm in size, per ASTM. Molds of any size, ranging from 5 cm × 5 cm × 5 cm to 15 cm × 15 cm × 15 cm, can thus be utilized. The perforated wooden sheets were used to create wooden molds for the research that were 10 cm × 10 cm × 10 cm in size as shown in Fig. 1. Electrical switchboards are made of wood sheets. For the purpose of creating POF cubes with varied percentages, perforated hardwood sheets with various numbers of drilled holes were affixed to the molds as illustrated in Fig. 2 [20]. The amount of fiber in the cube has an impact on both the diameter and spacing of the holes. The sheet's purpose is to make it possible to place POF in the concrete properly as depicted in Fig. 3.

In order to fit plastic optical fibers (POF) inside, wooden molds with holes were created with sheets at two parallel faces. At the top, the mold is open. POF was introduced into the mold through the holes punched in the sheet before pouring concrete as shown in Fig. 4. In order to avoid the development of pores in the concrete cube, after establishing POF, concrete was poured into the mold while it was being vibrated mechanically. Concrete was used to create 10 cm × 10 cm × 10 cm cubes in four distinct mix proportions: A_1, A_2, A_5, and A_6. To evaluate the strength, durability, and light transmission properties of POF, 4.0% of plastic optical fibers were utilized. After the cubes have been unmolded, the specimens are put in the curing tank to begin the curing process as illustrated in Fig. 5.

Fig. 1 Sheets with perforations

Fig. 2 Wooden molds

3.1 Experimental Properties Studied on Light Transmitting Concrete

For each of the series A_1, A_2, A_5, and A_6, samples of POF volume of 4% are cast in order to assess the light-directing property of light transmitting concrete [20, 21]. The photometer (also known as a lux meter), which has a range of 0.1–1,00,000 lx,

Fig. 3 Placing of fibers in the mold

Fig. 4 Light transmittance cube mold

Fig. 5 Unmolded cubes

was used to test the transmittance. As a light source, an incandescent bulb with a 1100 lx light output was selected [5–9]. The amount of light that entered the box from the sample was measured using a wooden box with a light source fitted on one face and a photometer attached to the other face as shown in Fig. 6. To ensure accuracy, the photometer's box was carefully inspected and securely attached so that all transmitted light entered the box correctly, as illustrated in Fig. 7.

To investigate the properties of light transmittance, a light transmittance test was conducted. Given that transparent concrete's primary function is to transmit light, this test must be done. By measuring the intensity of incoming and transmitted light, the transmittance ratio may be calculated. The photometer is used to measure light

Fig. 6 Experimental arrangement with a wooden box

Fig. 7 Light transmitting concrete experiment set

Fig. 8 Light transmittance through cube

intensity. Lumens are the units used to measure light intensity by photometers. Transmitted light measurements were made with a photometer. To prevent measurement errors, the photometer's box was securely fastened, ensuring that all transmitted light was directed into it, as depicted in Fig. 8.

4 Results

4.1 Test for Light Transmittance

The light transmittance test was conducted to measure the amount of light passing through the plastic optical fiber from one end of the specimen to the other, determining the light transmittance ratio. The experiment utilizes a halogen light source with an intensity of 1100 lux.

The transmittance values for different mix proportions are presented in Table 1. Table 1 shows the light transmittance (in lux) for different concrete mixes, including the control mix and those incorporating marble dust and marble aggregates. The results indicate that the mix prepared with marble dust and marble aggregates (A_6) improves it by 44% when compared to the control mix (made with standard coarse aggregate) as illustrated in Fig. 9 . It could be because the control mix specimen had the lowest light absorption while waste marble dust had the greatest. The quantity of light that may be transmitted steadily increases from 8.33% for the control mix to 12% when the proportion of marble dust changes.

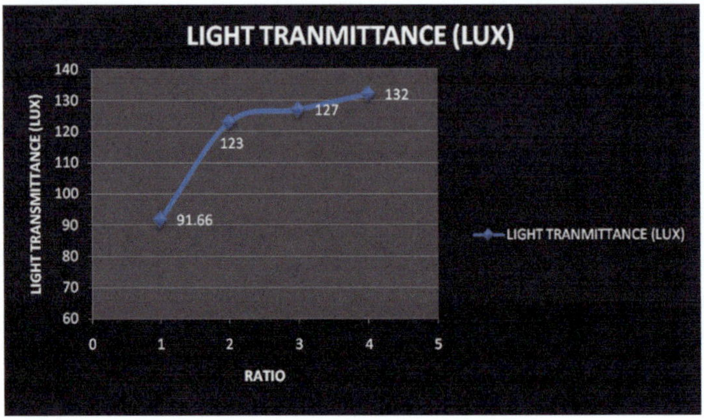

Fig. 9 Light transmittance (LUX) versus mix proportion

Table 1 Light transmittance (LUX)

Nomination	Stating	Mix	Cube 1	Cube 2	Cube 3	Average
A_1	QD/CA	Control mix	92	91	92	91.67
A_2	QD/MA	Control mix	123	125	122	123
A_5	MD75/MA	Control mix	125	129	127	127
A_6	MD100/MA	Control mix	135	131	130	132

4.2 Comparison of Compressive Strength of Normal Concretes and Transmit Light Concrete

Furthermore, the compressive strength of normal concrete and light-transmitting concrete was compared, as shown in Fig. 10. The results indicate that the strength of the control mix decreased by 7.6% when plastic optical fiber was introduced into the specimen. However, when marble dust was used as a complete replacement for fine aggregate, the reduction in strength was only 2.2%. This suggests that mix proportion A_6 exhibited the best performance in terms of both strength retention and light transmittance.

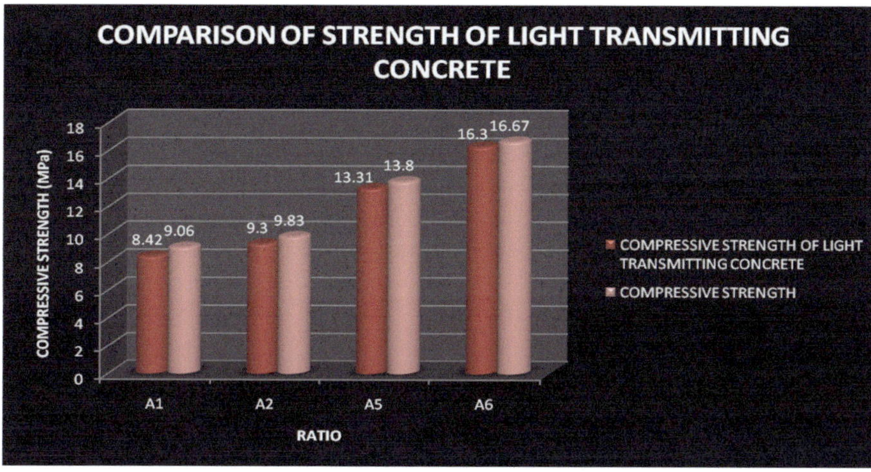

Fig. 10 Compressive strength measurements with and without optical fiber

5 Conclusion

Light transmitting concrete is a useful building material, according to the results of this experimental investigation. To improve the structure's energy efficiency, it can be employed in green buildings. It is unquestionably the material of choice for civil engineering projects in the future, and as time goes on, its application in building will only grow. The outcomes of this experimental effort allow for the following conclusions to be made. Increased marble powder replacement of fine aggregate results in an increase in compressive strength. The greatest compressive strength of 16.67 MPa is found after 28 days when employing a combination with a 100% substitution of fine aggregate with marble powder and coarse aggregate with marble. As the proportion rises, flexural strength likewise does. Marble powder is used in place of fine aggregate. After 28 days, maximum flexural strength of 3.09 MPa is obtained when marble totally replaces coarse and fine material. According to the results of the ultrasonic pulse velocity test, as the amount of marble powder in the mixture increases, the mixture gets denser and the time it takes for the ultrasonic wave to travel through the specimen decreases. In order to completely replace marble powder, ultrasonic pulse velocity must be more than 4.5 Kmps. So, the hardened concrete that is produced is of very high quality.

The results of the test show that replacement of marble powder increases while water penetration depth decreases. It is 10.6 cm when the fine aggregate is quarry dust and the coarse aggregate is regular, while it is 7.6 cm when the fine aggregate is marble powder and the coarse aggregate is marble. For a light transmittance ratio of up to 12%, a 4% Plastic Optical Fiber ratio was used. This illustrates how well light transmitting concrete is used in green constructions. The buildings' interiors will always have natural light. There will be reduced reliance on artificial energy sources as a result. Compared to control concrete, the light transmitting concrete's strength was not significantly affected by the compression strength tests. It was found that at the same mix percentage, using the same coarse aggregate as quarry dust, as well as the same fine aggregate as quarry dust, the concrete's 28-day compressive strength decreased from 9.06 N/mm^2 for control concrete to 8.42 N/mm^2 for concrete with 4% POF. A concrete mix percentage of coarse aggregate in marble powder had a compressive strength of 16.3 N/mm^2 for 28 days, whereas concrete with 4% POF had a compressive strength of 16.67 N/mm^2.

References

1. Aliabdo AA, Abd Elmoaty M, Auda EM (2014) Re-use of waste marble dust in the production of cement and concrete. Constr Build Mater 50:28–41
2. Ahuja A, Mosalam KM, Zohdi TI (2015) Computational modeling of translucent concrete panels. J Archit Eng 21(2):B4014008
3. Alyamaç KE, Ince R (2009) A preliminary concrete mix design for SCC with marble powders. Constr Build Mater 23(3):1201–1210
4. Bharti MK, Sharma M, Islam N (2020) Study on the dam & reservoir and analysis of dam failures: a data base approach. Int Res J Eng Technol 7(5):1661–1669
5. Sharma M, Ansari MI, Islam N (2022) Study of concrete filled unplasticized poly-vinyl chloride tubes as columns under axial loading. J Mech Mater Mech Res 5(1):10–17
6. Andre A, de Brito J, Rosa A, Pedro D (2014) Durability performance of concrete incorporating coarse aggregates from marble industry waste. J Clean Prod 65:389–396
7. Bashbash BF, Hajrus RM, Wafi DF, Alqedra MA (2013) Basics of light transmitting concrete. Glob Adv Res J Eng Technol Innov 2(3):76–83
8. Kashiyani BK, Raina V, Pitroda J, Shah BK (2013) A study on transparent concrete: a novel architectural material to explore construction sector. Int J Eng Innov Technol 2(8):83–87
9. Belachia M, Hebhoub H (2011, May) Use of the marble wastes in the hydraulic concrete. In: 6th International advanced technologies symposium (IATS'11), pp 16–18
10. Belaidi ASE, Azzouz L, Kadri E, Kenai S (2012) Effect of natural pozzolana and marble powder on the properties of self-compacting concrete. Constr Build Mater 31:251–257
11. Binici H, Kaplan H, Yilmaz S (2007) Influence of marble and limestone dusts as additives on some mechanical properties of concrete. Sci Res Essay 2(9):372–379
12. Corinaldesi V, Moriconi G, Naik TR (2010) Characterization of marble powder for its use in mortar and concrete. Constr Build Mater 24(1):113–117
13. Demirel B (2010) The effect of the using waste marble dust as fine sand on the mechanical properties of the concrete. Int J Phys Sci 5(9):1372–1380
14. Gameiro F, De Brito J, da Silva DC (2014) Durability performance of structural concrete containing fine aggregates from waste generated by marble quarrying industry. Eng Struct 59:654–662
15. Hebhoub H, Aoun H, Belachia M, Houari H, Ghorbel E (2011) Use of waste marble aggregates in concrete. Constr Build Mater 25(3):1167–1171
16. Hameed MS, Sekar ASS (2009) Properties of green concrete containing quarry rock dust and marble sludge powder as fine aggregate. ARPN J Eng Appl Sci 4(4):83–89
17. Rawat S, Pal S, Yadav SS, Krishna S, Manish S (2020) Comparative performance evaluation of fibre reinforced concrete & conventional concrete. Int J Innov Res Sci Eng Technol 5(9):3309–3313
18. Momin AA, Kadiranaikar RB, Jagirdar V, Inamdar AA (2014) Study on light transmittance of concrete using optical fibers and glass rods. IOSR J Mech Civil Eng 2(2):67–72

19. Chiew SM, Ibrahim IS, Ariffin MAM, Lee HS, Singh JK (2021) Development and properties of light-transmitting concrete (LTC)–A review. J Clean Prod 284:124780
20. Juan S, Zhi Z (2013) Some progress on smart transparent concrete. Pac Sci Rev 15(1):51–55
21. Jiménez-Muñoz E, Fernández-Martínez F (2014) Translucent concrete. Research with glass, optical fiber and glass fiber. In: Construction and building research. Springer, Netherlands, pp 111–114

Tailoring the Mechanical Strength of Porous Fly Ash-GGBS Alkalinized Geopolymer via Prefoaming for Thermal and Acoustic Characteristics

Manivannan Muthusamy, N. Vanitha, Janani Karuppaiyan, S. Surendhar, and R. Jeyalakshmi

Abstract This study investigates the development of porous geopolymeric structures with good thermal insulation and noise reduction capabilities. Pre-made foams stabilized with 2% Sodium Lauryl Sulfate (SLS) were added at varying dosages (10–15%) into fly ash-GGBS blend in the ratio of 60:40, with the binder activated using sodium silicate hardener. The impact of dosage of foam, curing temperature, and curing days on strength and density was evaluated. The binder phase present in the products was verified from XRD patterns and FTIR analysis. The resulting matrix displayed a compressive strength ranging from 3.13 to 10.91 MPa, decreasing with increasing foam content. Pore size distribution was examined using an image processing technique from the SEM micrograph and found to be in the range of 100–400 μm. The uniform distribution of pores facilitates a decrease in thermal conductivity (λ) as low as 0.023 W/mK by maintaining the density of the matrix at 620 kg/m^3. To produce a thin panel of a matrix to determine their acoustic properties, cylindrical specimens of two different sizes of 25 $\times$ 29 mm and 25 $\times$ 99 mm cast and sound impedance data were collected. The acoustic absorption coefficients were observed in the range of 0.01 to 0.78, at 100–6300 Hz frequency surpassing the control OPC matrix as 0.02. The study results demonstrate the suitability of a porous geopolymeric matrix featuring larger voids and a stronger gel network as promising thermal insulation and noise reduction.

Keywords Porous structure · Thermal conductivity · Sound absorption · Fly ash-GGBS geopolymer

---

M. Muthusamy · N. Vanitha · J. Karuppaiyan · S. Surendhar · R. Jeyalakshmi (✉)
Department of Chemistry, College of Engineering and Technology, SRM Institute of Science and Technology, Kattankulathur, Tamil Nadu, India
e-mail: jeyalakr@srmist.edu.in

Y. K. Guruprasad et al. (eds.), *Recent Advances in Infrastructure Development and Management—Volume 2*, Lecture Notes in Civil Engineering 566,
https://doi.org/10.1007/978-981-96-1558-2_12

1 Introduction

The growing importance of sustainability and the pursuit of enhanced quality of life have led to increased attention on energy efficiency within the construction industry and the overall convenience of indoor living space environments [1]. Notably, the building industry is a major contributor, accounting for nearly 40% of all energy used and CO_2 emissions in developed nations [2, 3]. Traditional cement-based materials have a high carbon footprint and often do not meet the evolving needs for thermal comfort and noise reduction in indoor spaces. Noise pollution in urban areas has become a major concern due to increasing traffic, industrial activities, and densely populated living conditions. This noise pollution can lead to various health issues such as stress, sleep disturbances, and cardiovascular diseases. Effective noise reduction strategies are essential for improving the quality of life in urban environments. Traditional building materials often fall short of providing adequate sound insulation, necessitating the development of new materials with enhanced acoustic properties [4].

Researchers are showing interest in geopolymer composites due to their low carbon footprint with lower warming potential, as well as their superior performance for broader applications against traditional cement. Construction industries, particularly for indoor use, are actively investigating lighter materials and structures to accomplish thermal comfort and noise reduction. Generally, lightweight concrete is produced by the usage of lightweight aggregates or porous structured matrices, typically by introducing gas into a slurry/mortar or adding pre-made foam. Due to the inherent porosity and high surface area, the porous or cellular matrix is utilized in the construction and infrastructure sector, ceramics, and as adsorbents in wastewater treatment [5, 6]. Considering the above, tailoring the porosity in the most popular geopolymer matrix, namely a binary blend of fly ash, GGBS aimed by using a pore-forming agent. This type of geopolymer has considerably lower embodied energy during production, exhibiting enhanced strength and chemical and thermal resistance.

The prevalent approach for foaming in geopolymer production involves mechanical or chemical foaming. In chemical foaming, a blowing agent (such as fine metallic powders or hydrogen peroxide) is introduced into the geopolymer paste. This agent reacts within the alkaline environment, generating gas that becomes trapped within the slurry [7]. Whereas, mechanical foaming integrates a pre-made foam into the geopolymer paste or incorporates surfactants to capture bubbles through air entrainment during mixing. Comparatively, pre-made foaming is able to control pore size and distribution than the chemical foaming method, thereby achieving desirable density. The reported literature available in this field is summarized in Table 1.

The objectives of this study are to develop and characterize an ambient-cured, alkali-activated geopolymer foam composite using a binary blend of fly ash and GGBS, with controlled porosity by incorporating protein-based prefoam and foam stabilizing agents; evaluate the mechanical, physical, and acoustic properties of the resultant product through a series of standard tests; identify the optimal conditions for

Table 1 Properties of alkali-activated foam concrete by mechanical foaming method

Binder	Foaming agent	Alkali activator	Density (kg/m^3)	Pore size (μm)	Compressive strength (MPa) (28 days)	Thermal conductivity (W/m K)	Sound absorption	References
GGBFS	Protein base	NaOH (12 M) and Na_2SiO_3 (Ms = 2.5)	550–1500	< 100–1000	2.4–13	n.d.	Acoustic absorption coefficients higher than 0.5 in the medium to high-frequency regions	[8]
GBFS/FA	Commercial surfactant with water	Anhydrous sodium metasilicate	600–1200	n.d.	1.6–13	0.18–0.937	n.d.	[9]
GGBFS/FA	Protein-based agent	(i) $Ca(OH)_2$ and Na_2SiO_3 (ii) $Ca(OH)_2$ and $Mg(NO_3)_2$	325–492	< 0.01–100	0.5–2	0.088–0.121	n.d.	[10]
BFS/CRS powders	Aqueous solution of a foaming agent diluted with water	NaOH and Na_2SiO_3 (Ms = 2.87)	700–1000	0.1–1 mm	2–5	n.d.	Transmission loss = 30–50 dB at 100–5000 Hz	[11]
MK/BFS	Aqueous solution of a foaming agent diluted with water	NaOH and Na_2SiO_3	400–1000	0.1–0.5 mm	0.4–11	n.d.	0.15–0.9 at 100–4000 Hz	[12]

GGBFS—ground granulated blast furnace slag; n.d.—not detected; GBFS—granulated blast furnace slag; FA—fly ash; solution; MK—metakaolin; BFS—blast furnace slag; CRS—calcined reservoir sludge; SDS—sodium dodecyl sulfate; Ms.—SiO_2/Na_2O or K_2

Table 2 Chemical composition of the fly ash and GGBS in Wt% (XRF analysis)

Materials	SiO_2	Al_2O_3	CaO	Fe_2O_3	K_2O	TiO_2	k_2O	SO_3	MgO	Na_2O
Fly ash	60.949	27.533	1.569	5.974	2.181	1.659	–	0.134	–	–
GGBS	21.58	14.88	55.25	1.78	0.48	–	0.48	–	2.63	0.015

producing geopolymer foams with specific densities and compressive strengths suitable for enhancing energy efficiency and comfort in indoor living environments; and assess the noise reduction capabilities of these geopolymer foams to determine their effectiveness in mitigating noise pollution in urban areas. By addressing these objectives, the study seeks to contribute to the development of sustainable, high-performance building materials that align with the growing emphasis on sustainability, energy efficiency, and noise reduction in the construction industry.

2 Materials and Methods

2.1 Materials

The porous geopolymer synthesis entails the use of primary raw materials: Class F Fly ash (FA) sourced from the Thermal Power Plant in Gummidipoondi, and ground granulated blast furnace slag (GGBS) acquired from the Jindal Steel Plant in Bellary. The source material's chemical composition was analyzed using the XRF techniques listed in Table 2. The alkali activator solution is meticulously prepared by combining sodium hydroxide and sodium silicates to arrive at the silica modulus, SiO_2/Na_2O of 1.1, and stored for 24 h before use. The sodium silicate was procured from M/S Kuttuva silicates, with a silica modulus (Ms) of 2.2, consisting of 34% SiO_2, 15% Na_2O, and 51% H_2O. Stable preformed air bubbles were generated through the consistent pressurization of an aqueous solution containing a foaming agent with compressed air. A protein-based foaming agent consisting of animal protein, gelatin, and sodium dodecyl polyoxyethylene ether sulfate was sourced from Astra Chemicals in Chennai. This foaming agent underwent dilution with water at a 1:20 weight ratio and was subsequently aerated, resulting in a density of 35 kg/m^3. Additionally, Sodium lauryl sulfate (SLS, C12H25SO4Na) is incorporated as a foam-stabilizing agent.

2.2 Preparation

Fabrication of geopolymer specimens involved two stages, firstly homogeneous geopolymer slurry prepared with precursors FA: GGBS in the weight % 3:2 dry mixed and then an alkali activator at an L/S ratio of 0.45 was added. Homogenization is

Table 3 Material mix proportions of the prefoam fly ash: GGBS binder (wt% = 60:40) (L/S = 0.45) (weight of SLS = 2 wt%)

Sample ID	FG	FG-10PR	FG-10PO	FG-15PR	FG-15PO
Weight of foam	0	10	10	15	15
Curing temperature	30 °C ± 1 °C	30 °C ± 1 °C	60 °C ± 2 °C	30 °C ± 1 °C	60 °C ± 2 °C

ensured by thorough mixing for 5 min. To this slurry, different doses of prefoam were added to arrive at the final mix design proportions. Subsequently, the constant 2 wt% of surfactant was introduced in the geopolymer slurry and mixed for an additional 2 min. The pre-made foam was prepared by mixing protein base foamer with water at a weight ratio of 1:20. Subsequently, this mixture was introduced into the foaming generator, where it was combined with compressed air. To enhance the workability of the paste, 20% of the required pre-made foam was added to the geopolymer mix, with the remaining 80% gently blended afterward. The resulting wet foam mixtures were poured into 30 × 35 mm cylindrical molds. Two sets of samples were prepared by curing one set under ambient conditions (FG-10PR, FG-15PR) after proper sealing, while the other set was initially kept at ambient temperature for 24 h and then cured in an oven at 60 °C for 24 h (FG-10PO, FG-15PO) to expedite the polycondensation reaction and investigate the impact of setting temperature. Following the curing process, the geopolymer matrix was left to stabilize at ambient temperature for a duration of 28 d before undergoing testing. The mix proportions for the preparation of different foam dosage (10–15 wt.%) are detailed in Table 3. Studies have shown that a binary blend of GGBS and FA can offer a balance between strength and workability in geopolymer composites. FA generally contributes to higher workability, while GGBS improves short- and long-term strength [13]. The specific ratio will depend on the target compressive strength of the final geopolymer foam. The synthesized samples underwent further characterization studies.

2.3 Characterization Studies

FT-IR spectra were obtained using a Bruker IFS 66v/S FT-IR spectrometer equipped with ATR accessories and a KRS-5 lens. A total of 32 scans were performed with a resolution of 2 cm^{-1} and a scanning speed of 5 kHz to record the absorbance spectra spanning the range of 4000 to 400 cm^{-1}. The XRD pattern was obtained using a Phillips diffractometer (model X-pert) equipped with Cu-Kα radiation. At a speed of 2°/min and a step size of 0.02°, the scanning range covered 10° to 80° (2θ). Analysis of the results unveiled crystallographic phases, identified through JCPDS cards and subsequently confirmed using High Score Plus software. The SEM and microscopic sample specimens were prepared in pellets of 4 mm diameter with 2 mm thickness. Every sample had a minimum of 100 pores measured and SEM pictures were used to assess the cell size, side distribution, and cell window size using the image analysis software of ImageJ.

Compressive strength was measured on 30 mm diameter × 35 mm height cylinders according to ASTM C39-20 at a loading rate of 2 kN/s (average of 3 samples). Water absorption, a key permeability measure, was tested on 40 × 40 × 80 mm specimens following 28 days of room temperature curing. As per ASTM C642-06, samples were oven-dried at 100–110 °C for 24 h, then submerged in water for 24 h. After removing surface moisture, total water absorption (%) was calculated. The thermal conductivity of the specimens was determined through the hot wire technique, in accordance with ASTM standard code C177, using specimens of 2 mm × 4 mm pellet size. The samples were cleaned using compressed air after being polished to have parallel, smooth surfaces. The assessment of porous geopolymer's acoustic absorption utilized a modified impedance tube following ASTM E 1050 procedures as detailed elsewhere [14]. The experimental setup included a vacuum pump positioned behind the sample to measure sound absorption characteristics for acoustic insulating materials. Testing occurred in an ambient air environment, with each geopolymer variant subjected to trials with three specimens. All specimens shared a consistent thickness of 25 mm, with diameters of 29 mm and 99 mm. The testing tube, equipped with a specimen holder at one end and a sound source at the other, featured two strategically placed microphones along its length. This arrangement facilitated the distinction between incident and reflected waves, enabling the calculation of the sound pressure reflection coefficient (R) across frequencies from 40 to 6400 Hz. The sound absorption coefficient (α) was subsequently determined using Eq. (1).

$$\alpha = 1 - |R^2| \tag{1}$$

3 Result and Discussion

3.1 *Density and Compressive Strength*

The compressive strength and dry density results for geopolymers subjected to varying foam dosages (10–15 wt%) and different curing conditions are depicted in Fig. 1. When the rigid FG group mix is added with a 10–15 wt% of prefoam dosage and the 2 wt% of SLS, the measured D changed from 1020 to 620 kg/m^3. Under ambient curing conditions, the rigid sample (FG) exhibited a compressive strength of 39.31 MPa at a density of 1933 kg/m^3. On the other hand, the foam geopolymer samples exhibited a range of compressive strengths of 10–5 MPa under the same ambient conditions. The highest recorded strength was 10.91 MPa at a density of 1020 kg/m^3, while the lowest was 4.28 MPa at 750 kg/m^3. The observed decrease in compressive strength with increasing foam content (10–15 wt%) aligns with theoretical expectations and existing research on geopolymer foams [15, 16].

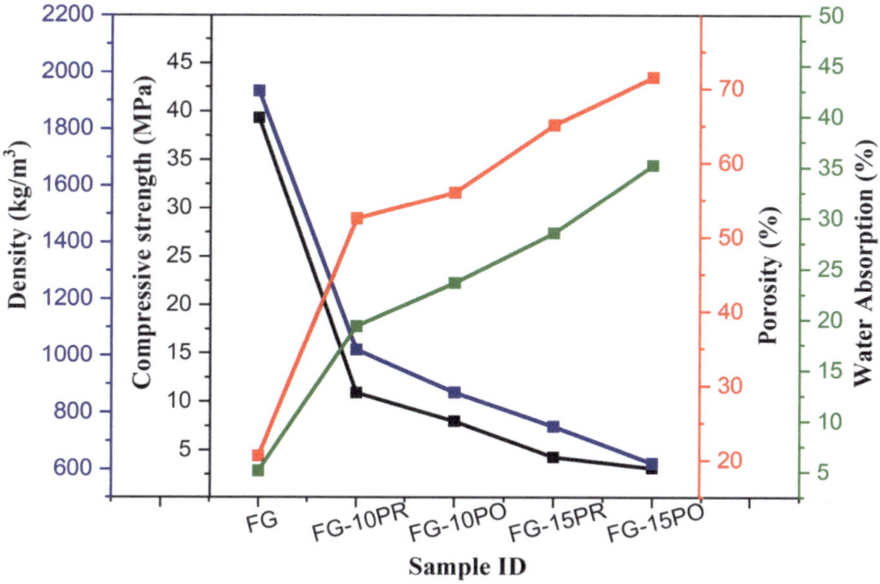

Fig. 1 Compressive strength, density, porosity, and water absorption analysis of the control and prefoam geopolymers

When the foam geopolymer samples were subjected to a slightly increased temperature of 60 °C, a noticeable decrease in compressive strength was observed in comparison to the samples cured under ambient conditions. Raising the curing temperature diminishes the overall strength of the specimens. This is attributed to the gas released above ambient temperature, consequently leading to an increase in the porosity of the samples. This reduction in compressive strength under higher temperature conditions suggests that the foam geopolymer material may be sensitive to thermal influences during the curing process. The change in temperature could potentially affect the material's internal structure, leading to a decrease in its mechanical strength. Understanding the impact of temperature variations on compressive strength is vital for predicting the material's behavior in different curing environments.

3.2 *Porosity and Pore Size Distribution*

The porosity in the matrix is determined by the volume of gas bubbles, which is similar to the density of the geopolymer foam matrix and exhibits a positive connection with rising porosity as the bubble volume increases [17]. The total porosity of different geopolymer samples is shown in Fig. 1. For comparison purposes, the densities of various mixes are also plotted on the same graphs. For example, the total porosity of FG-10PR was determined to be 52%, whereas FG-10PO exhibited a porosity of 56%. Similarly, the porosity of FG-15PR was 65%, while FG-15PO observed with a porosity value of 71%. This indicates that the total porosity of FG-10PF, FG-10PO, FG-15PR, and FG-15PO increased by 62%, 64%, 69%, and 72%, respectively, compared to their corresponding reference mixes of FG. The FG-15PO mix exhibits a maximum porosity of 71% at a density of 620 kg/m^3. The porosity increases with increasing prefoam dosage, this can be seen by comparing the total porosity with different foam dosages. However, using more than 15 wt% of prefoam results in the collapse of the pores, significantly impacting the mechanical properties. Higher proportions of air voids in the foamed concrete corresponded to increased water penetration. Consequently, the percentage of water absorption exhibited a rise from rigid samples to porous samples, aligning with expectations. Additionally, heat curing augmented the air void content, resulting in slightly higher values for the hot-cured samples compared to their counterparts cured under standard conditions. Water absorption values are correlated directly with compressive strength, porosity, and density.

Figure 2 displays the SEM micrograph of the porous structure along with the corresponding pore size distribution. The different percentage of prefoam mixes reveals a notable distinction in pore sizes, with FG-10PR and FG-10PO displaying fine pores and FG-15PR and FG-10PO containing coarse pores. The comparison of pore size distribution between different curing temperatures of foam geopolymer mixes for the same prefoam dosage reveals that the 60 °C curing samples have a large number of pores with coarse bubbles. The foaming agent plays a more significant role in foam geopolymer; increasing the foam dosage from 10 to 15 wt% results in an increase in pore size distribution from 291 to 336 μm at ambient temperature. The pore diameter of the rigid geopolymer (FG) is 141 μm, which is smaller than that of all others. Additionally, heat cured geopolymers exhibit a pore size distribution ranging from 336 to 395 μm. Figure 3a–e are the microscopic images from the scanning electron microscope.

3.3 XRD Analysis

To characterize the phase assemblages, XRD analysis was carried out for raw feedstock as well as geopolymer samples. XRD patterns are depicted in Fig. 4. From the XRD patterns of FA, crystalline phases of mullite (Al_2O_3, PDF#84-1205), α-quartz (SiO_2, PDF#89-8934), and hematite (Fe_2O_3, PDF#89-8104) were identified. In raw GGBS, an observable amorphous alumina silicate hump is present approximately at 2θ value ranging from 22 to 30◦. Additionally, peaks corresponding to anhydrite ($CaSO_4$, PDF #01–072-0916), akermanite (Ca_2MgSi_2O, PDF #01-074-0990), and gehlenite ($Ca_2Al_2SiO_7$, PDF #00-035-0755) are evident [18]. The alkali excited samples exhibit distinct characteristics, such as a diffuse peak at a 2θ value of 29°–30°, corresponding to the presence of the reaction product C–S-H ($Ca1.5SiO_3.5H_2O$, PDF #00-033-0306), with partial substitution of Si by Al, commonly known as the

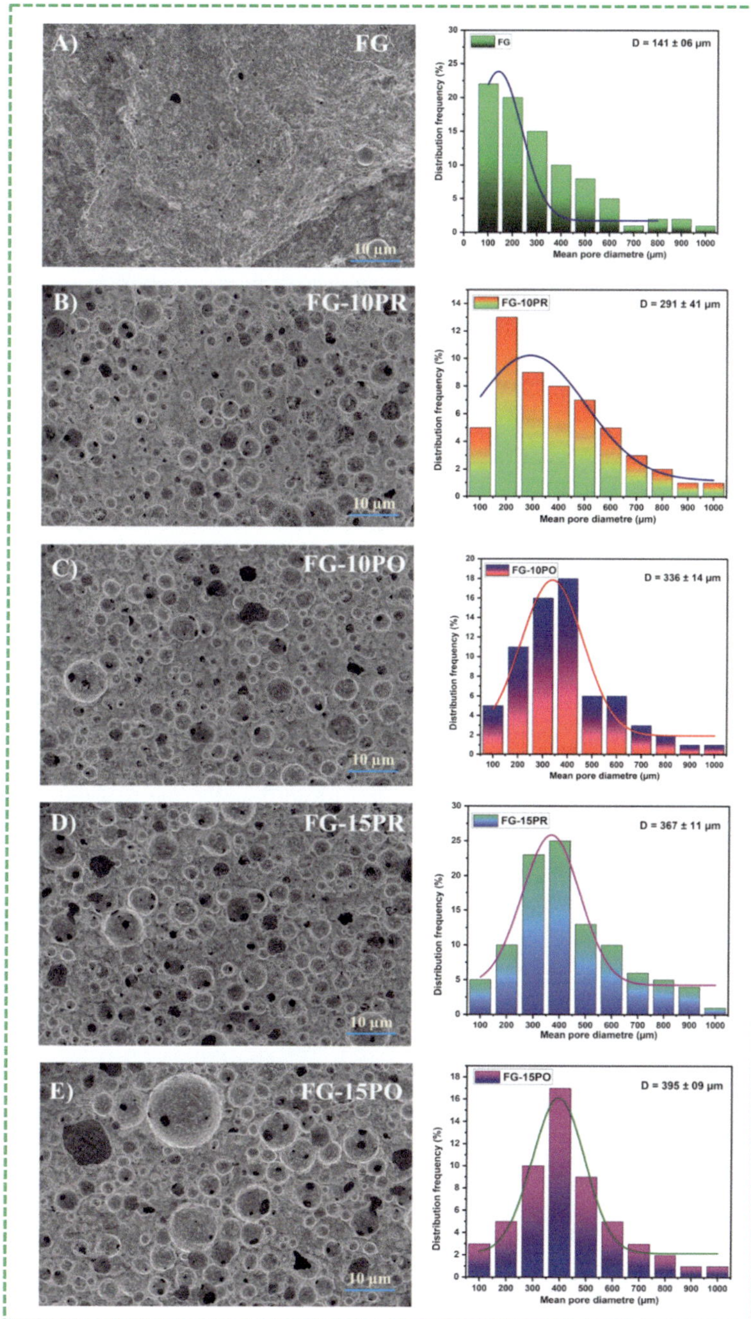

Fig. 2 SEM images of as-fabricated geopolymers containing different dosage of prefoam contents and the respective cell size analysis

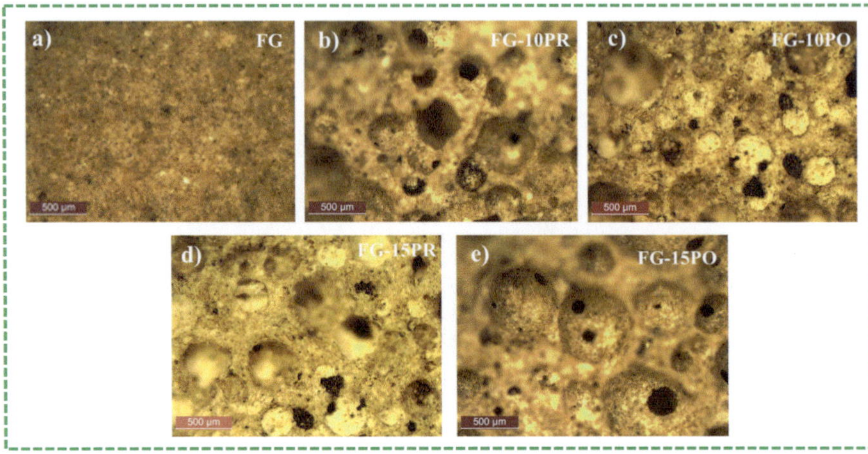

Fig. 3 Optical microscope image analysis of control and prefoam geopolymers

'C-A-S-H' gel. Additionally, reflections at a 2θ value between 20° and 28° are associated with N-A-S-H gel resulting from the alkali activation of FA. The presence of insoluble quantities of quartz and mullite constitutes the other crystalline phases in the geopolymer matrix [19]. The crystalline peaks observed in both FG and foam geopolymer samples are the same, but they appear with lower intensity in the foamed sample. Thus, it is clear to infer that the alkali activation reaction predominantly targets the amorphous phases of feedstock fly ash, GGBS leading to the formation of primarily amorphous products of gel type. Additionally, the foaming agent has minimal impact on the phases of geopolymer products.

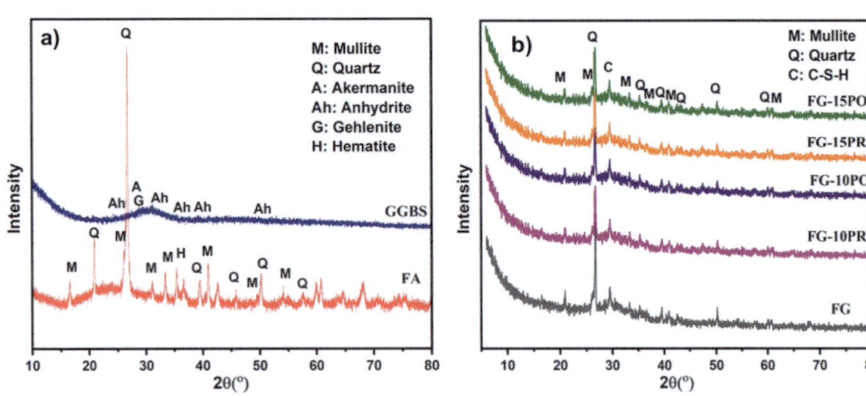

Fig. 4 XRD analysis of **a** Raw materials (FA & GGBS) and **b** Control (FG) and prefoam geopolymers (FG, FG-10PR, FG-10PO, FG-15PR & FG-15PO)

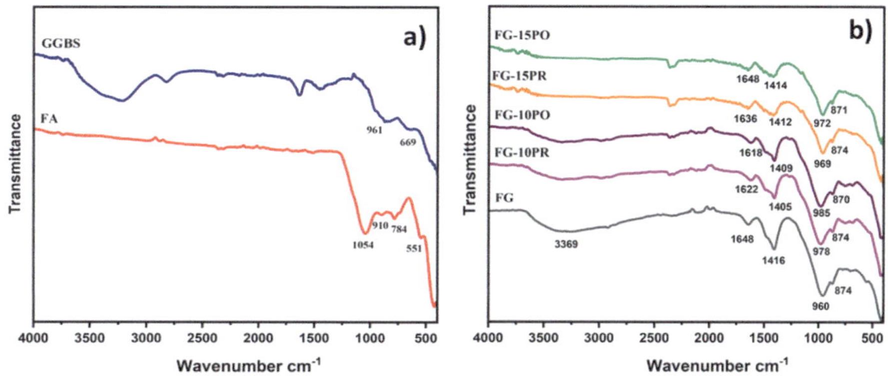

Fig. 5 FTIR analysis of **a** Raw materials (FA & GGBS) and **b** Control (FG) and Prefoam geopolymers (FG, FG-10PR, FG-10PO, FG-15PR & FG-15PO)

3.4 FTIR Analysis

FTIR results for the raw materials of GGBS, FA, and the geopolymer, both with and without a foaming agent, are illustrated in Fig. 5. In the geopolymer network, Si-O-Al and Si-O-Si connections constitute the primary tetrahedral coordination. These aluminosilicate networks, acting as network modifiers in the alkali hardener, can incorporate either sodium or calcium. The substitution of Al^{3+} for Si^{4+} during reactions indicates geopolymerization, evident in the shifting asymmetric vibration of the tetrahedral unit of T-O-Si (T-Al/Si) band within the wavenumber range of 900–1100 cm^{-1}, as observed in the raw material spectrum of FA and GGBS. This absorption band is shifted to a lower wavenumber in all geopolymer samples in the range of 985 to 960 cm^{-1} which provides confirmation of the network [18]. This distinct band is essential for condensing $Si(OH)_4$ and $Al(OH)_3$, thereby enhancing the precipitation of cementitious binders. Additionally, the vibration band observed at 1416–1405 cm^{-1} results from the –C–O stretching of carbonates [20]. Excessive alkali activators under atmospheric conditions lead to the formation of sodium carbonates. The vibrational bands at 3400^1 and 1648–1618 cm^{-1} correspond to the symmetric stretching vibration and bending vibration of hydroxyl groups originating from various water molecules, including free water, water within pores, and chemically bound water molecules [21].

3.5 Thermal Conductivity

The rigid sample exhibits the highest λ with 0.482 W/mK. The pre-made foam geopolymer samples (FG-10PR, FG-10PO, FG-15PR, and FG-15PO) have a λ of 0.14, 0.08, 0.04, and 0.02 W/mK, respectively. As the foam percentage increases, a decrease in thermal conductivity is observed in geopolymer samples. Specifically, the geopolymer with 15 wt% pre-made foam exhibits the lowest thermal insulation, registering 0.021 W/mK at a D of 620 kg/m^3. This suggests that, as expected, the increased water absorption can lead to a decrease in the thermal conductivity of foam geopolymer samples. The significant reduction in thermal conductivity with increasing foam content (0.482 W/mK for a rigid sample to 0.02 W/mK for 15 wt% foam) demonstrates the effectiveness of the pre-made foam approach for achieving excellent thermal insulation. Huiskes et al. [22] synthesized a foam geopolymer with a matrix similar to ours using a blend of FA-GGBS. They reported a slightly higher λ of 0.27 W/mK, having D of 720 kg/m^3. The achieved thermal conductivity of 0.021 W/mK at 620 kg/m^3 density is lower than the value reported by Huiskes et al. (0.27 W/mK), highlighting the potential advantage of the pre-made foam method for superior insulation at comparable densities.

3.6 Sound Absorption

Figure 6 illustrates the geopolymer's sound absorption coefficient (α) at different foam dosages and curing temperatures. The rigid sample exhibits a relatively low acoustic absorption coefficient within the range of 0.01 to 0.14 at 40–6300 Hz. The specimens prepared with pre-made foam dosages ranging from 10 to 1.5% and raising the setting temperature from the ambient temperature to 60 °C demonstrate a noteworthy acoustic absorption rate within the frequency range, with values ranging from 0.14 to 0.78. Similar reports by Zhang et al. [8] on the α of prefoamed FA/GGBS geopolymer concrete exhibited a maximum coefficient of 0.8. Nevertheless, this coefficient was linked to a highly concentrated absorption peak within the range of 40–150 Hz. In a contrasting approach, Neithalath et al. [23] created foamed concrete samples, reporting α ranging from 0.20 to 0.30, accompanied by a D range of 400–700 kg/m^3. Thus, the developed samples in this study showed a remarkable reduction of sound absorption. Further, the noise reduction values are computed.

The Noise Reduction Coefficient values for FG, FG-10PR, FG-10PO, FG-15PR, and FG-15PO are 0.09 ± 0.01, 0.33 ± 0.03, 0.48 ± 0.02, 0.53 ± 0.01, and 0.61 ± 0.02, respectively. The analysis indicates that NRC rises proportionally with increasing porosity. Generally, there is a close relationship between the structure of porous materials and their acoustic characteristics. This association arises from the direct link between sound absorption and the energy loss caused by friction in the pore walls of porous materials. Nonetheless, the porous geopolymer featured a closed pore network, limiting sound propagation and decreasing sound absorption. Hence, optimizing processing parameters, such as the ideal quantities of surfactant and foaming agent, is crucial to achieving porous geopolymer with significantly higher open porosity when aiming for enhanced acoustic absorption properties. The foam concrete produced displayed higher porosity, leading to increased sound absorption. Therefore, optimizing processing parameters, such as determining the ideal quantities of foaming agent and surfactant, is crucial for achieving foam concrete with

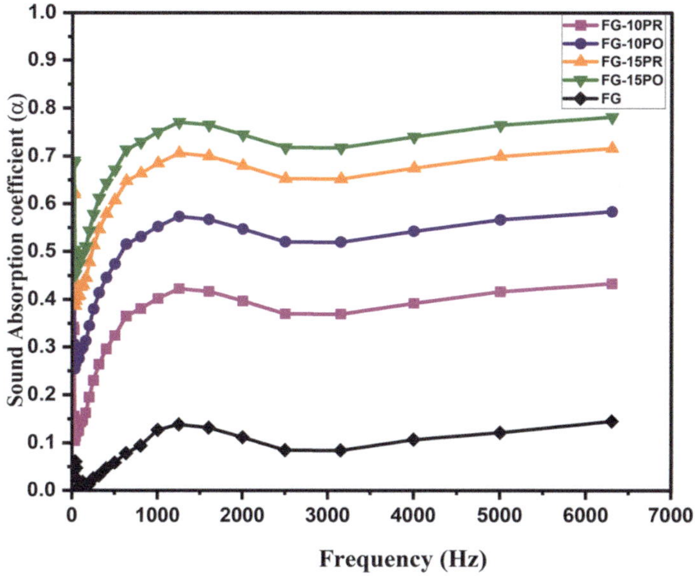

Fig. 6 Sound absorption analysis of control and prefoam geopolymers

significantly enhanced porosity when seeking improved sound absorption properties. The developed foam geopolymer attains a relatively high level of sound absorption performance, surpassing that of other foam plasters, thanks to its exceptional acoustic absorption properties.

4 Conclusion

In the current study, a porous matrix of fly ash-GGBS alkali excited geopolymers fabricated using the optimized dose of pre-made foam, showcasing a diverse range of densities. The 40% slag-infused foam geopolymer demonstrated 28 d compressive strengths that varied between 3.13 and 10.91 MPa, contingent upon the density. The thermal conductivity of the synthesized foam geopolymer system exhibited intriguing behavior, transitioning from 0.14 W/mK at a density of 1020 kg/m^3 to a significantly lower value at 0.02 W/mK at 620 kg/m^3. This noteworthy variation underscores the potential for tailoring the thermal properties of foam geopolymer based on specific density requirements. Additionally, thin lightweight geopolymer specimens with a thickness of 25 mm were subjected to acoustic absorption characteristics. Impressively, in the frequency range of 40–6300 Hz, these specimens demonstrated notable acoustic absorption rates (α) ranging from 0.14 to 0.78. This outcome suggests the viability of foam geopolymer in applications where acoustic performance is a critical consideration. Higher stability and lower thermal conductivity are the merits of lightweight geopolymers as eco-friendly construction materials. Future studies will be focused on the development of various building products and to find their performance in field conditions.

Contribution Statement

Manivannan Muthusamy, N. Vanitha, Janani Karuppaiyan, S. Surendhar: Investigation, Methodology, Data curation, Writing original draft.

R. Jeyalakshmi: Conceptualization, Formal analysis, Supervision.

5 Declaration of Competing Interest

The authors declare that they have no known competing financial interests or personal relationships that could have appeared to influence the work reported in this paper.

Acknowledgements This work was supported by the Department of Science and Technology, Government of. India (GOI), under the grant DST/TDT/WMT/2017 14/03/18, GOI. Authors greatly acknowledge the utilization of facilities provided by NRC, Department of Chemistry, SRM Institute of Science and Technology, Kattankulathur.

References

1. Davidovits J (2008) Geopolymer. In: Chemistry and applications. Institute Geopolymere, Saint-Quentin, France
2. Nejat P, Jomehzadeh F, Taheri MM, Gohari M, Majid MZA (2015) A global review of energy consumption, CO_2 emissions and policy in the residential sector (with an overview of the top ten CO_2 emitting countries). Renew Sustain Energy Rev 43:843–862. https://doi.org/10.1016/j.rser.2014.11.066
3. Hashim M, Tantray M (2021) Comparative study on the performance of protein and synthetic-based foaming agents used in foamed concrete. Case Stud Constr Mater 14:e00524. https://doi.org/10.1016/j.cscm.2021.e00524
4. Fediuk R, Amran M, Vatin N, Vasilev Y, Lesovik V, Ozbakkaloglu T (2021) Acoustic properties of innovative concretes: a review. Materials 14(2):398. https://doi.org/10.3390/ma14020398
5. Zhang Z, Provis JL, Reid A, Wang H (2014) Geopolymer foam concrete: an emerging material for sustainable construction. Constr Build Mater 56:113–127. https://doi.org/10.1016/j.conbuildmat.2014.01.081
6. Narayanan A, Shanmugasundaram P (2017) An experimental investigation on flyash-based geopolymer mortar under different curing regime for thermal analysis. Energy Build 138:539–545. https://doi.org/10.1016/j.enbuild.2016.12.079
7. Kočí V, Černý R (2022) Directly foamed geopolymers: a review of recent studies. Cement Concr Compos 130:104530. https://doi.org/10.1016/j.cemconcomp.2022.104530
8. Zhang Z, Provis JL, Reid A, Wang H (2015) Mechanical, thermal insulation, thermal resistance and acoustic absorption properties of geopolymer foam concrete. Cement Concr Compos 62:97–105. https://doi.org/10.1016/j.cemconcomp.2015.03.013
9. Hajimohammadi A, Ngo T, Kashani A (2018) Sustainable one-part geopolymer foams with glass fines versus sand as aggregates. Constr Build Mater 171:223–231. https://doi.org/10.1016/j.conbuildmat.2018.03.120

10. Yang K-H, Lee K-H, Song J-K, Gong M-H (2014) Properties and sustainability of alkali-activated slag foamed concrete. J Clean Prod 68:226–233. https://doi.org/10.1016/j.jclepro.2013.12.068
11. Yang K-H, Lo C-W, Huang J-S (2013) Production and properties of foamed reservoir sludge inorganic polymers. Cement Concr Compos 38:50–56. https://doi.org/10.1016/j.cemconcomp.2013.03.017
12. Hung T-C, Huang J-S, Wang Y-W, Lin K-Y (2014) Inorganic polymeric foam as a sound absorbing and insulating material. Constr Build Mater 50:328–334. https://doi.org/10.1016/j.conbuildmat.2013.09.042
13. Provis JL, Bernal SJ, van Deventer JSJ (2003) Alkali-activated materials: chemistry, microstructure and application. Cem Concr Res 33:1451–1460. https://doi.org/10.1007/978-94-007-7672-2
14. Choy YS, Huang L, Wang C (2009) Sound propagation in and low frequency noise absorption by helium-filled porous material. J Acoust Soc Am 126(6):3008–3019. https://doi.org/10.1121/1.3257182
15. Pacheco-Torgal F, Medeiros MA, Vieira CMF, Monteiro PMS, Labrincha JA (2010) Effect of the incorporation of metallic powders on the properties of fly ash-based geopolymer composites. Compos Part B Eng 41:303–311. https://doi.org/10.1016/j.measurement.2023.112616
16. Zhang Z, Zhu H, Zhou Y, Chen X (2017) Preparation and properties of fly ash-based geopolymer foams. Constr Build Mater 132:542–549. https://doi.org/10.1016/j.jclepro.2022.132043
17. Kitnasamy D, Pasupathy K, Ramakrishnan S, Sanjayan J (2020) Progress, current thinking and challenges in geopolymer foam concrete technology. Cem Concr Compos 103886. https://doi.org/10.1016/j.cemconcomp.2020.103886
18. Revathi T, Jeyalakshmi R (2020) Fly ash–GGBS geopolymer in boron environment: a study on rheology and microstructure by ATR FT-IR and MAS NMR. Constr Build Mater 120965. https://doi.org/10.1016/j.conbuildmat.2020.120965
19. Poornima N, Sivasakthi M, Jeyalakshmi R (2022) Microstructure investigation of the Na/Ca aluminosilicate hydrate gels and its thermal compatibility in fly ash–GGBS cementitious binder. J Build Eng 50:104168
20. Dlapa P, Bodí MB, Mataix-Solera J, Cerdà A, Doerr SH (2013) FT-IR spectroscopy reveals that ash water repellency is highly dependent on ash chemical composition. CATENA 108:35–43. https://doi.org/10.1016/j.catena.2012.02.011
21. Yan S, Zhang F, Feng X, Kong J, Wang B, Yang J (2021) Effect of high temperature on the mechanical properties of hierarchical porous cenosphere/geopolymer composite foams. Int J Appl Ceram Technol 18(3):817–829. https://doi.org/10.1111/ijac.13681
22. Huiskes DMA, Keulen A, Yu QL, Brouwers HJH (2016) Design and performance evaluation of ultra-lightweight geopolymer concrete. Mater Des 89:516–526. https://doi.org/10.1016/j.matdes.2015.09.167
23. Neithalath N, Weiss J, Olek J (2005) Acoustically efficient concretes through engineered pore structure. In: 2003 Autoclaved aerated concrete-properties and structural design. American Concrete Institute, pp 135–148

Seismic Stress Analysis of High-Speed Railway Track Embankment Under Saturated Pore Pressure Conditions

Pavan Kumar Emani and S. P. Mohan Swamy

Abstract The seismic stability of high-speed railway track embankment is of great importance for the safety and operational reliability of modern transportation networks, especially in regions subjected to seismic activity. The present study deals with a critical aspect of embankment performance and its behaviour under saturated pore pressure conditions during seismic events. By employing advanced geotechnical analysis and numerical simulations, this research provides a rigorous assessment of the gravity analysis, seismic response and stress analysis of railway track embankments under saturated pore pressure conditions. During seismic loading, the generation and dissipation of pore pressures are central to embankment stability. This research characterises the dynamics of pore pressure development within the embankment materials, focusing on the very complex interplay between excess pore water pressure and effective stress. By taking advantage of the advanced numerical simulations, the study thoroughly investigates the embankment's response to seismic forces, with careful consideration of the influence of saturated pore pressures. This analysis offers valuable insights into potential deformations, settlements, stress and strain developments that occur during seismic events. The results of this study significantly advance our understanding of the complex interplay between seismic forces, saturated pore pressures and embankment responses. Such insights are vital for the development of engineering solutions aimed at ensuring the long-term safety and high performance of high-speed railway systems in regions susceptible to seismic activity.

Keywords High-speed railway track · Pore pressure development · Seismic simulation · OpenSeesPy modelling

P. K. Emani (✉) · S. P. Mohan Swamy
Siddaganga Institute of Technology, Tumkur, India
e-mail: pavankumar@sit.ac.in

Y. K. Guruprasad et al. (eds.), *Recent Advances in Infrastructure Development and Management—Volume 2*, Lecture Notes in Civil Engineering 566,
https://doi.org/10.1007/978-981-96-1558-2_13

1 Introduction

A high-speed railway track is a specially designed railroad track, built to facilitate the operation of trains that can travel at much higher speeds than traditional trains. These tracks are constructed to enable high-speed train operation, often exceeding 155 miles per hour (250 kms per hour) and, in some instances, even faster. Analysing the connection between saturated pore pressure and deformation in high-speed railway track embankments is crucial due to the dynamic loading they experience from passing trains affecting both pore pressure and deformation. This understanding is essential for assessing the embankments stability and overall performance. Hence, this analysis helps to quantify how these embankments respond including deformations, lateral displacements and vertical strains.

Reference [1] stated that elevated water table levels lead to a reduction in the stiffness of the geological formation and an increase in dynamic responses. When train speed exceeds the Rayleigh wave velocity in the ground, there is a significant risk of elevated ground vibration levels [2] added. In the study done by [3, 4], complex interactions between load speed, soil permeability and dynamic pore water pressure generation in saturated soil beneath a track are demonstrated. Vibration levels vary based on the earthwork type with embankments producing the lowest and cuttings the highest [5, 6] stated that dynamic soil stresses and vibrations rise with higher water content or water table levels. Reference [1] stated that factors like permeability and porosity play a crucial role in sub-grade deformation as higher permeability reduces the potential for excess pore pressure buildup.

2 Methodology

The modelling and analysis of high-speed railway track embankments are conducted using OpenSeesPy, an open-source Python library. The process of numerical simulation using OpenSeesPy involves estimation of the geostatic stresses using gravity analysis. The vertical displacement at the base is restrained to simulate a vertically fixed foundation. This geostatic stress analysis is very precious for engineers as it provides insights into how a high-speed railway track structure reacts to the force of gravity, which is fundamental for evaluating its stability and performance under real-world conditions. In OpenSeesPy, the gravity analysis is performed as a dynamic analysis with very large time steps and a large value for the permeability of soils. This is done to achieve hydrostatic pressures at all points rapidly.

The next part of the analysis is seismic analysis. This involves analysing how the HSRT embankment responds to seismic forces such as ground shaking scenarios during earthquakes. This assessment is crucial to ensure their stability and safety within railway infrastructure. This process involves conducting a dynamic analysis, unlike the static analysis for gravity loads. For seismic analysis, a dynamic approach is adopted. This step involves introduction of ground motion records or accelerograms which simulate the time-dependent earthquake induced forces acting on the embankment. This dynamic analysis captures the structure's response over time as seismic waves pass through it. Boundary conditions are then established to represent the constraints imposed on the embankment during seismic events. This includes fixing the base of an embankment in vertical direction and giving equal degree of freedom to the nodes of the embankment along left and right faces. Once these conditions are set, the model is ready to execute the dynamic analysis using the accelerograms as input. This dynamic finite element analysis computes the embankment's response including deformation and stress distribution over time as seismic waves impact it. Upon completing the dynamic analysis, post-processing is carried out to extract critical data such as displacements, stresses and strains at various locations within the embankment.

The overall study is carried out to analyse the behaviour of HSRT embankment and its response to gravitational and seismic loading conditions under saturated pore pressure. The embankment is modelled and numerical results and analysed by plotting the deformations, stress, strain and pore pressure contours and the same plots are briefly discussed in results.

3 Details of the Numerical Model

The HSRT embankment model was created in OpenSeesPy software using the 'basic model builder' of two dimensions and three degrees of freedom. The model involves studying the HSRT embankment under saturated pore pressure conditions. The slope of the HSRT embankment modelled is 1.5:1. The width of the concrete layer is 3 m and thickness or depth of 0.3 m. A coarse fill of 0.4 m depth is modelled. A sub-ballast of 2.4 m depth and a sub-grade with 5 m depth is modelled. After creating an embankment with this geometry, the embankment model is discretised into finite number of 9-noded Quad $u - p$ elements. The meshing method used in the modelling is structured finite element meshing. The HSRT embankment is then divided into four parts namely: sub-grade, sub-ballast, coarse fill and concrete layer, each having different material properties. The total number of nodes formed during the modelling is 1965 nodes connecting 465 elements. The material used for the modelling of HSRT embankment is PressureIndependMultiYield material. The material modelling having slightly varying properties in four different layers is used. Finite element meshing, commonly referred to as meshing, plays a critical role in the accuracy of the finite element analysis (FEA) results. It involves dividing complex geometric domains or structures into smaller interconnected subdomains or elements. These elements are typically simple geometric shapes such as triangles or quadrilaterals in two dimensions (2D). Meshing is an essential step in addressing and solving complex engineering problems numerically. For the modelling of an embankment using the Nine Four Node Quad $u - p$ element, the sub-grade is divided into a total of 330 mesh elements. This is achieved by using 33 elements along the

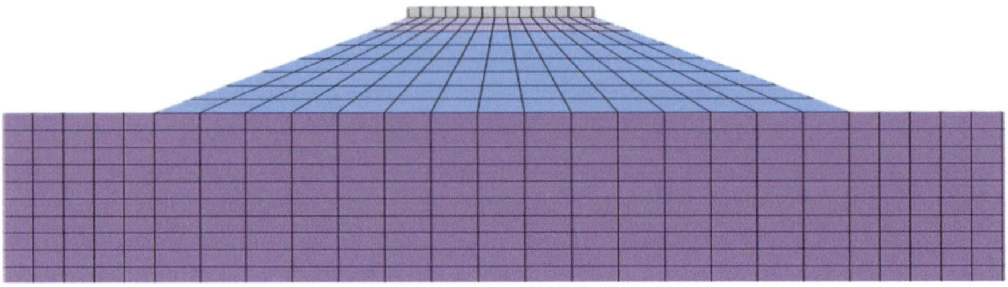

Fig. 1 Finite element meshing of HSRT embankment

horizontal x-direction and 10 elements along the horizontal y-direction. The sub-ballast portion is composed of 90 elements, distributed with 15 elements along the horizontal x-direction and 6 elements along the vertical y-direction.

The coarse fill, also known as the roadbed, is discretised into 30 elements with 15 elements along the horizontal x-direction and 2 elements along the vertical y-direction. And the topmost layer, the concrete layer, is discretised into 15 elements, with the distribution of 15 elements along the horizontal x-direction and only single element along the vertical y-direction. The meshing of the HSRT embankment is shown in Fig. 1. The element used in the modelling process is Nine Four Node Quad $u - p$ element. Nine Four Node Quad $u - p$ element has been specifically developed for the simulation of dynamic responses in solid–fluid fully coupled materials.

To ensure the overall stability of the HSRT embankment, the base is rigidly fixed to prevent vertical movement. This is achieved by using the 'fix' command in the analysis and fixing the nodes in the required direction. Therefore, nodes at the base of the embankment are immobilised in vertical direction. In gravity analysis of the embankment, the base nodes are fixed vertically and for the side phase, the rollers are horizontally constrained and dynamic axle loading is applied for the analysis. Whereas, in seismic analysis, the constraints on the side phase are removed and seismic excitation file, which stores acceleration values to be given to the base and side faces, is imposed. This excitation can be either a defined seismic load or recorded acceleration data from the specific site. The acceleration values provided by one-dimensional ground response analysis represent the ground motion experienced by the earthen slope. These acceleration values applied at the base and at the soil-structure interface form the loading for seismic analysis of the HSRT embankment.

4 Data Collection and Analysis

The material properties used for the analysis such as permeability, density, etc. are specified and the particular properties are given as inputs through material definition through input command file. The seismic data used as an input for the seismic analysis is recorded data from El-centro earthquake station located at the Imperial valley. In the analysis part, gravity analysis of high-speed railway track embankment using

OpenSeesPy involves simulating the embankment's response to gravitational forces. The time step size of 0.01 s used in analysis is compatible with the shear wave velocity of 175.74 m/s, damping ratio of 0.05 and integration using Newmark integrator method. The total time steps used are 2000. When creating the excitation pattern, ground motion records that are responsible for the seismic hazard in the region of seismic affected area are to be selected. These records are usually obtained from the established seismic stations like the one mentioned earlier and should be compatible with the analysis requirements. While preparing the selected ground motion record for analysis, they were pre-processed by adjusting the time step, filtering out high-frequency noise and ensuring that they are properly scaled to match the target seismic intensity. The processing of seismic inputs was carried out using the method of Fourier transformation reported in [7] and is shown in Fig. 2.

To simulate seismic excitation, the ground motion records from one-dimensional ground response analysis are applied at various levels of the HSRT embankment model. using OpenSeesPy commands, namely, Ground-motion, imposed ground motion and multiple support excitation pattern to apply the selected ground motion record as nodal accelerations. The intensity of the applied seismic data was ascended from base to the top node of the embankment to achieve the actual seismic environment. After this analysis, procedures are defined, which included setting up the load

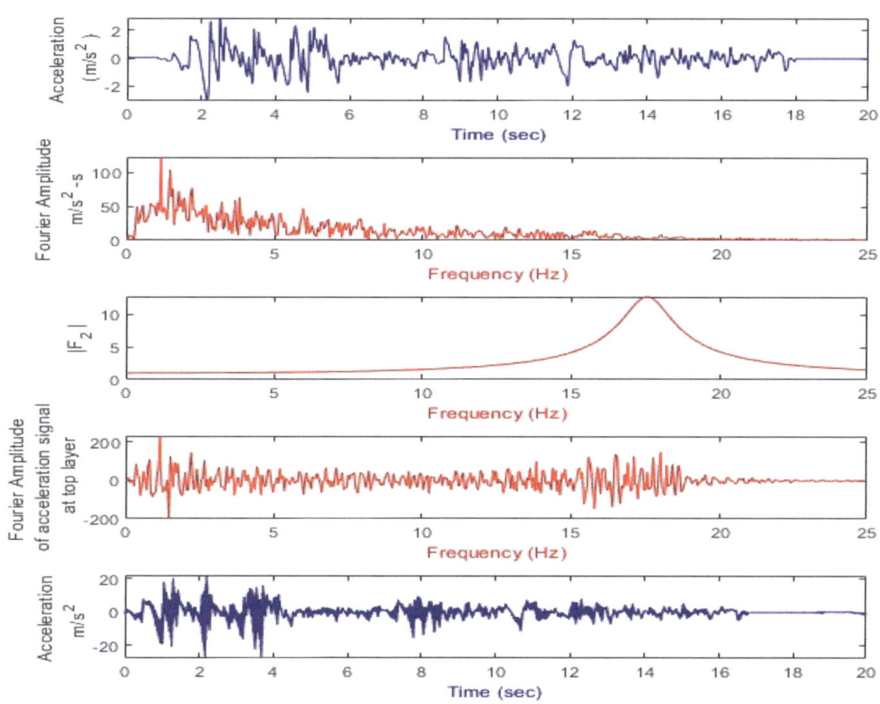

Fig. 2 One-dimensional ground response analysis method using Fourier transform

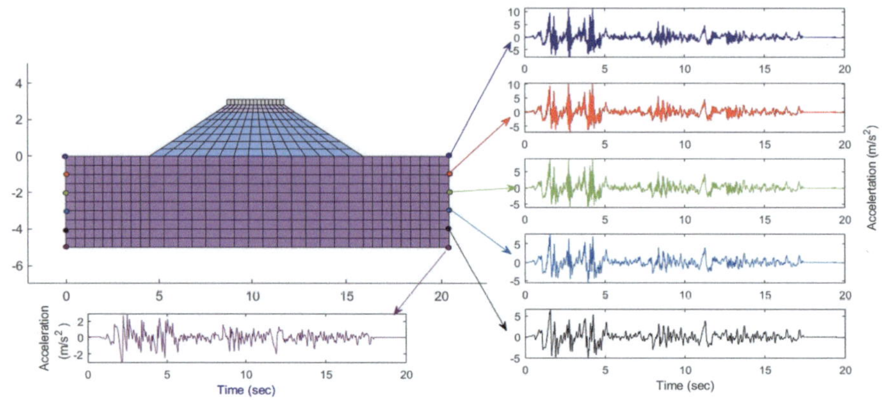

Fig. 3 Ground response inputs at various levels of the subsoil

pattern. In this research work, plain load pattern is used for the seismic loading and also other necessary procedures like convergence criteria are defined. After running the seismic analysis using OpenSeesPy commands, the MATLAB-based post-processing commands are used to analyse the numerical response printed in the output files. This involves obtaining the plots of deformations, displacements, stresses, strains, time histories and pore pressure contours. After completing the analysis, post-processing of the results to visualise and interpret the embankment's behaviour is done.

Figure 3 represents the input acceleration versus time plots at various depths of an embankment. It depicts the dynamic input that an embankment or any structure experiences due to the seismic waves generated by the earthquake. The acceleration values which are to be given as inputs were modified or amplified using Fourier transformation and then the acceleration values are stored in text format. The embankment from base to top face consists of 21 nodal points and so 21 modified seismic inputs are given at different nodes to simulate the seismic condition and the analysis was run up to 2000 time steps. In the figure, only the selected nodal plots are depicted for your reference.

5 Results

In the results section, the results obtained from the gravity analysis of the HSRT embankment are discussed first. Figure 4a represents the deformed shape. The deformed shape shows a downward settlement of the HSRT embankment due to the self-weight of an embankment. In the figure, the dotted lines, as indicated in the legend, represent the actual shape or geometry of an embankment and the solid lines represent the settled or deformed shape of an embankment under the action of gravity. The pore pressure analysis is shown in Fig. 4b, the pore pressure at the bottom of an embankment is much higher and it diminishes as we move towards the top of an embankment, which is exactly similar to the hydrostatic conditions. The vertical stress contours of an embankment generally refer to the distribution of vertical stresses within the embankment soil due to the applied weights or loads. When constructing the earthen or any embankment, the weight of the embankment material from which it is being constructed and the additional loads such as traffic loads create stresses within the soil layers. Here, the vertical stress contours demonstrate the stresses which might vary at different depths and various locations within an embankment. The vertical stresses are high at the base of the embankment and decrease with the depth as shown in Fig. 4c. These contours help very much to understand the stress distribution, which is so important for designing a safe and stable embankment. Vertical stress contours are of utmost importance as they are used to ensure that the embankment can support the loads applied without undergoing any settlement or failure. Shear stress measures the force applied parallel to a surface per unit area. The main effect of the shear stress is that it causes one layer of material to slide or to deform relative to an adjacent layer.

Figure 5 represents the deformed shape of the HSRT embankment at various time steps. The time steps selected for plotting the deformed shapes are 500, 1000, 1500 and 2000 and the plots of the same are represented consecutively in the figure. The magnitude of the seismic values is high at the top of an embankment and it gets reduced as we move towards the base. Hence, the deformation is higher at the top end of an embankment and it is minimal towards the base. Also, it can be seen that as the duration of seismic activity increases, the deformation will only get worse and finally lead to the shear collapse of an embankment. Figure 6 represents the displacement histories of nodes in the sub-base layer along the y-direction. The x-displacement graph indicates the displacement that has undergone by the embankment in the form of the sideways movement during the seismic event. The overall lateral displacement

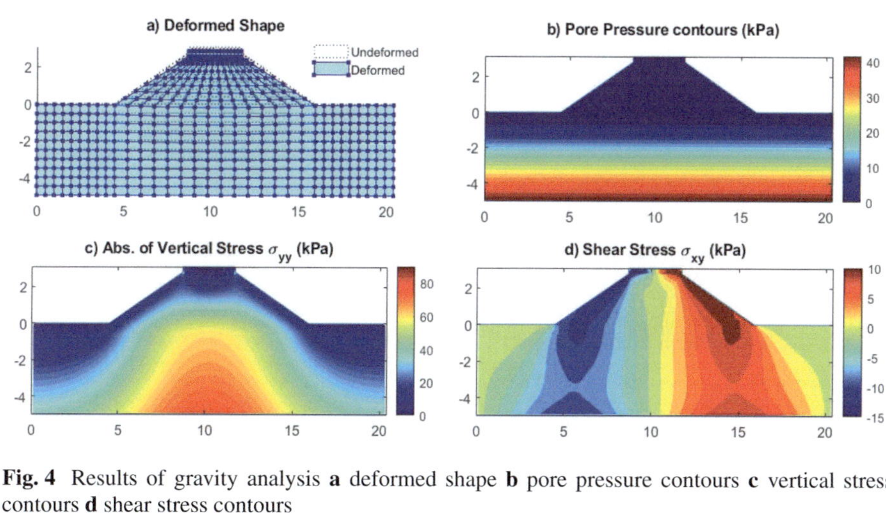

Fig. 4 Results of gravity analysis **a** deformed shape **b** pore pressure contours **c** vertical stress contours **d** shear stress contours

of an embankment is 600 mm in the span of 20 s. *Y*-displacement time history graph indicates the displacement of the selected nodes. The displacement of the 395 node is more as it is present in the upper side of the selected node number in the embankment and also due to the presence of higher lateral displacement. The displacement of the 389 and 383 node is relatively less as compared to node 395. Whereas the displacement of the 379 node is nil or zero as it is fixed according to the boundary condition. Figure 7 indicates the element number and its location along the embankment which are selected as the critical elements for stress–strain study. Figures 8, 9, 10 and 11 represent the stress–strain histories of the respective elements under seismic loading. As we move from top to bottom, the values of shear stress and shear strain keep on increasing. The most common thing which is observed in the stress–strain study of the critical elements is that the ratio of shear stress to yield stress is a linear graph which indicates the elastic property of the element. Also, it can be noted that the deviatoric stress is directly proportional to the amount of confinement, i.e., the deviatoric stress keeps on increasing with the amount of confinement. From the results of all the elements presented in Figs. 8, 9, 10 and 11, it can be observed that the material of the embankment remained in elastic condition during the seismic event. A possible reason is the lack of due amplification factor due to which a very weak PGA is applied to the embankment leading to very elastic results. A maximum shear stress value of 50 kPa is recorded in the whole of embankment, which is too small. So, also a maximum shear strain of 0.1% is too small. Stronger ground motion can initiate yielding, development of high excess pore pressures and instabilities.

Figure 12 compares the shear stress and shear strain time histories at the centre of the integration point of the critical elements. This comparative study easily points out that the shear strain and shear stress values go on increasing as we move towards the base.

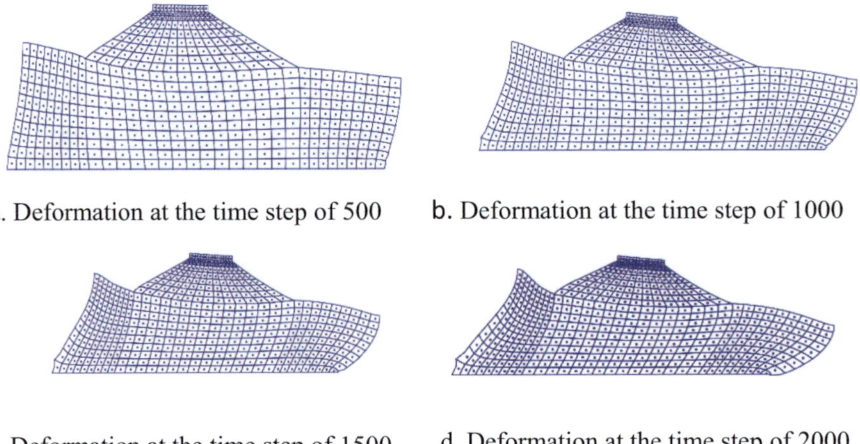

a. Deformation at the time step of 500

b. Deformation at the time step of 1000

c. Deformation at the time step of 1500

d. Deformation at the time step of 2000

Fig. 5 Deformation of the HSRT embankment due to seismic loads

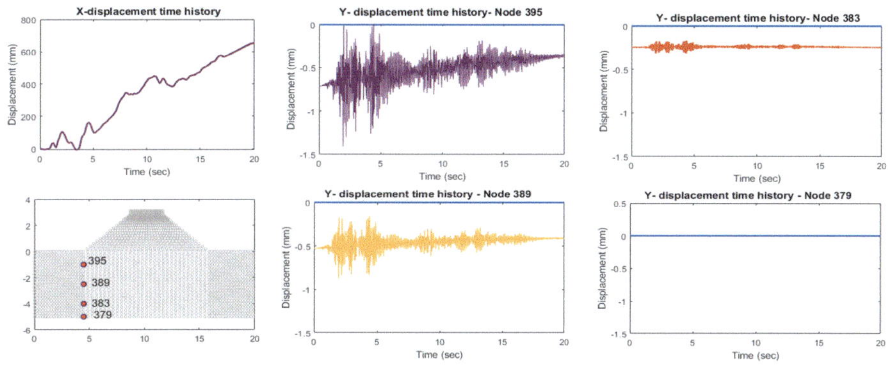

Fig. 6 *Y*-displacement histories of nodes in the sub-base layer

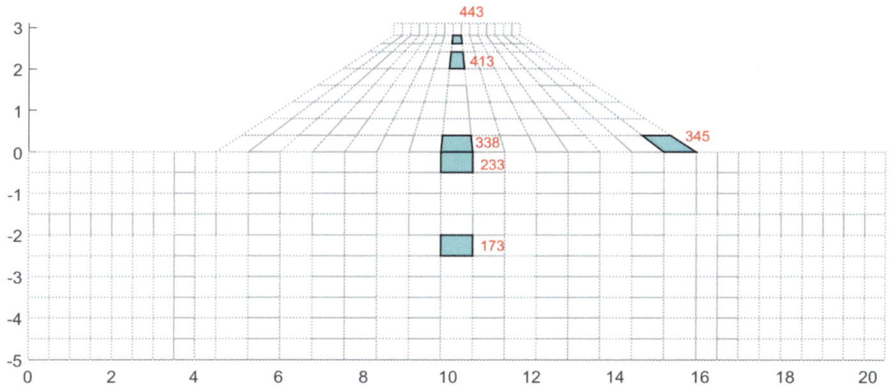

Fig. 7 Critical elements for stress–strain study

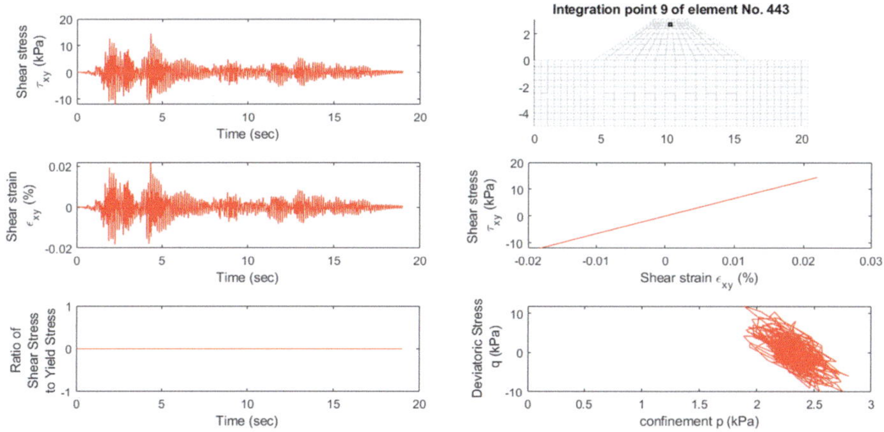

Fig. 8 Stress–strain histories of element 443 under seismic loading

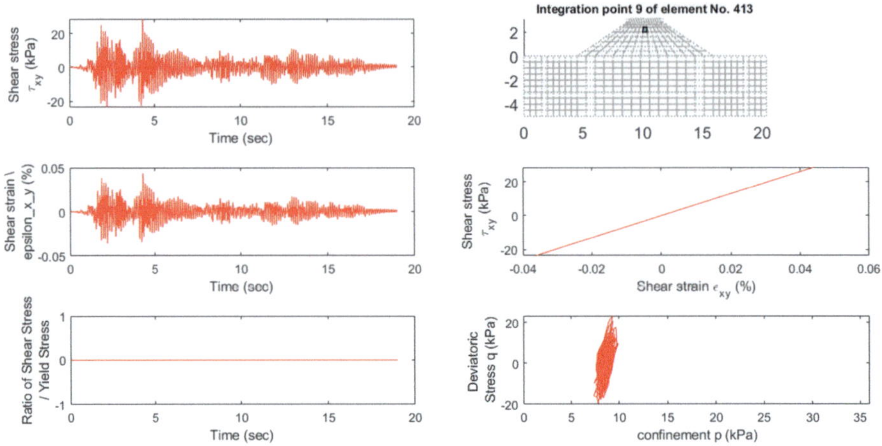

Fig. 9 Stress–strain histories of element 413 under seismic loading

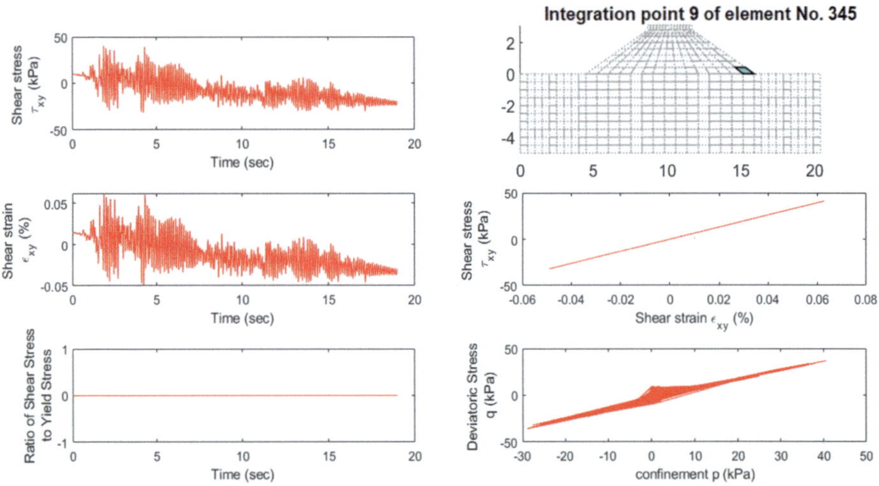

Fig. 10 Stress–strain histories of element 345 under seismic loading

Figure 13 represents the contour plots of an embankment subjected to the seismic analysis at the time step of 251, i.e., 2.51 s. The distribution of shear strain which is the measure of the deformation due to the application of shear stress that acts parallel to the surface can be shown. The contour plot of shear stress to yield stress is almost zero as the material is elastic. The plot of the absolute value of vertical stress has some value greater than 0.8 due to the self-weight and dynamic loading experienced by the embankment. The shear stress contours indicate the values around 400 within an embankment. Pore pressure in soils can experience significant changes due to the dynamic loading caused by seismic inputs. While looking at the pore pressure

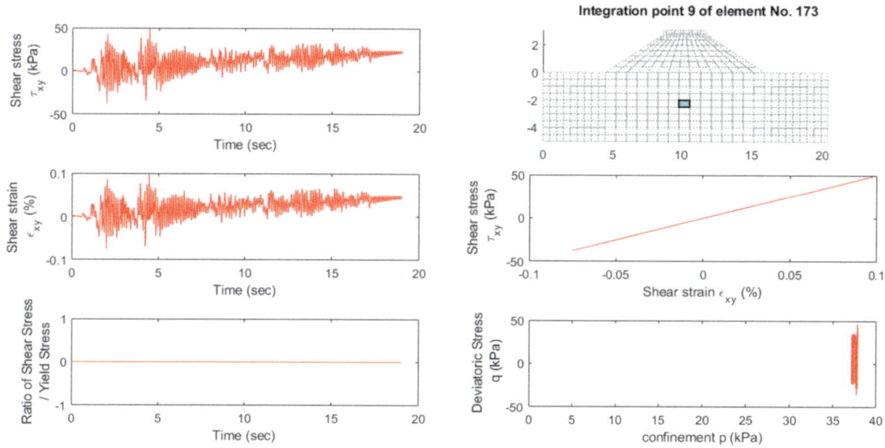

Fig. 11 Stress–strain histories of element 173 under seismic loading

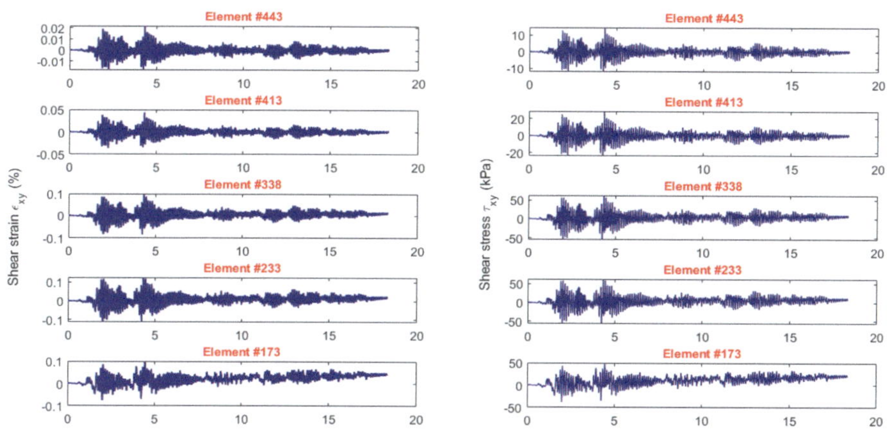

Fig. 12 Comparative time histories of shear stress and shear strains at the centre of the critical elements

contours, we can say that the pore pressure is almost dissipated due to the action of seismic loading and there is only some pore pressure left at some bottom most corner of an embankment.

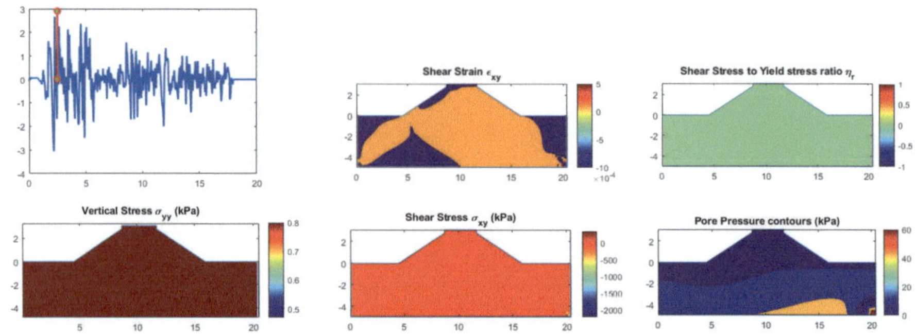

Fig. 13 Contour plots of the shear stresses and strains at times step 251 = 2.51 s

6 Conclusions

- The embankment analysed using OpensSeesPy simulates the embankment in the most realistic way. It shows downward deformation or displacement during the gravity loading and more sideways displacement during the seismic or dynamic loading.
- The pore pressure effect generated using Nine Four Node Quad $u - p$ element worked well in the process of simulation giving more pore pressure at the base and almost dissipating at the top.
- Saturated conditions lead to the pore pressure buildup within the embankment affecting the effective stress and shear strength of the soil. The ratio of shear stress to yield stress is almost zero, which indicates the elastic property of the material modelled. This can be due to a lack of due scaling of the seismic motion.
- Stronger seismic loading can induce increased stress on an embankment, affecting its stability.

References

1. Jiang H, Bian X, Jiang J, Chen Y (2016) Dynamic performance of high-speed railway formation with the rise of water table. Eng Geol 206:18–32
2. Krylov VV (1995) Generation of ground vibrations by superfast trains. Appl Acoust 44(2):149–164
3. Bian X, Hu J, Thompson D, Powrie W (2016) Pore pressure generation in a poro-elastic soil under moving train loads. Soil Dyn Earthq Eng 125:105711
4. Patchamatla JRR, Emani PK, Kumar R (2024) Dynamic axle load effects on high-speed railway embankment: pore pressure and deformation in varied subsoils. Geotech Geol Eng 42(5):3561–3578. https://doi.org/10.1007/s10706-024-02744-1

5. Connolly DP, Kouroussis G, Woodward PK, Costa PA, Verlinden O, Forde MC (2014) Field testing and analysis of high-speed rail vibrations. Soil Dyn Earthq Eng 67:102–118
6. Jiang H, Bian X, Cheng C, Chen Y, Chen R (2016) Simulating train moving loads in physical model testing of railway infrastructure and its numerical calibration. Acta Geotech 11:231–242
7. Kramer SL (1996) Geotechnical earthquake engineering. Prentice-Hall, New Jersey

Effective Precoating Treatment for Recycled Concrete Aggregates to Improve the Compressive Strength of Concrete

C. Chella Gifta, M. Lavanya, R. Sridharan, V. Kalimuthu, and M. Chandhru

Abstract The rapid depletion of natural aggregate resources and the growing environmental concerns associated with construction waste have prompted a shift toward sustainable building practices. Hence, the utilization of recycled aggregates obtained from buildings and demolition waste has become a perfect solution. However, the high water absorption, poor interfacial zone, and presence of adhered mortar are the major drawbacks that hinder the wide application of recycled aggregates. Therefore, this study aimed to improve the performance of recycled aggregates by using a suitable precoating technique. Supplementary cementitious materials such as fly ash, silica fume, and rice husk ash were chosen as precoating materials for the development of treated RCAs and specimens were prepared with this pretreated recycled aggregate. Water absorption studies were conducted on the test specimens with natural and recycled aggregates with and without precoating treatment. Among the various pozzolanic materials, the silica fume-coated aggregates had a positive effect on the RA and reduced the water absorption by 46.49% compared with the untreated RA. Compressive strength is almost similar to the natural aggregate concrete. This approach not only addresses the technical improvements but also aligns the study with broader sustainability goals in construction, making the research more relevant to current industry challenges.

Keywords Recycled aggregate · Water absorption · Compressive strength

1 Introduction

Concrete sustainability is a major problem in the construction industry because we use a lot of natural resources to produce concrete, which is the most important building material. One such sustainability challenge that we must address is the depletion of

C. C. Gifta (✉) · M. Lavanya · R. Sridharan · V. Kalimuthu · M. Chandhru
Department of Civil Engineering, National Engineering College, Kovilpatti, Tamil Nadu, India
e-mail: ccgcivil@nec.edu.in

Y. K. Guruprasad et al. (eds.), *Recent Advances in Infrastructure Development and Management—Volume 2*, Lecture Notes in Civil Engineering 566,
https://doi.org/10.1007/978-981-96-1558-2_14

natural resources. Construction industry recently adopted a trend of using alternative sources of construction materials in the place of virgin sources. Owing to the construction of new buildings and the demolition and maintenance of old buildings, excessive construction and demolition (C&D) waste is being created worldwide. Their improper disposal causes flooding and other negative effects on human health, environmental problems such as landfill shortages, etc. Utilization of recycled aggregates, derived from concrete, ceramics, and brick masonry, addresses natural resource depletion. Shortage in raw materials puts us at high risk because they may increase the market price of the concrete and due to this reason, the building industry is looking for alternative environmentally friendly cheap aggregate materials. Therefore, construction and demolished waste, called recycled concrete aggregate (RCA), is recycled and regenerated as a new material. C&D waste is recycled into smaller particles, and the properties of the resulting RCA fully depend upon the number of recycling stages and the methods involved in the crushing process. A few buildings like Shangai ecological house and Pelican Self Storage building are made from recycled aggregate and 80% to 100% of the concrete is produced from the demolished buildings, reducing the demand for the non-renewable virgin materials. With advancements in the field of concrete technology, the impact of recycled aggregates on the development of mechanical, shrinkage, permeability, creep deformation, and durability properties of concrete has been explored in many ways. However, due to their adverse effects on concrete, it is considered a low-value product and is recommended for low-level replacement in practical applications. The inferior performance of recycled aggregates has been reported because of their excessive water absorption, high crushing capacity, lower apparent density, and porosity, etc. The weak transition zone developed between the old mortar and natural aggregate hinders its application, and many attempts are being made to improve the quality of recycled aggregate. Ball mill grinding, acid washing, microbial carbonate precipitation, precoating, or slurry pretreatment are a few examples. Among the various methods, the precoating technique is very easy to perform and inexpensive. On the other hand, the presence of acidic ions in the coating materials sometimes reduces the pH and slightly improves the strength of the concrete. However, surface pretreatment with some supplementary cementitious materials like zeolite, as mentioned in the literature, is very expensive and not locally available.

The use of C&D waste as road base and subbases in pavement construction was reported by Rabin Tuladhar. K&C-RCA's high clay content made it suitable for use as a selective fill to replace or enhance subgrades [1]. A laboratory investigation on the precoating technique of recycled concrete aggregates by Veda Vardhan Reddy et al. summarized the cost-effectiveness of recycled aggregates in concrete, which constitutes between 25 and 35% of the total aggregate production [2]. MS Pourtahmasb presented a practical way of reusing recycled aggregate in a special type of asphalt mix called stone mastic asphalt (SMA) to partially replace coarse and fine aggregates [3]. According to Priyadharshini Perumal and Mamdouh Omran, the water absorption of recycled aggregate was decreased by 10% by the carbonation process. Moreover, carbonation significantly improves the interfacial transition zone of the concrete made with recycled aggregate when applied after pozzolanic coating, increasing the compressive strength [4]. Jun Lu Zhedong et al. experimented with recycled aggregates from crushed hydrated cement mortar pretreated with CO_2 curing. Three treatment methods were adopted: standard CO_2-curing, CO_2-curing with presaturated $Ca(OH)_2$ immersion, and CO_2-curing with cement slurry prewrapping [5]. Olonade et al. [6] reported that the problem of forming a weak zone or inadequate bonding within the cement paste matrix is most common in RCA, provided that the aggregates are adequately and effectively pretreated [6]. Bio-deposition method is microbially-induced carbonate precipitation and during the process, sodium alginate (SA) was utilized to increase the homogeneous precipitation on the aggregate surfaces. When SA is applied for bio-deposition, aggregate strength and recycled concrete durability can be increased [7]. Bruno Wenzel et al. reported the greatest improvement achieved in the properties of recycled aggregate precoated with zeolite powder cement (ZP-C) slurry [8].

Using statistical analysis of the findings from previous literature, Ouyang et al. examined the effects of several pretreatment techniques (removal, polymer impregnation, pozzolanic slurry immersion, accelerated carbonation curing, and bio-deposition) on the intrinsic properties of RCAs and compared their efficacy. Pretreatment techniques are very appealing, according to the authors, as they improve mechanical strength and durability and reduce drying shrinkage [9]. Ning Li et al. developed a high-impact resistance self-compacting concrete by precoating coarse particles with an energy-absorbing layer made of asphalt [10]. Yongjae Kim et al. compared the various pretreatment techniques, viz-a-viz HCl and Na_2SO_4 pretreatment reported the most efficient and feasible method of pretreatment [11]. Manu Garg et al. used epoxy-coated recycled aggregate by brushing it on the surface and added glass fibers to overcome the inferior problems associated with recycled waste aggregates [12]. Dong Lu et al. reported the results of concrete containing graphene oxide @RCA, strongly indicating the locality at the ITZ when compared with non-coated RCA specimens [13]. Graphene oxide used in the precoating of recycled aggregate presents superior performance in terms of mechanical and durability characteristics of the concrete and the local cement hydration at the ITZ was effectively promoted by the hydrophilic functional groups.

From the literature review, it is clear that the performance of RCAs mainly depends on the C&D waste source, grade of concrete, exposure to environmental conditions, etc. Therefore, this study attempts to evaluate effective precoating materials, which are locally available by analyzing the performance of four different pozzolanic materials: fly ash, silica fume, rice husk, ash and cement. It mainly focused on the performance of precoated recycled aggregate by measuring the water absorption, and the best-performing SCM coating was further considered for the compressive strength comparison. The influence of precoating materials on the aggregate properties was analyzed for all four materials, and an efficient precoating method was identified. Furthermore, the strength of the concrete cube specimens was made with the best pretreated RCA. The use of waste and demolished concrete as coarse aggregates provides several benefits, such as the effective recycling of waste and disposal, economic benefits, reduced landfills, the replacement of natural aggregates, resource conservation, and a sustainable environment.

Therefore, the objectives of this study are as follows:

- To evaluate the effectiveness of precoating pozzolanic materials for recycled aggregates based on water absorption properties;
- To explore the compressive strength of the concrete prepared with the best pretreated recycled concrete aggregate.

2 Experimental Program

2.1 Natural and Recycled Aggregates

Natural coarse aggregates of silicious origin were obtained from a local crusher, confirming that the IS383 specification was used. The specific gravity of the 20 mm and 12.5 mm aggregates under SSD conditions is 2.8 and 2.6, respectively. The crushing strength, water absorption, and impact resistance of the natural aggregate (NA) were 24.58%, 0.3%, and 18.91%, respectively. Manufactured sand with a specific gravity 2.65 was used as the fine aggregate. This experimental study mainly focused on the efficient precoating materials required to enhance the compressive strength of recycled concrete aggregate (RCA) obtained from Construction & Demolition waste. The basic properties of the crushed RCAs were measured, and the results were compared with those of the NA. The specific gravity is much lower than the natural aggregates, and the crushing strength, water absorption, and impact resistance of recycled aggregates are 2.10%, 26.6%, 3.85% and 13.58%, respectively. Figure 1 shows the characterization of the natural and recycled aggregates before surface treatment.

2.2 Material for Precoating

This study explored the strengthening of recycled aggregates with four locally available cementitious materials: fly ash, silica fume, rice husk ash, and cement. Then, the results are compared with those of non-coated RCAs and natural aggregates. The recycled aggregate precoated with SCM enhances the performance of recycled aggregate and also provides a solution for industrial waste disposal problems. Fly ash is a noncombustible mineral obtained from the coal burning process, and a small quantity of carbon remains after incomplete combustion and produces fine fly ash powder. Silica fume is a byproduct of making elemental silicon or alloys containing silicon in electric arc furnaces. At about 200 °C, high-purity quartz is transformed into silicon, resulting in silicon dioxide vapor, which at low temperatures oxidizes and condenses to form silica fume. Rice husk ash is vegetable waste ash obtained from charred rice husk burnt at 800 °C and then ground to few micron level as a fine powder. In the present study, locally available 53-grade OPC was used as a precoating

Fig. 1 Testing of natural and recycled

material for surface treatment and for casting the test specimens. Figure 2 shows the SCMs used in the precoating technique.

Fig. 2 SCM used for precoating aggregates

(a) Flyash (b) Silica fume (c) Cememt

Fig. 3 a–c Precoating of aggregate

2.3 Two-Stage Coating Process

Two stages are adopted in the treatment of recycled concrete aggregates: (i) Aggregate Preparation Process and (ii) Precoating Treatment of RCA. In the first stage, the waste concrete is collected from the disposal site, and it is crushed in a Los Angeles abrasion machine with six balls at 100 rpm. Then, it is further crushed by a manual hammering process and graded to the required size between 10 and 20 mm. The crushed particles were removed, and the fine mortar adhered to the surface of the RCA was cleaned by thorough washing in clean water before the precoating process. The second treatment stage aimed to enhance the inferior qualities of the recycled aggregate by modifying its surface with a coating of fine supplementary cementitious materials. After the completion of the first stage, two kg of the RCA sample was taken and coated with fly ash paste, and the water binder ratio was maintained between 0.6 and 0.7. These pretreated aggregates were continuously immersed in paste medium for 48 h. Three trials were conducted for each precoating material, and Fig. 3 shows the precoated aggregate treatment of the recycled aggregates. This procedure is repeated for other SCMs also.

2.4 Water Absorption and Compressive Test on Cubes

The water absorption property is one of the main parameters used to determine the internal structure of aggregates. High water absorption values indicate that aggregates are with greater absorption and more porous. High water absorption harms the compressive strength of concrete due to the excess amount of free water in the mixture. Furthermore, this aggregate becomes unsuitable and unfit for strength development, impact resistance, and hardness properties. To measure the percentage of

Fig. 4 Water absorption test

water absorption, the test was conducted as per the guidelines mentioned in IS 2386 (Part 3), 1986, and Fig. 4 shows the test setup for the water absorption test.

Control concrete cubes of size 150 mm × 150 mm × 150 mm were cast based on the design mix proportions, and the specimens were tested as per the IS 516:2016 (Part 1) guidelines. The specimens taken from the curing tank after 28 days of curing are immediately tested for the compression test in wet stage itself. The test was performed in a 2000 kN capacity hydraulically operated compression testing machine and the load was gradually applied without any shocks. It is continued at a rate of 140 kg/sq.cm until the crushing failure of the specimen. The failure load at crushing is noted and the compressive strength of the specimens is calculated. The average of the three specimens shall then be recorded as the compressive strength of concrete. Figure 5 shows the test setup for the compression test. At the time of failure, its appearance is also observed and any unusual features in the type of failure shall be noted. Figure 6 shows the failure of the cube specimen.

Fig. 5 Compressive strength test

Fig. 6 Crushing failure

3 Results and Discussion

3.1 Water Absorption Properties

Because of the weak ITZ and porous surface of the recycled aggregate, the water absorption of recycled aggregate is 12.8 times greater than natural aggregate. Precoating with SCM reduces the water absorption to some extent and exceeds by 5.86 times, which confirms that the recycled aggregate performance is improved. This enhancement is due to removal of porous adherent mortar, which strengthens recycled aggregate's intrinsic qualities and is the weakest component of RCA. Among the five series, the water absorption of aggregates pretreated with silica fume was better than that of the other series. Among various SCMs, the pozzolanic action and filling activity of silica fume are the key factors to close the pores and microcracks exist on the surface of the precoated aggregate and reduce the water absorption percentage of recycled aggregate. Figure 7 shows a comparison of the water absorption results. High levels of water absorption in recycled aggregates without treatment adversely affect the mechanical properties of concrete. When RCA is immersed in SCM for pretreatment, the slurry penetrates the pores of the adhered mortar until the voids are filled. Therefore, the surface of the recycled aggregate is modified. Among many SCMs, silica fume particles perform better due to their high reactivity; moreover, smaller silica fume particles can enhance water absorption by 46.49%. A thinner layer of cement slurry formed on the surface of the RCA; hence, the old and new mortar bonding was greatly enhanced.

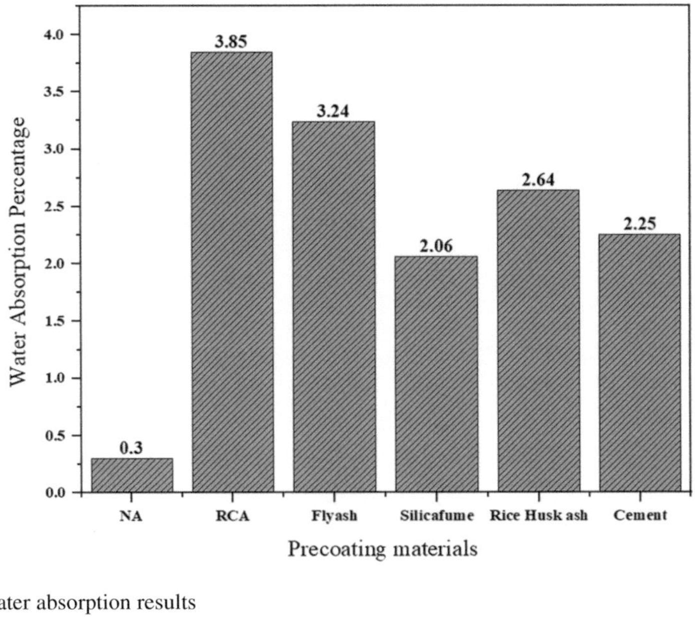

Fig. 7 Water absorption results

3.2 Crushing Strength

Based on the comparison studies on water absorption and specific gravity values among the four precoating materials, silica fume performed best according to both criteria. Hence, concrete cube specimens are made with silica fume-coated recycled aggregate for M30 grade concrete. Around 30% of the natural aggregate was replaced with treated and untreated recycled aggregate, and three series, namely CC, RCA 30 NC, and RCA 30 SFC, were made. For each series, three cubes were tested. After 28 days of curing, the concrete cube specimens were tested in a compression testing machine with a 200 T capacity. The crushing failure strength values for the CC, RCA 30 NC, and RCA 30 SFC series are 39.18 MPa, 31.23 MPa, and 39.01 MPa, respectively. Replacement of 30% recycled aggregate reduces the crushing strength, which is 20.29% lesser than corresponding control concrete (zero recycled aggregate). But the impregnation of silica fume improves the inferior ITZ of recycled aggregate and strengthens the attached mortar on the surface of RCA. Free lime (CH) and silica fume combine to generate C–S–H, which alters the microstructure and produces compressive strength almost similar to concrete produced with natural aggregate.

However, the loss in the compressive strength was compensated by the silica fume precoating on the RCA, which improved the compressive strength. Figure 8 shows the comparison among the three series of compressive strength results. Silica fume is a good material that provides resistance to chemical reactions and promotes good adhesive behavior. Additionally, it offers very dense surface on the coarse aggregate and it allows very less capillarity and exceptional sustainability [14]. The presence of active silicate and aluminate compounds has the potential advantage of bonding characteristics and it acts as a filler to produce pozzolanic strength. [15]

Precoating with silica fume provides the best results due to the pozzolanic reactivity of the silica content in the ash particles, which results in the formation of an improved bond between the old mortar and the precoated aggregate. It is worth

Fig. 8 Compressive strength results

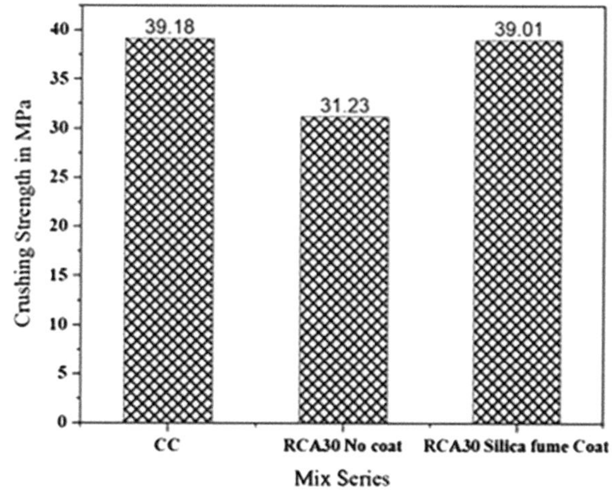

mentioning that after the pretreatment process, the width of the cracks in the ITZ and cement matrix decreases because the pores and microcracks were that filled with hydrated products, enhancing the compressive strength.

4 Conclusions

Due to the porous nature of the adhered mortar on the surface of the untreated recycled aggregate, the waterabsorption amount was six times greater than the natural aggregate. It makes a huge impact on the workability properties and the crushing resistance was significantly reduced. Therefore, small adjustment in water cement ratio is essential to achieve the desired workability.

- Due to the finer particle of silica fume, it could fill all the pores on the surface of recycled aggregate and its filler effect could reduce the water absorption by 46.49% while compared with the untreated recycled aggregate. But no matter the water absorption percentage of RA remains higher than the natural aggregate.
- The replacement of 30% recycled aggregate decreases the compressive strength by 20.29% when compared to the control concrete prepared with untreated aggregate after 28 days of hardening due to the inherent characteristics of waste concrete.
- Precoating of the RCAs with silica fume increased the compressive strength by 19.94% in comparison with the untreated RCAs. The same when compared with the natural aggregate shows only a meager loss of about 0.43%.

With the limited investigation, it is understood that the pretreatment with silica fume slurry could bring beneficial effect on compressive strength. Though it increases the cost of men, materials, and machinery, this pretreatment method is the appropriate one to meet the future demand of the natural aggregate.

Literature reports that 100% replacement of natural aggregate with RA in one cubic meter of concrete could approximately save 49% mineral resources depletion and reduce 2.175×10^5 kg CO_2 equivalent in 15,000 sq.m floor area. Hence, the utilization of waste materials not only offers environmental and economic benefits but also significantly improves the RCA's performance compared to its untreated counterpart. The mechanical and durability properties of SCM-coated recycled aggregate can be explored further to increase the percentage of replacement in the development of new concrete.

Acknowledgements The author wishes to express his gratitude to the Department of Civil Engineering National Engineering College, Kovilpatti Tamilnadu for providing laboratory facilities to carry out this investigation.

References

1. Tuladhar R, Marshall A, Sivakugan N (2020) Advances in construction and demolition waste recycling, pp 181–197 https://doi.org/10.1016/B978-0-12-819055-5.00010-3
2. Reddy VV, Reddy RA, Shetty R, Kishore S, Reshma EK (2020) Laboratory investigation on pre-coating technique on recycled concrete aggregate. Int Res J Eng Technol
3. MS Pourtahmasb (2014) Utilization of recycled concrete aggregates stone asphalt mixtures. Adv Mater Sci Eng https://doi.org/10.1155/2014/902307
4. Perumal P, Omran M (2021) Sustainable treatment methods for recycled concrete aggregate. Young Concrete Researchers. Engineers and Technologist
5. Lu J, Cai Z, Gao Y, Yin Y, Ma Z, Liang C (2023) Effects of pretreatment methods on the properties of recycled aggregates and prepared concrete under CO_2-curing. Case Stud Constr Mater 18(5). https://doi.org/10.1016/j.cscm.2023.e01826.
6. Olonade KA, Akindahunsi AA, Ajagbe WO, Adebanjo AU, Tijani MA (2023) Pretreatment of recycle aggregates, multifunctional concrete with recycled aggregates. Woodhead Publishing Series in Civil and Structural Engineering, pp 17–39 https://doi.org/10.1016/B978-0-323-89838-6.00005-0
7. Zhang R, Xie D, Wu K, Wang J (2023) Optimization of sodium alginate aided bio deposition treatment of recycled aggregates and its application in concrete. Cem Concr Compos 139:105031. https://doi.org/10.1016/j.cemconcomp.2023.105031
8. Wenzel B, Bustamante M, Muñoz P, Ortega JM, Loyola E, Letelier V (2023) Physical and mechanical behavior of concrete specimens using recycled aggregate coated using recycled cement paste. J Clean Prod 393:132015. https://doi.org/10.1016/j.conbuildmat.2023.132015
9. Ouyang K, Shi C, Chu H, Guo H, Song B, Ding Y, Guan X, Zhu J, Zhang H, Wang Y, Zheng J (2020) An overview on the efficiency of different pretreatment techniques for recycled concrete aggregate. J Clean Prod 263:121264. https://doi.org/10.1016/j.jclepro.2020.121264
10. Li N, Long G, Li W, Ma C, Zeng X, Xie Y, Li H, Yang H (2020) Designing high impact-resistance self-compacting concrete by addition of asphalt-coated coarse aggregate. Constr Build Mater 253:118. https://doi.org/10.1016/j.conbuildmat.2020.118758
11. Kim Y, Hanif A, Kazmi SM, Munir MJ, Park C (2018) Properties enhancement of recycled aggregate concrete through pretreatment of coarse aggregates—comparative assessment of assorted techniques. J Clean Prod 191:339–349. https://doi.org/10.1016/j.jclepro.2018.04.192
12. Garg M, Verma N (2023) Experimental investigation of concrete incorporated with waste aggregates with coating of epoxy and glass fiber as replacement of coarse aggregates. Mater Today Proc 46(2). https://doi.org/10.1016/j.matpr.2020.12.456
13. Lu D, Wang D, Wang Y, Zhong J (2023) Nanoengineering the interfacial transition zone between recycled concrete aggregates and fresh paste with their graphene oxide. Constr Build Mater 384:131244. https://doi.org/10.1016/j.conbuildmat.2023.131244
14. Malayali AB, Venkatesh R, Alharbi SA, Kalam MA (2024) Hybridization of concrete by the inclusions of kaolin, alumina and silica fume: performance evaluation. Heliyon 10(9):0674. https://doi.org/10.1016/j.heliyon.2024.e30674
15. Olivia M, Maulidi MA, Wibisono G (2024) Characteristics of palm oil fuel ash concrete admixed with precipitated silica and silica fume. Clean Eng Technol Clean Eng Technol 19:100738. https://doi.org/10.1016/j.clet.2024.100738

Research and Practice on Progressive Collapse and Robustness of Reinforced Concrete Frame Structures by Linear Static Analysis

Prashant Sunagar, Rushikesh Ghodekar, J. Sumalatha, Piyush Dhadge, Vedant Nawale, and Prasad Ahire

Abstract Progressive collapse is a significant issue in the field of structural engineering. It refers to the chain reaction of localized structural failure, which results in the subsequent destruction of nearby elements. This phenomenon becomes especially apparent when vertical load-bearing elements, such as columns, give way to unintentional loads, whether caused by human-made accidents or natural calamities. The current study explores the complexities of progressive collapse by doing a thorough analysis of a building utilizing the advanced FEM software. The main objective of this research is to model and assess the progressive collapse of a structure through the selective removal of columns. Multiple scenarios involving the removal of columns are analysed in order to gain a thorough understanding of the vulnerabilities and consequences connected with these changes to the structure. The Demand Capacity Ratio (DCR) of beams is computed in accordance with the rules set by the General Services Administration (GSA) for each specific scenario. This essential measurement acts as a pivotal signal in evaluating the stability of a structure against increasing collapse. The computed DCR values offer valuable information about the structural robustness in various loading scenarios and allow for a precise assessment of the building's ability to endure localized breakdowns. The study highlights the importance of comprehending gradual collapse in relation to seismic events. This research enhances the ongoing discussion on structural integrity by providing a thorough investigation of progressive collapse using advanced computer analysis. It emphasizes the importance of designing structures that demonstrate resilience against unexpected events.

P. Sunagar (✉) · R. Ghodekar · P. Dhadge · V. Nawale · P. Ahire
Department of Structural Engineering, Sanjivani College of Engineering, Savitribai Phule Pune University, Kopargaon, Maharashtra, India
e-mail: prashant.sjce@gmail.com

J. Sumalatha
Ramaiah Institute of Technology, Bangalore, India

P. Sunagar
Department of Civil Engineering, Sandip Institute of Technology and Research, Nashik, India

Y. K. Guruprasad et al. (eds.), *Recent Advances in Infrastructure Development and Management—Volume 2*, Lecture Notes in Civil Engineering 566,
https://doi.org/10.1007/978-981-96-1558-2_15

Keywords Progressive collapse · Linear static analysis · Structural vulnerability · Collapse mechanism

1 Introduction

Progressive collapse occurs when structural components such as columns, beams, and slabs are failed due to abnormal loads [1, 2]. In many buildings, column failure is the main reason for progressive collapse [3–7]. When column fails, then the beam connected to it gets damaged. Then adjacent components connected to beam are also got damaged. If damage can spread, finally it leads to collapse of building. Bracings are provided to minimize the damage occurred due to abnormal loads. These bracings can effectively transfer the loads [8, 9]. During the building collapse due to the accidental loads, there is no time for the people to leave the buildings which leads to loss of life [8, 9]. Some authorities made guidelines for design of structures to resist progressive collapse analysis. In India, we can accept the guidelines given by GSA and DOD guidelines by UFC, New York. These guidelines give a clear methodology step to avoid progressive collapse. Alternative Path Method (APM) is used in both GSA and DOD guidelines. When one of the columns gets damaged or fails, then load is transferred to the ground through an alternate path.

The study conducted by [10] examined the ability of steel moment-resistant frames to withstand progressive collapse. The research explored several strategies recommended by the GSA and DOD guidelines. The study focused on examining how catenary movement affects the vulnerability of steel moment frames to progressive collapse. A separate investigation conducted by [11] examined the ability of inclined building structures to withstand gradual collapse. The study focused on a specific scenario where columns were intentionally removed.

The study conducted by [12] examined the ability of reinforced concrete (RC) buildings to withstand seismic events. The analysis was performed using linear static analysis, following the recommendations of the GSA. The study conducted by [13] focused on analysing the progressive collapse of a reinforced concrete (RC) structure that included external non-structural walls. The main objective was to determine the influence of these walls on the propensity for collapse. The study primarily examined the analytical load and dynamic rise elements and compared the formulas for the Lateral Impact Factor (LIF) and Dynamic Impact Factor (DIF) with the guidelines provided by the UFC. In 2011, Liu [14] utilized structural optimization methods to create seismic steel moment frames that have improved resistance to progressive collapse, specifically in situations where key columns are abruptly eliminated. [15] proposed an empirical method to calculate the dynamic increase factor (DIF), which amplifies the gravity loads on affected sections during nonlinear static alternate path analysis. Furthermore, [16] introduced an innovative nonlinear static analytic approach for estimating the highest structural responses that occur when a column is suddenly removed. The study conducted a comprehensive assessment of the progressive collapse resistance of a pre-existing steel frame building, with a specific emphasis on its performance in such scenarios. Traditional design guidelines for progressive collapse frequently recommend using simplified analytical procedures that involve the prompt removal of specific key columns in a structure.

The study conducted by [17] analysed the effectiveness of commonly employed methods for assessing and developing progressive collapse. This investigation utilized both numerical simulations and experimental data. [18] has made substantial contributions to this field by assessing the vulnerability of current steel frame buildings to progressive collapse. [19] conducted a thorough analysis of different structural systems and scenarios including the removal of columns. The study conducted by [20] utilized three-dimensional finite element modelling to analyse the progressive collapse of multi-storey structures that have composite steel frames. The researchers conducted experiments on a 1/3 scale model to study the collapse process of steel frames with CFST columns in a column removal scenario. The introduction of fragility functions for low-rise RC framed buildings, switching from deterministic to probabilistic, was pioneered by [21]. The concept of pushdown analysis was introduced by [22] as a method for constructing fragility models. In this study, the researchers conducted a numerical analysis on framed buildings with PE beams to evaluate their ability to handle vertical loads after the removal of columns. In their study, researchers [23] conducted a comparative analysis of the progressive collapse-resisting capacity of a newly developed system and the standard diagrid system. The study specifically examined the local failure of structural elements in the above-ground narrative.

The sensitivity of candidate structures to progressive collapse is evaluated by a sensitivity index, which measures the structures' susceptibility to the dynamic effects caused by progressive collapse [7]. In their study, [1] propose a methodology that is recommended for the computation of the dynamic increase factor (DIF). This factor is crucial in accounting for the dynamic impacts in nonlinear static analysis. The list of influential works comprises the publications by [8]. The study conducted by [9] focuses on the energy-based evaluation of the dynamic amplification factor in the context of linear static analysis for the design of ductile reinforced concrete frame structures in progressive collapse scenarios. The analytical formulation of the Dissolved Air Flotation (DAF) is derived by applying the principle of energy conservation. The issue of gradual collapse in building structures has garnered significant attention from engineers and researchers since the occurrence of the "911" incident. In this paper, [2] evaluate the existing research on this matter through the examination of experimental studies, numerical simulations, and theoretical analyses. Consequently, there is a potential violation of the seismic design concept known as "strong-column-weak-beam," which might result in undesirable failure modes and compromise the overall seismic performance. The work conducted by [24] explores a new approach to enhance the seismic and progressive collapse capabilities of reinforced concrete (RC) frames through the implementation of innovative structural detailing techniques. In order to mitigate the detrimental impacts of progressive collapse design on the seismic resilience of reinforced concrete (RC) frames, an innovative approach to structural detailing is put forth.

The objective of this research work is to understand the behaviour of building under progressive collapse by performing linear static analysis for evaluation of progressive collapse resistance of the building. It is proposed to check the Demand Capacity Ratio (DCR) value for flexure for the beams under the GSA guidelines and determine which column removal case is more critical to progressive collapse of building. It was also proposed to study the behaviour of building for progressive collapse when bracings are fixed on building.

2 Methodology

In this study, a rectangular G + 8 storey building is considered (Fig. 1). It has five bays along *X*-direction and three bays along *Y*-direction. Column to column distance is 6 m along both the directions.

2.1 Detailed Data of Rectangular Building

Number of floors = G + 8
Floor to floor height = 3 m (constant)
Column size = 300 X 450 mm
Column spacing = 6 m

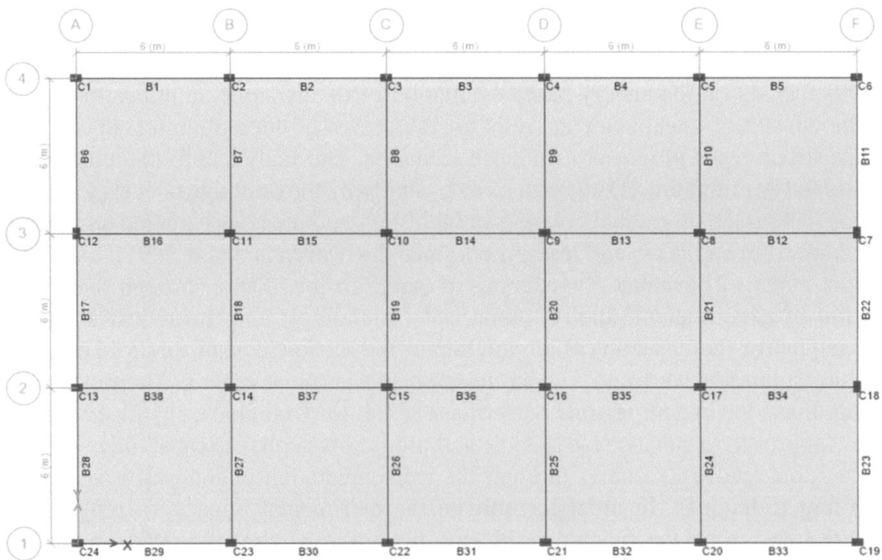

Fig. 1 Plan of rectangular building

Beam size = 300 X 450 mm
Slab depth = 150 mm
Grade of concrete = M30
Grade of steel = Fe 500

2.2 Loads on the Building

Live load = 3 kN/m
Seismic load: As per IS 1893–2013 [25]
Seismic zone: Zone II.
Response reduction factor = 5
Importance factor (I) = 1
Fundamental period (Ta): as per the IS 1893:2013
Bracings used: ISMB 400

2.3 Analysis of the Model

Linear static method is simpler than other methods. So, linear static method is considered for this analysis. The column failure is dynamic but, in this study, it is considered static. After column removal, the building is analysed using ETAB software as per IS codes. The steps that are followed in the analysis of the model in ETABS are listed below:

Step 1: Structure is modelled in ETAB. Loads are applied on the model including seismic load. Then the model is analysed in linear static approach.
Step 2: DCR values of the beams are determined without any column removal.
Step 3: Columns are removed at different locations.
Step 4: DCR values of the beams are determined for each case of column removal.
Step 5: If DCR is > 1, then the demand is more than the capacity; if the DCR is > 2, then there will be progressive collapse.
Step 6: Bracings are applied on the face on which column is removed, then compare the DCR values of beams.

2.4 Permissible Limit

As per GSA, Demand Capacity Ratio (DCR) = $Q_{\mathrm{ud}}/Q_{\mathrm{ue}}$.

Allowable DCR is < 2 for symmetrical structure and DCR < 1.5, for asymmetrical structure.

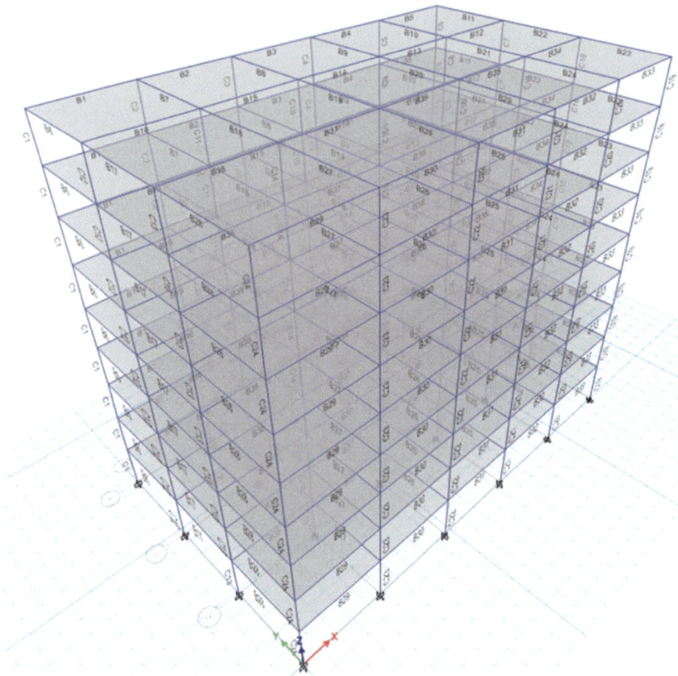

Fig. 2 3D model of rectangular building

2.5 Cases Considered for Removal of Column

Case 1: Removal of corner column—In rectangular building, column C24 is removed. In L-shaped building, column C16 is removed (Fig. 2a).

Case 2: Removal of exterior column—In rectangular building, column C22 is removed. In L-shaped building, column C13 is removed (Fig. 2b).

Case 3: Removal of interior column—In rectangular building, column C15 is removed. In L-shaped building, column C11 is removed (Fig. 3).

The bracings fixed at corner bay and the bays adjacent to removed columns are also examined in this study (Fig. 4).

3 Results and Discussion

The estimated DCR values of beams without removal of any column are given in Table 1. The DCR values of beams after the removal of corner column C24 are given in Table 2. The DCR values of beams after fixing the bracings at corner bay are given

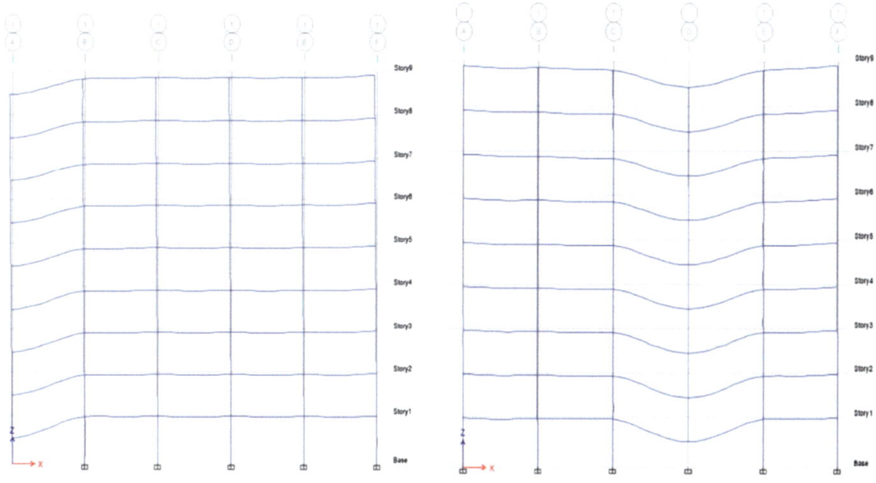

(a) (b)

Fig. 3 **a** Deflection of rectangular building when corner column is removed and **b** Deflection of rectangular building when exterior column is removed

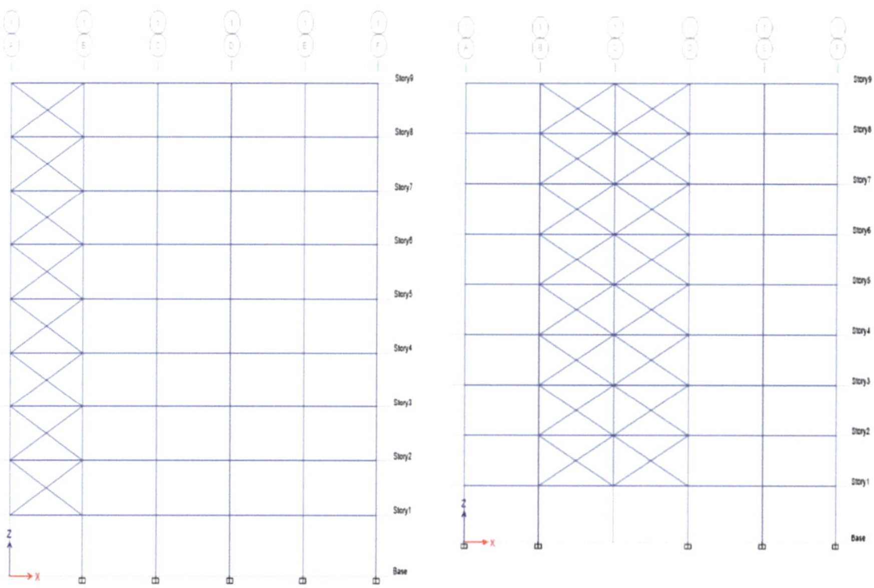

(a) (b)

Fig. 4 **a** Bracings are fixed at corner bay and **b** Bracings are fixed at bays adjacent to removed column in Table 3. The comparison of DCR values among different floors (without bracings) after the removal of column C24 is shown in Fig. 4. The DCR values of beams at different floors are compared when bracings are fixed (Fig. 5). The comparison of DCR values of beam B29 with and without bracings is shown in Fig. 6.

Table 1 presents the DCR (Design Code Ratio) values for beams that have not undergone any column removal on various floors. The range of DCR values observed in this study spans from 0.32 to 0.53, suggesting that the structural integrity of the system remains within acceptable thresholds during typical operating conditions. The DCR values exhibit a downward trend as one ascends the floors, with the lowest values seen on the highest floor (Floor 9). Conversely, the highest DCR values are found on the lower floors, indicating a greater demand on the structural elements at lower levels owing to the transfer of loads.

Table 2 displays the Dissipation Factor values subsequent to the exclusion of column C24. Significantly, the DCR (Damage Control Ratio) values exhibit a

Table 1 DCR values of beams without removal of any column

Floor	DCR value of the beams				
	Beam 29	B30	B31	B32	B33
Floor 9	0.34	0.42	0.43	0.43	0.32
Floor 8	0.38	0.43	0.45	0.45	0.34
Floor 7	0.40	0.47	0.48	0.48	0.38
Floor 6	0.43	0.50	0.51	0.53	0.43
Floor 5	0.42	0.52	0.51	0.51	0.40
Floor 4	0.44	0.52	0.52	0.52	0.44
Floor 3	0.45	0.53	0.54	0.54	0.44
Floor 2	0.47	0.53	0.54	0.54	0.47
Floor 1	0.47	0.53	0.53	0.53	0.48

Table 2 DCR values of beams after the removal of column C24

Floor	Column removed	DCR value of the beams				
		Beam 29	B30	B31	B32	B33
F 9	C24	1.37	0.50	0.47	0.46	0.36
F 8		1.54	0.54	0.52	0.50	0.37
F 7		1.50	0.60	0.55	0.53	0.36
F 6		1.52	0.63	0.57	0.56	0.39
F 5		1.60	0.71	0.59	0.58	0.45
F 4		1.56	0.63	0.58	0.58	0.46
F 3		1.68	0.74	0.59	0.59	0.49
F 2		**1.79**	0.79	0.60	0.60	0.52
F 1		1.74	0.84	0.55	0.55	0.53

Table 3 DCR values of beams after fixing of bracings at corner bay

Floor	Column removal	DCR value of the beams				
		Beam 29	B30	B31	B32	B33
F 9	C24	0.44	0.46	0.56	0.52	0.36
F 8		0.44	0.45	0.61	0.59	0.40
F 7		0.44	0.45	0.63	0.61	0.44
F 6		0.44	0.53	0.66	0.64	0.48
F 5		0.46	0.59	0.69	0.68	0.52
F 4		0.48	0.61	0.72	0.72	0.62
F 3		0.50	0.73	0.78	0.77	0.66
F 2		0.54	**0.84**	**0.84**	0.83	0.75
F 1		0.70	0.79	0.71	0.71	0.67

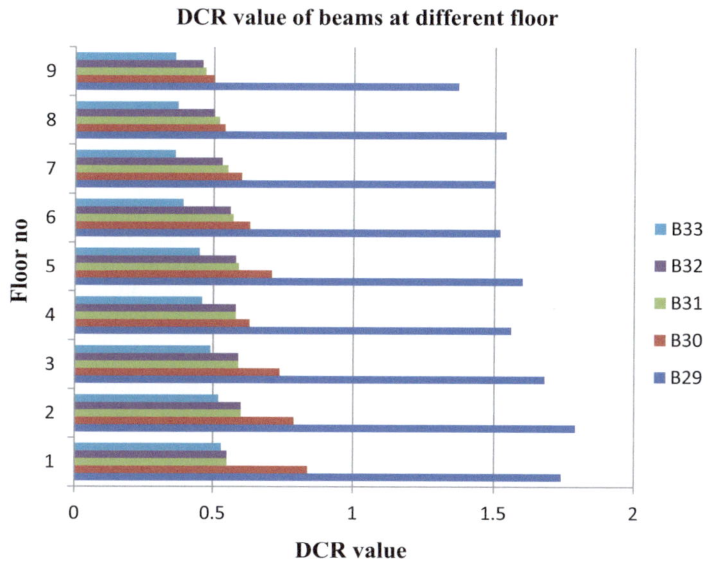

Fig. 5 DCR value of beams at different floors after the removal of column C24

substantial rise, particularly for beam 29, where values beyond 1.5 are observed in the majority of instances, hence indicating an elevated susceptibility to failure.

The floor with the greatest DCR value, specifically Floor 2 with a value of 1.79, indicates that the elimination of a column significantly burdens neighbouring beams, particularly on lower levels.

Table 3 presents a comparison of the DCR (Displacement Control Ratio) values subsequent to the implementation of bracings at the corner bay, where column C24 had been previously removed. The implementation of bracings significantly

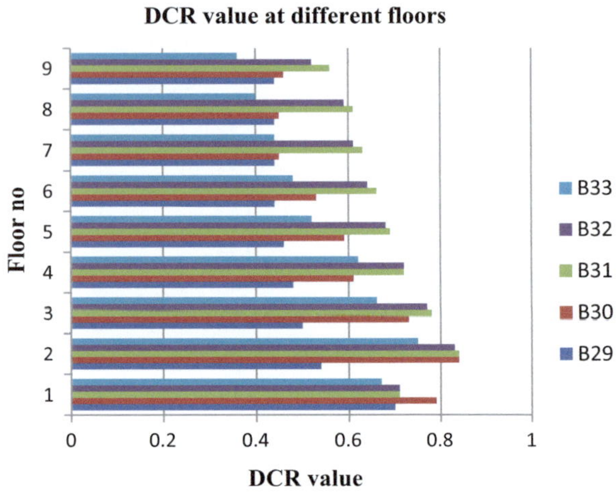

Fig. 6 DCR value of beams at different floors when bracings are fixed

decreases the values of DCR, so demonstrating an efficient mitigation of the risk of progressive collapse. The DCR (Damage Control Ratio) value for beam 29, which exhibited a consistently high value in the majority of cases following the removal of the column, experiences a dramatic reduction to a level well below 1 subsequent to the installation of bracing.

Figures 5 and 6 present a graphical representation that facilitates the comparison of DCR (Damage Control Ratios) among various floors in scenarios involving both braced and unbraced conditions. The provided figures depict the significant impact of bracings in mitigating the structural load imposed on beams.

Figure 7 provides a direct comparison of the DCR (Damage Control Ratio) values for beam B29, both with and without the inclusion of bracings. This comparison serves to demonstrate the significant impact of bracing on improving the overall structural safety of the beam.

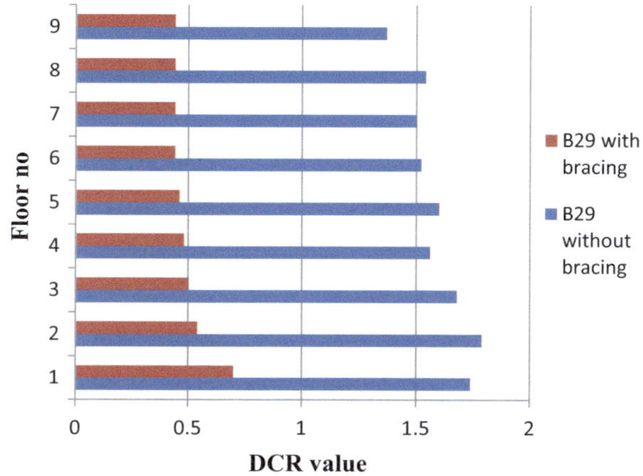

Fig. 7 Comparison of DCR values of beam B29 with and without bracings

4 Conclusion

From the obtained DCR values, it was found that DCR values of beams surrounding the removed column are higher than far beams. In this rectangular building, DCR values of surrounding beams are less than 2. But these surrounding beams have DCR value of nearly 1.7. These surrounding beams simply fail but not collapse. In this model of the building, maximum DCR value is obtained when interior column is removed. In almost all the cases, DCR values at the bottom stories are more as compared to the top stories. When bracings are fixed on the bay of the removed column, DCR value of the beams is reduced considerably. From bar graph, it was observed that, beam B29 has DCR value of more than 1.5 in almost all the floors. But after fixing the bracings, DCR value considerably reduced to value within 1. This shows that fixing bracings is most effective in reducing the progressive collapse. The study's findings suggest that the DCR values in the immediate vicinity of the removed column exhibit larger values compared to those observed at greater distances, implying the presence of localized stress concentration. Nevertheless, it is usually observed that these values are typically below 2, indicating that although the beams may experience failure, they do not result in a complete structural collapse. The highest value of DCR is obtained when an interior column is eliminated, highlighting the susceptibility of internal supports.

The findings additionally indicate that the values of DCR exhibit an upward trend as the elevation decreases, which signifies the heightened transmission of loads to these particular levels. The usefulness of bracing as a mitigation approach against progressive collapse is highlighted by the large reduction in DCR values resulting from the placement of bracings on the bay of the removed column. The research findings indicate that the implementation of bracing is an effective measure in reducing the DCR (Damage Control Ratio) values to levels that are considered safe. Consequently, this practice contributes to the improvement of the structural integrity of the building in situations where column removal is necessary.

References

1. Sunagar P, Bhashyam A, Shashikant M, Sreekeshava KS, Chaurasiya, AK (2021) Effect of different base isolation techniques in multistoried RC regular and irregular building. In: Narasimhan MC, George V, Udayakumar G, Kumar A (eds) Trends in civil engineering and challenges for sustainability. Lecture notes in civil engineering, vol 99. Springer, Singapore. https://doi.org/10.1007/978-981-15-6828-2_29
2. Sunagar P, Nayak SG, Geethakumari TG, Mahesh Kumar CL, Kiran BM, Sanjith J (2022) Progressive collapse analysis of T shape RCC building. In: Abstract IOP Conf Ser: Earth Environ Sci 1125(1):012017. https://doi.org/10.1088/1755-1315/1125/1/012017
3. Tsai MH (2012) Assessment of analytical load and dynamic increase factors for progressive collapse analysis of building frames. Adv Struct Eng 15(1):41–54
4. Song BI, Sezen H (2009) Evaluation of an existing steel frame building against progressive collapse. In: Structures congress 2009: don't mess with structural engineers: expanding our role, pp 1–8
5. Xu M, Gao S, Guo L, Fu F, Zhang S (2018) Study on collapse mechanism of steel frame with CFST-columns under column-removal scenario. J Constr Steel Res 141:275–286
6. Mucedero G, Brunesi E, Parisi F (2021) Progressive collapse resistance of framed buildings with partially encased composite beams. J Build Eng 38:102228
7. Mashhadiali N, Gholhaki M, Kheyroddin A, Zahiri-Hashemi R (2016) Analytical evaluation of the vulnerability of framed tall buildings with steel plate shear wall to progressive collapse. Int J Civ Eng 14:595–608
8. Sunagar P, Bhashyam A, Dharek MS, Sreekeshava KS, Ramegowda RS, Lakshmi HS (2020) Blast resistance of steel plate shear walls designed for seismic loading. In: Abstract IOP Conf Ser: Mater Sci Eng 814(1):012041. https://doi.org/10.1088/1757-899X/814/1/012041
9. Sreekeshava KS, Dharek MS, Sunagar P (2020) Non-linear seismic analysis of steel plate shear wall subjected to blast loading. In: Abstract IOP Conf Ser: Mater Sci Eng 955(1):012025. https://doi.org/10.1088/1757-899X/955/1/012025
10. Kim J, Kim T (2009) Assessment of progressive collapse-resisting capacity of steel moment frames. J Constr Steel Res 65(1):169–179
11. Kim J, An D (2009) Evaluation of progressive collapse potential of steel moment frames considering catenary action. Struct Design Tall Spec Build 18(4):455–465
12. Tsai MH, Lin BH (2008) Investigation of progressive collapse resistance and inelastic response for an earthquake-resistant RC building subjected to column failure. Eng Struct 30(12):3619–3628
13. Tsai MH, Huang TC (2011) Progressive collapse analysis of an RC building with exterior non-structural walls. Procedia Eng 14:377–384
14. Liu M (2011) Progressive collapse design of seismic steel frames using structural optimization. J Constr Steel Res 67(3):322–332
15. Liu M (2013) A new dynamic increase factor for nonlinear static alternate path analysis of building frames against progressive collapse. Eng Struct 48:666–673
16. Liu MM, Pirmoz A (2016) Energy-based pulldown analysis for assessing the progressive collapse potential of steel frame buildings. Eng Struct 123:372–378
17. Song BI, Sezen H (2013) Experimental and analytical progressive collapse assessment of a steel frame building. Eng Struct 56:664–672
18. Song BI, Sezen H, Giriunas KA (2012) Collapse performance evaluation of steel buildings after loss of columns. In: Structures congress 2012, pp 213–224
19. Fu F (2009) Progressive collapse analysis of high-rise building with 3-D finite element modeling method. J Constr Steel Res 65(6):1269–1278
20. Fu F (2012) Response of a multi-storey steel composite building with concentric bracing under consecutive column removal scenarios. J Constr Steel Res 70:115–126
21. Brunesi E, Nascimbene R, Parisi F, Augenti N (2015) Progressive collapse fragility of reinforced concrete framed structures through incremental dynamic analysis. Eng Struct 104:65–79

22. Brunesi E, Parisi F (2017) Progressive collapse fragility models of European reinforced concrete framed buildings based on pushdown analysis. Eng Struct 152:579–596
23. Mashhadiali N, Kheyroddin A (2014) Progressive collapse assessment of new hexagrid structural system for tall buildings. Struct Design Tall Spec Build 23(12):947–961

Advances in Structural Systems for Tall Buildings: Emerging Developments Using Exterior Bracing and Diagrid Systems

Prashant Sunagar, Bhagure Aditya, J. Sumalatha, Mahesh Yele, Saurabh Jadhav, and Shrutika Pawar

Abstract Diagrid systems possess the key characteristic of substantially decreasing the quantity of columns typically used in building projects. A reduced number of columns leads to a floor plan that is more spacious and flexible, which is a crucial factor in contemporary architectural design, particularly for tall commercial and residential buildings. This research study examines the analysis and design of diagrid systems for tall buildings, specifically focused on storey heights ranging from 30 to 66 floors. An essential component of this study involves investigating several height-to-base (*h*/*b*) ratios, specifically 2.58, 3.1, 3.616, 4.13, 4.65, 5.16, and 5.68, as specified by the Indian Standard Code 800-2007. This code functions as a blueprint for the architectural layout of steel structures in India, guaranteeing that the diagrid systems comply with national safety and construction regulations. The findings of this study are extensive, including a range of structural factors. The axial force of columns, shear force of columns, and axial force of diagonals are thoroughly contrasted and analysed. Furthermore, the study focuses specifically on deflection, which is a crucial aspect in ensuring the structural stability and safety of tall buildings. Through the analysis of these factors, the study offers valuable perspectives on the efficacy of various diagrid layouts and their influence on building performance. In summary, this paper makes a substantial contribution to the field of architectural engineering by providing a comprehensive analysis of diagrid systems in high-rise buildings. The findings not only improve the diagrid system design but also impact the design of skyscrapers.

Keywords Diagrid · Triangular framing · High-rise structures · Structural efficiency

P. Sunagar (✉) · B. Aditya · M. Yele
Department of Civil Engineering, Sandip Institute of Technology and Research, Nashik, Maharashtra, India
e-mail: prashant.sjce@gmail.com

J. Sumalatha
Ramaiah Institute of Technology, Bangalore, India

M. Yele · S. Jadhav · S. Pawar
Department of Structural Engineering, Sanjivani College of Engineering, Kopargaon, India

Y. K. Guruprasad et al. (eds.), *Recent Advances in Infrastructure Development and Management—Volume 2*, Lecture Notes in Civil Engineering 566,
https://doi.org/10.1007/978-981-96-1558-2_16

1 Introduction

The speedy growths of urban populace and consequent strain on constrained area have drastically motivated the residential improvement of city [1]. The high value of land, the preference to keep away from non-stop urban sprawl, and the need to preserve essential agricultural production have all contributed to force residential homes upward. As the peak of building increase, the lateral load-resisting gadget turns into extra vital than the structural system that resists the gravitational hundreds. Rigid body, shear wall, braced tube device, outrigger device, and tubular system are lateral load-resisting systems used in tall structure [2–4]. Diagrids, the present system of study, are considered a mutation of tubular structural system. The present work aims to study and understand diagrid structural system, analyse the parameters impacting the behaviour of the structure, and provide preliminary design procedure considering these parameters. Diagrid is an external structural configuration. The main feature of this system is a grid of diagonals kept at narrow spaces along the external perimeter to carry both dead loads as well as lateral loads which are also called exo-diagonal system. Application of steel members in diagonal manner for strength and stiffness efficiency is just the old natural principle of triangulation which is re-established in a new fashion [5].

The structural designers of tall homes shifted their interest to the braced or trussed tube by employing mega-diagonals instead of the narrow grid of diagonal contributors' feature of diagrids. It turned into other version of diagonalized structural system. Now, again the interest is shifted to a huge range of utility of diagrids. Use of more than one and variegated triangulations which outline the brash aesthetic nature of a constructing gave delivery to a new fashion in tall building's structural system [6] (Figs. 1 and 2).

In a study conducted by Moon [7], an analysis is conducted on the structural characteristics of tall buildings with intricate shapes, such as those that are twisted, tilted, tapering, or freeform in nature. The primary findings indicate that twisted towers have enhanced dynamic reactivity in crosswind scenarios as a result of vertex shedding. Both diagrids and braced tubes experience a reduction in lateral stiffness as the rate of twist increases. However, it is worth noting that braced tubes exhibit a higher degree of sensitivity to twisting, particularly in taller constructions [8]. When

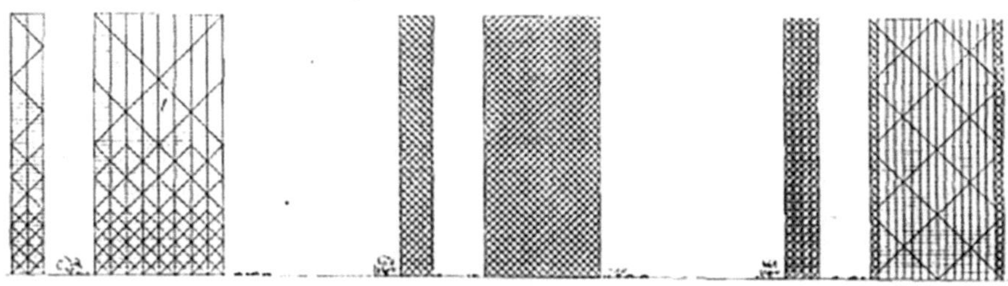

Fig.1 Diagonalized structure solutions for tall buildings proposed by Goldsmith

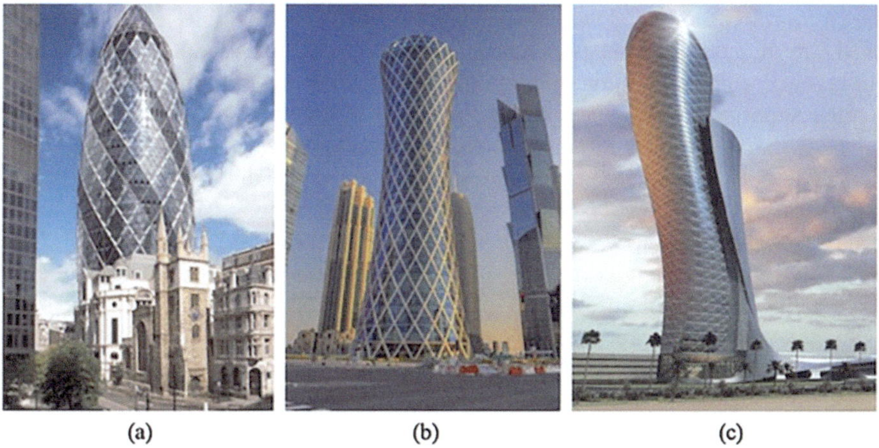

Fig. 2 Diagrid structural systems: **a** Swiss Re Tower, London (from http://larryspeck.com); **b** Tornado Tower, Doha (from http://www.asergeev.com); **c** Capital Gate, Abu Dhabi (from https://www.rmjm.com)

outriggers are inclined, they offer enhanced stability, and as the degree of inclination increases, there is an observable increase in lateral stiffness. Tapered forms have superior performance in terms of lateral stiffness and a reduced vertex shedding effect when compared to prismatic forms. Additionally, the research revealed that freeform shapes are well-suited for diagrid structures, with the ideal angle of the diagrid adjusting as the divergence from the original form becomes more pronounced. In a study conducted by Sree Harsha et al. [1], a comparative analysis was performed on diagrid structures with varying angles. The analysis was carried out using the ETABS software, specifically for non-linear analysis [9]. The research findings indicate that diagrid structures exhibit best performance when the angle between their components falls within the range of 60° to 80°. These structures demonstrate satisfactory behaviour in dynamic analysis, remaining within acceptable bounds.

According to Yadav et al. [10], this study examines the structural behaviour of regular and diagrid steel buildings with a total of 15 storeys in a high seismic region. The implementation of diagrid structures resulted in a notable decrease in axial forces, bending moments in columns, and shear forces in beams as compared to traditional buildings [11]. The decrease of military personnel resulted in the utilization of smaller beam and column sections within diagrid structures, resulting in an approximate 12% improvement in total cost-effectiveness. In a study conducted by Giulia Milana and colleagues [5], a comparative analysis of the sustainability aspects of outrigger and diagrid structures was conducted, with a specific focus on the structural performance of tall buildings [12]. The utilization of a diagrid structure with a 60° angle exhibited superior performance and material efficiency compared to alternative structural designs. In their work, Montuori et al. [13] conducted a detailed examination of diagrid patterns in a 90-storey building. The researchers assessed several patterns, such as variable angle and variable density diagrids. The research put up a proposal for the implementation of a secondary bracing system (SBS) with the objective of mitigating local deformations. The findings indicated that utilizing variable angles yielded superior performance. However, it was observed that combining variable angles with moderate variations resulted in a more optimum solution. In their study, Korsavi et al. [6] examined the evolutionary aspects of the diagrid structural system by analysing data obtained from a sample of 40 constructions. The study conducted indicates that the majority of diagrids exhibit heights ranging from 120 to 150 m. Additionally, it was noted that there is a growing prevalence of non-uniform angle constructions as time progresses.

In their study, Devaraja et al. [14] examined the performance of curved perimeter diagrid lateral systems in a 30-storey steel building with a core shear wall [15, 14]. The findings indicate that both static and dynamic results were within acceptable limits, suggesting that diagrids are successful in such situations. The study conducted by Nishith B. Panchal and colleagues in 2014 The findings of this study pertaining to the ideal angle of diagrid constructions indicate that angles ranging from 65° to 75° are found to be optimal. The study employed various storey heights and diagrid angles in order to analyse and compare the responses, displacements, and material usage. According to the study conducted by Singh et al. [11], the investigation of a typical five-storey reinforced concrete building situated in a region with significant seismic activity revealed that diagrid structures had superior performance and cost-effectiveness when compared to conventional buildings, exhibiting reduced forces and displacements. The study conducted by Jani et al. [15] primarily centred on the examination, formulation, and specification of steel components within diagrid frameworks. The study determined that diagrid structures provide effective distribution of gravity loads and exhibit strong resistance to lateral loads, particularly in the context of larger buildings. According to Boake [12], this research focused on the theoretical features of diagrid structural systems, emphasizing the significance of node design and constructability. It suggested shop fabrication as a potential approach to address issues related to erection and site work. According to the study conducted by Kim et al. (2012), the research primarily investigates the seismic performance of diagrid structures and their effectiveness in mitigating shear lag effects. The findings of this study indicate that diagrids have notable efficacy in resisting shear lag effects, however, their efficiency is significantly influenced by the angle at which the diagrids are confined. In a series of studies conducted by Moon [16], an investigation was carried out to analyse the impact of bending and shear deformations on diagrid designs. The findings of these studies indicated that the utilization of variable angles in diagrid structures is an effective approach for taller buildings with high aspect ratios. Conversely, the implementation of uniform angles is deemed more appropriate for structures with lower aspect ratios.

With research and advance of technology in the field of engineering, it became much simple to obtain structure's behaviour through finite element modelling and analysis. The finite element representation of a structure includes nodes connecting structural elements along with material properties, boundary conditions, loading, and all other characteristics of the physical system. With all the complexities included and the most accurate solutions as results, finite element analysis became more of an engineer's choice of analysis. Hence, the parametric study considered is also performed using finite element method for more accurate results and conclusions. The ETABS software, which incorporates finite element modelling and analysis, is used to develop required physical features of the structures and understand the behaviour. It has a legacy of 40 years as software for professional structural design engineers. Its user-friendly graphic interface and accurate results made ETABS 2013 software to be considered above all other software present in the field of finite element modelling and analysis.

2 Methodology

2.1 Description of Present Work

Stage 1

Height of structure varying from 100 to 200 m.

We have to find optimal Height (H_{opt}) and Optimal Angle (A_{opt}) in each case mentioned above.

Procedure for optimal Height

- Started with a random Height (H), width (B), and Angle (A).
- Fixing B, A, and varying H.
- For different H/B, we get the results of stiffness w.r.t 1/(deflection).

Conditions considered fixing optimal height are

- As plan remains same (as B is fixed) with varying height, height is also increased in such a way that the modules are constant throughout the height (which helps in maintaining angle constant).
- Now optimized H is decided based on stiffness.
- But, only considering directly the stiffness will not give the solution to choose optimized H as H/B is varying.
- From studies and graphs we got, as H/B increases, K decreases, and vice-versa. Therefore, optimal H of the system is the one at which difference between max. deflection allowed at top and actual deflection at top is maximum.

Procedure for optimal Angle

By considering H = optimal H, B = optimal B, A is varying.

Condition considered fixing optimal angle.

- Here, since H and B are constant, the plan remains constant.
- Angle is varied by changing a number of modules.

- If stiffness is the only criterion considered, we cannot get an optimized solution as less module size gives a lower angle, which in turn gives more stiffness.
- Therefore, for optimal angle, stiffness per unit weight of steel consumed by diagonals is considered a deciding factor.

By this procedure, in two cases, optimal Height (*H*), optimal width (*B*), optimal angle (*A*) are found.

Stage 2

- Now, by using the above results, an equation is developed for both cases, which helps to predict *H*/*B* ratio and angle for the preferred case.
- These equations are based on the allowable deflection or limiting deflection or preferred deflection at the top of the building when subjected to loading conditions closer to those considered in the present study.

2.2 *Model Description*

For the above-mentioned work, the models have been made and analysed. Description of the models is provided in this segment.

2.2.1 Models of Stage 1 (For Height from 93 to 204.6 m)

Models of varying height (H)

In this case Width (*B*) = 36 m and Angle (*A*) = $64^{\circ}10'44''$, 6 storey module and height is varying.

Models of varying Angle (*A*)

In this case, Height $(H) = H_{2\text{opt}} = 111.6$ m and Width $(B) = B_{2\text{opt}} = 31.5$ m and Angle (*A*) is varying. The height and width considered here are based on the analysis and results obtained from procedure explained in Tables 1 and 2).

Table 1 Models of varying Height

Model name	No. of storey	Height (*H*)	*H*/*B*
Model2H.1	30	93	2.58
Model2H.2	36	111.6	3.1
Model2H.3	42	130.2	3.616
Model2H.4	48	148.8	4.13
Model2H.5	54	167.4	4.65
Model2H.6	60	186	5.16
Model2H.7	66	204.6	5.68

Table 2 Models of varying Angle (A) in Case 2

Model name	Angles, module
Model2A.1	49°44′35″, 3 storey
Model2A.2	57°34′52″, 4 storey
Model2A.3	67°3′10″, 6 storey
Model2A.4	73°58′53″, 8 storey
Model2A.5	77°50′55″, 12 storey

Table 3 Models of varying Width (B)

Model name	Width (B)	H/B
Model2HB.-1	13.5	8.26
Model2HB.1	18	6.2
Model2HB.2	22.5	4.96
Model2HB.3	27	4.13
Model2HB.4	31.5	3.54
Model2HB.5	36	3.1
Model2HB.6	40.5	2.75
Model2HB.7	45	2.48

2.2.2 Models of Stage 2

Models with constant optimal Height ($H_{2\mathrm{opt}}$) with varying Width (B) and Angle (A) (Table 3).

Storey height: Storey height was maintained at 3.1 m in every model considered for analysis (Tables 4 and 5).

Models' Section Properties

Size of column: All columns are steel box sections.

Size of beam and brace: All beams and braces are steel box sections (Table 6).

Table 4 Varying Angles (A) to be tested for each H/B w.r.t Width (B)

Model name	Angles, Module
Model2HB.-1	45°56′21″, 1.5 storey 54°1′40″, 2 storey 64°10′44″, 3 storey 70°3′15″, 4 storey 76°23′58″, 6 storey
Model2HB.1	
Model2HB.2	
Model2HB.3	
Model2HB.4	
Model2HB.5	
Model2HB.6	
Model2HB.7	

Table 5 Column sizes in Case 2

Column size (mm X mm)	Web and flange thickness (mm)
1000 × 1000	100, 250, 300, 400
1200 × 1200	400
1400 × 1400	600
2000 × 2000	700

Table 6 Beam and brace sizes in Case 2

Beam and brace size (mm X mm)	Web and flange thickness (mm)
500 × 500	40
1000 × 1000	300

Column, beam, and brace sizes were chosen depending on design requirements.

Thickness of slab: The slab in present study is 150 mm steel slab.

Material properties: The grade of steel was Fe 345, which was a default in ETABS 2013 software.

2.3 *Restraint Supports*

In all the models, the beam column joints are normal joints. But the braces are considered truss elements. The columns are taken as hinged as to be able to transfer the lateral loads to the external diagrid system. The diagonal elements are modelled as truss elements. But the diagonal elements starting from ground have one end fixed to the ground and the other end pinned. The diagonal elements were braced to the external beam members in order to prevent buckling of both beams and diagonal members.

2.4 *Loading Details and Types Considered*

- Self-weight (or) Dead load of the structure (DL) [17]
- Live load of 2.5 KN/m^2 (LL) [18]
- Wind loads (WL) are as per IS 875 (Part 3): 1987 for the conditions

Basic wind speed (V_b) = 50 m/s, Class B structure and Terrain category = 2.
The risk coefficient (K_1) = Topography coefficient (K_3) = 1.
Windward coefficient = 0.8.
Leeward coefficient = 0.5 (Figs. 3, 4 and 5).

Fig. 3 Plan of diagrid model

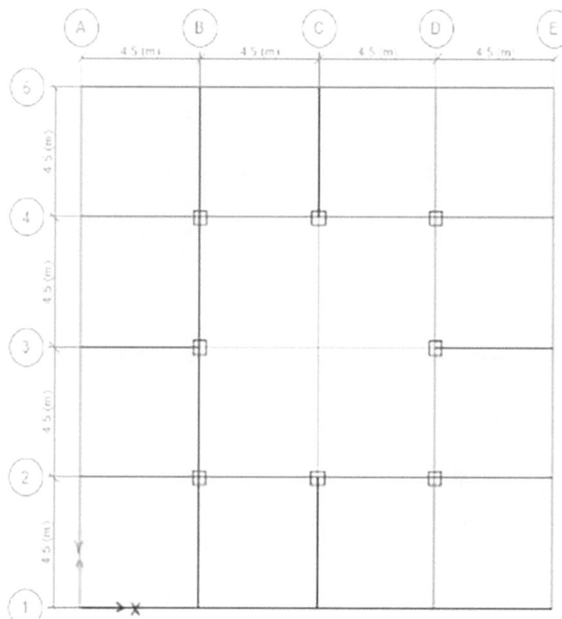

2.5 *Diagrid Model Example*

3 Results and Discussions

The study on the models has been carried out extensively, and behaviour of beams, columns, and braces in terms of bending moment (BB), axial (AFC) and shear forces (SFC), axial forces (AXBR), respectively, is discussed along with deflection (DEF) and the behaviour of the system as a whole. All the results of the study which are used in the context are provided here.

3.1 *Effect of Variation of Height*

Change in behaviour is clearly visible due to max and minimum humps on the graph at *H*/*B* 4.13. At 4.13 *H*/*B* ratio, axial forces of braces are max and all others (BB, SFC) are minimum values, with axial force and deflection values within limit.

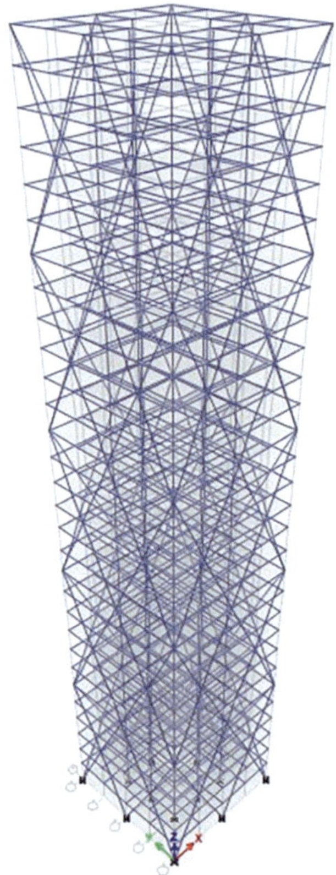

Fig. 4 Elevation and model diagrid building

3.2 *Effect of Variation of Angle*

At Approximately 60°, the axial force in the braces is minimal, i.e., with respect to the lower heights of Case 1, the angle shift occurred towards the higher angle side, which represents bending behavior. Deflection is minimum value at around 60°, but previously in Case 1, it is around 65°, which means the angle shifted to lower angle, reflecting the behaviour of bending and also need of shear strength. After 67°, there is not much change in the values of BB, AFC, and SFC, but AXBR remains high, implying that at 67°, diagrids' participation is more (Table 7).

4 Conclusions

From this research study, it was observed that in lower heights (Case 1), the behaviour is similar for low *H*/*B* ratios due to minimum effect of lateral forces. The change of deflection with respect to *H*/*B* is much steeper when *H*/*B* is achieved by height rather

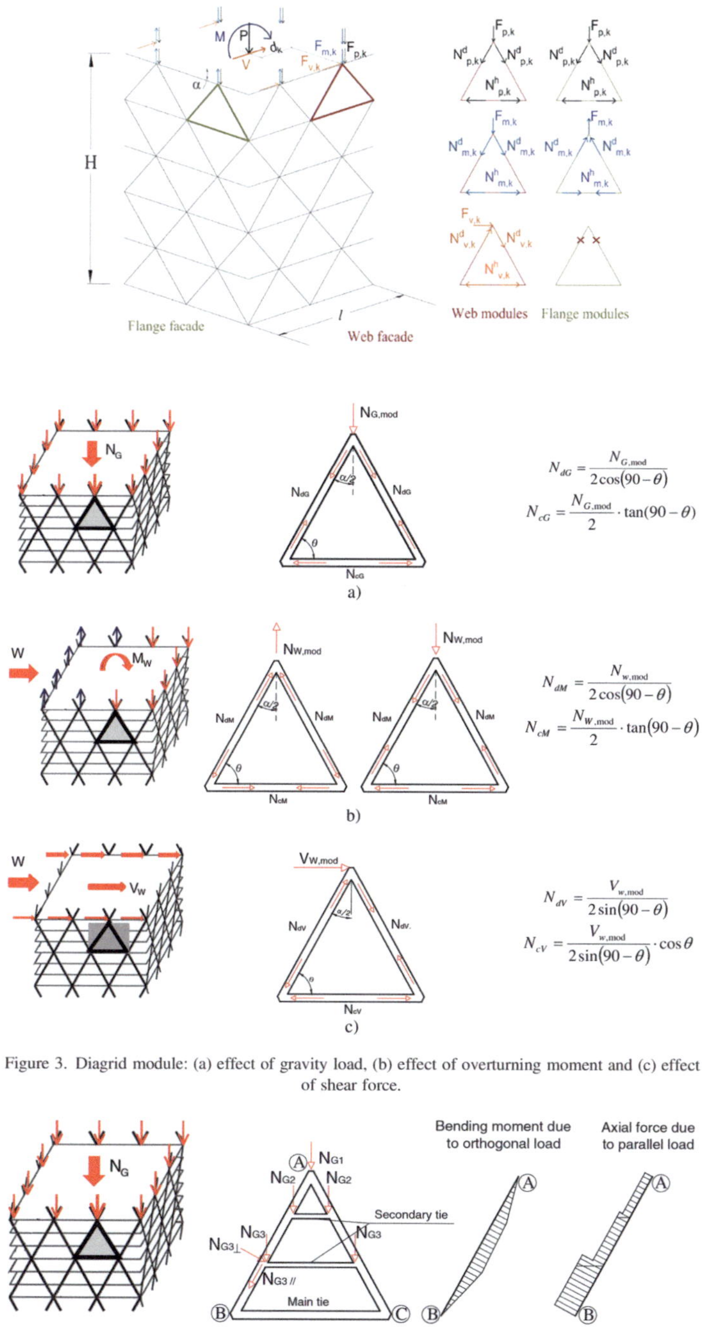

Fig. 5 Geometrical parameters and Load path distribution due to gravity and lateral loads in the diagrid structure ([13] and [16])

Table 7 Results of Case 2 varying angle

$B = 36$ m, $H = 111.6$ m, $A =$ varying					
Angle	BB (KNm)	AFC (KN)	SFC (KN)	AXBR (KN)	DEF (mm)
49.44	2298	60,803	1625	33,795	129.7
57.34	1738	54,128	1323	15,160	61.1
67.3	1434	51,534	1158	14,736	60
73.58	1317	50,100	1090	14,624	70.7
77.5	1302	49,802	1068	14,505	85.8

than width. When behaviour is studied by varying width, the graphs are unusual due to change in no. of columns with plan of the structure. In any case, deflections are minimum when angle is maintained closer to 65° for lower H/B, i.e., 2.48, 2.75, 3.13, 3.3, and 4.13. For higher H/B, i.e., 8.26, 8.23, 6.2, and 5.511, the deflections are minimum at angle closer to 70° (away from 65°). The angle change effect can be clearly seen with respect to forces in members of system, which helps to attain sustainability solution. The kinks of maximum or minimum values help to deduce the behaviour of member. It would help to minimize or maximize a particular member's participation with respect to requirement. When H/B is 4.13 and if we could provide an angle closer to 65°, it can give an optimal solution and also diagrid system's better participation. The max stressed elements behaviour helps to increase or decrease members participation under loads. This behaviour study helped to provide details on members to be focused when changes in initial chosen parameters happen. Mostly, the critical sections, i.e., maximum stressed elements, are column connecting beams, storey 1 columns, and storey 1 braces at the junction of adjacent faces. In diagonals, stress in members decreases when we move up from bottom storey (Figs. 6, 7, 8, 9, 10, 11, 12, 13, 14 and 15).

Structural diagrid systems have become a fundamental element in the construction of tall structures in the ever-changing field of modern architecture, representing both structural efficiency and architectural sophistication. These systems, distinguished by their unique line diagrams consisting of triangular beams, can have either a straight or curved shape and frequently include a horizontal ring system. This arrangement

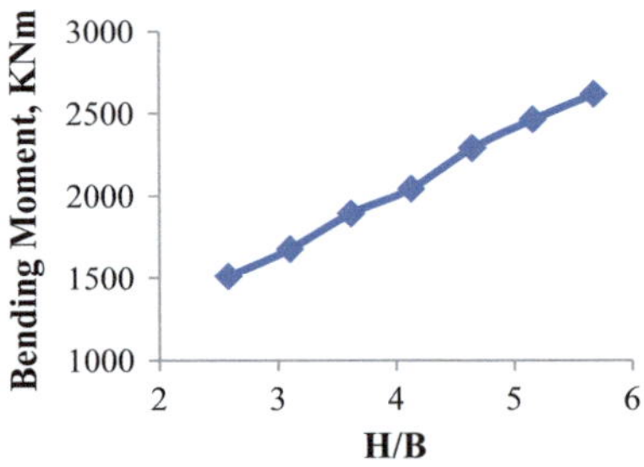

Fig. 6 H/B Versus Bending moment of beam

Fig. 7 *H/B* Versus Bending Moment

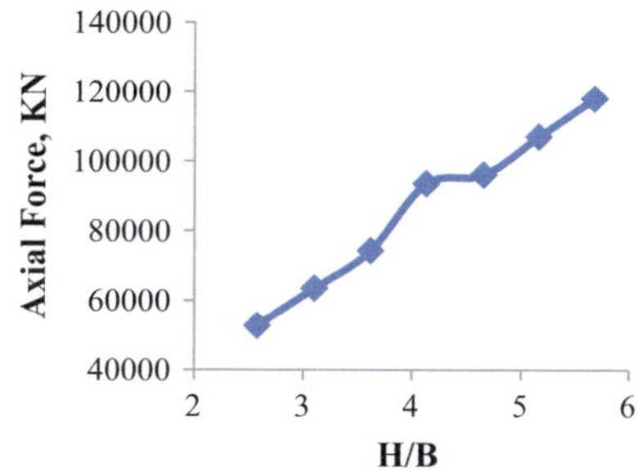

Fig. 8 *H/B* Versus Shear force of column

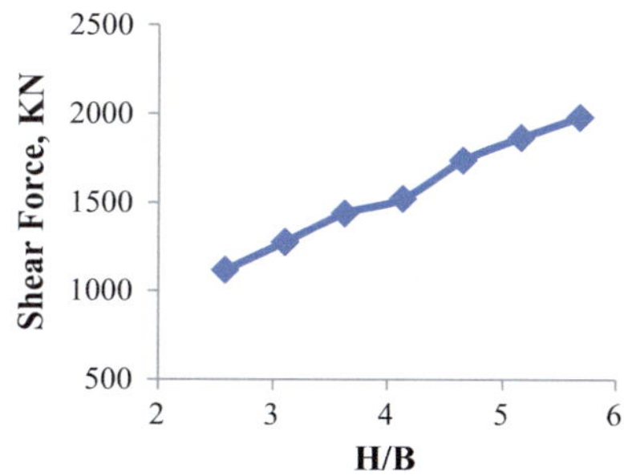

Fig. 9 *H/B* Versus Shear Force of diagonals

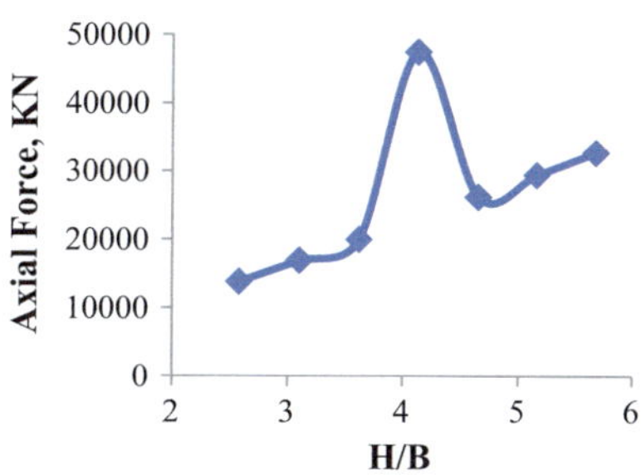

Fig. 10 *H/B* Versus Axial Force

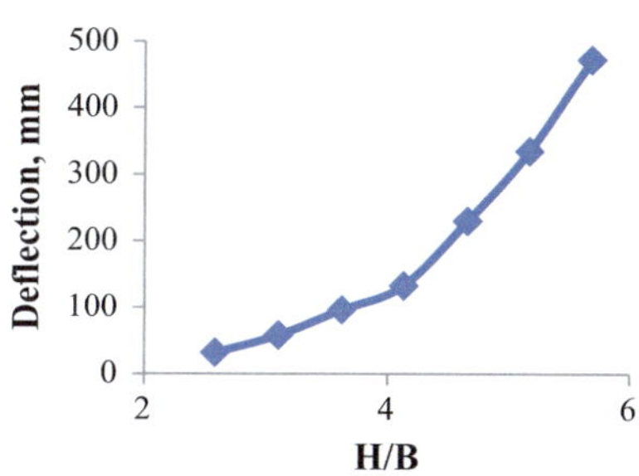

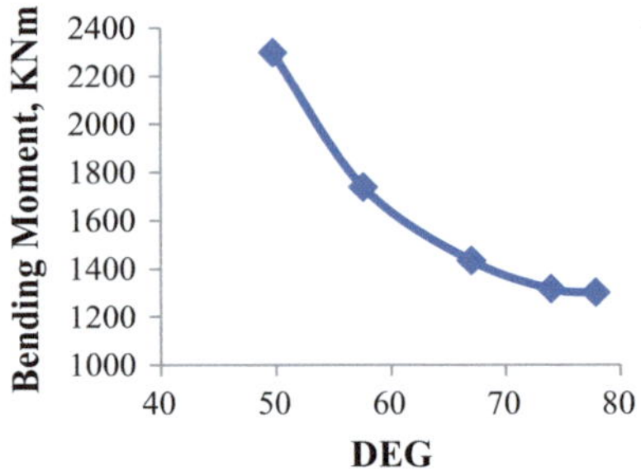

Fig. 11 *H*/*B* Versus Deflection

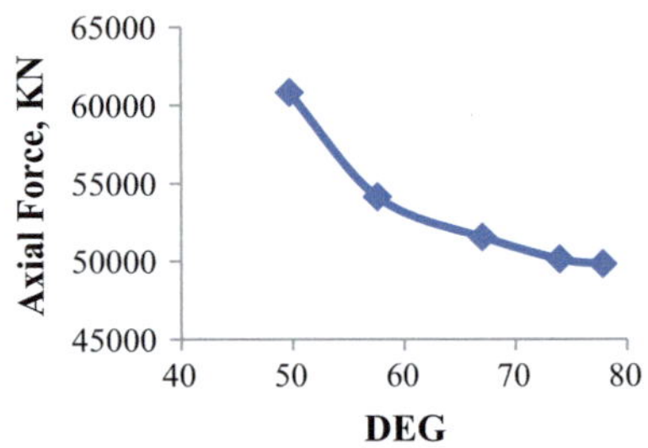

Fig. 12 *H*/*B* Versus Axial force of column

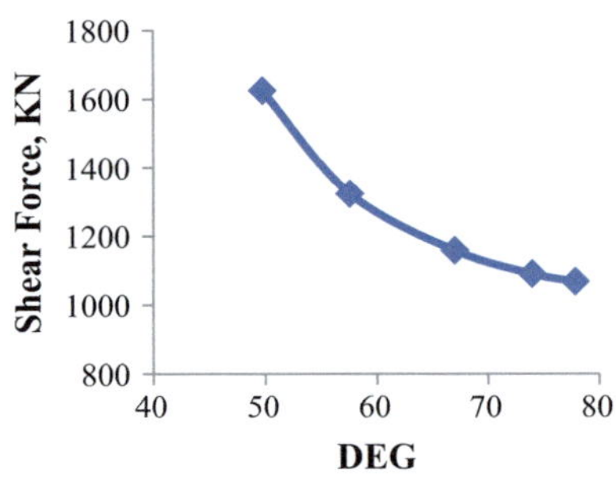

Fig. 13 *H*/*B* Versus Shear force of column

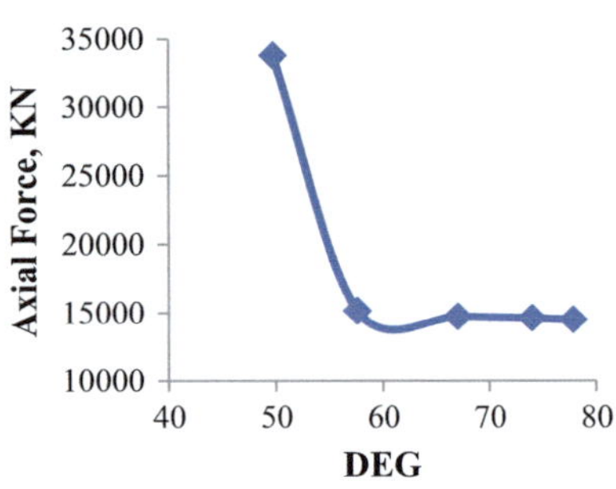

Fig. 14 *H*/*B* Versus Shear Force of diagonals

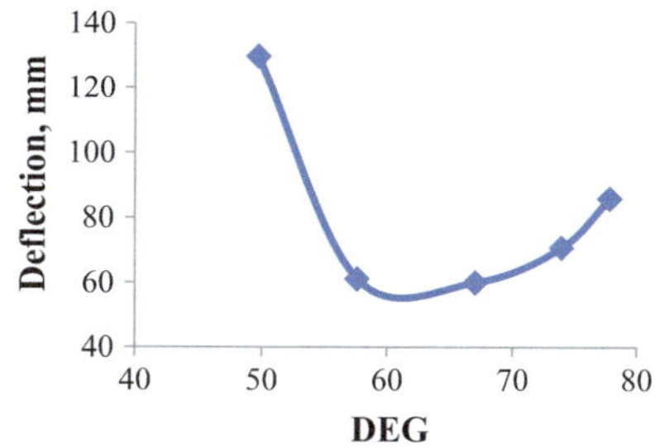

Fig. 15 *H*/*B* Versus Axial Force

is very widespread in the architectural composition of tall buildings, providing a distinct visual and practical benefit. Diagrid systems have the principal characteristic of substantially decreasing the quantity of columns typically necessary in building constructions. This decrease not only improves the design adaptability and usable area within the structure but also significantly enhances the overall structural efficiency. A reduced number of columns leads to a floor plan that is more spacious and flexible, which is a crucial factor in contemporary architectural design, particularly for tall commercial and residential buildings. This research study examines the analysis and design of diagrid systems for tall buildings, specifically focused on storey heights ranging from 30 to 66 floors. An essential component of this study involves investigating several height-to-base (*h*/*b*) ratios, specifically 2.58, 3.1, 3.616, 4.13, 4.65, 5.16, and 5.68, as specified by the Indian Standard Code 800-2007. This code functions as a blueprint for the architectural layout of steel structures in India, guaranteeing that the diagrid systems comply with national safety and construction regulations. The findings of this study are extensive, including a range of structural factors. The axial force of columns, shear force of columns, and axial force of diagonals are thoroughly contrasted and analysed. Furthermore, the study focuses specifically on deflection, which is a crucial aspect in ensuring the structural strength and safety of tall buildings. Through the analysis of these factors, the study offers valuable perspectives on the efficacy of various diagrid layouts and their influence on building performance.

In summary, this paper makes a substantial contribution to the field of architectural engineering by providing a thorough analysis of diagrid systems in high-rise buildings. The findings not only improve comprehension of diagrid system design but also serve as a valuable point of reference for future architectural projects, potentially impacting the design and construction of buildings worldwide.

References

1. Sree Harsha J, Raghu K, Narayana G (2015) Analysis of tall buildings for desired angle of Diagrids. IJRET: Int J Res Eng Technol 4(04)
2. Sunagar P, Bhashyam A, Shashikant M, Sreekeshava KS, Chaurasiya AK (2021) Effect of different base isolation techniques in multistoried RC regular and irregular building. In:

Narasimhan MC, George V, Udayakumar G, Kumar A (eds) Trends in civil engineering and challenges for sustainability. Lecture notes in civil engineering, vol 99. Springer, Singapore. https://doi.org/10.1007/978-981-15-6828-2_29
3. Sunagar P, Bhashyam A, Dharek MS, Sreekeshava KS, Ramegowda RS, Lakshmi HS (2020) Blast resistance of steel plate shear walls designed for seismic loading. IOP Conf Ser: Mater Sci Eng 814(1):012041. https://doi.org/10.1088/1757-899X/814/1/012041
4. Sunagar P, Dharek MS, Nruthya K, Sreekeshava KS, Nagashree B, Ramegowda RS (2020) Non-linear seismic analysis of steel plate shear wall subjected to blast loading. IOP Conf Ser: Mater Sci Eng 955(1):012025. https://doi.org/10.1088/1757-899X/955/1/012025
5. Milana G, Gkoumas K, Bontempi F (2014) Sustainability concepts in the design of high-rise buildings: the case of diagrid Systems. In: Proceedings of the 3rd international workshop on design in civil and environmental engineering. Mary Kathryn Thompson, pp 21–23
6. Korsavi S, Maqhareh MR (2014) The evolutionary process of diagrid structure towards architectural, structural and sustainability concepts: reviewing case studies. J Arch Eng Technol 3(2):1–11
7. Moon K (2009) Design and construction of steel diagrid structures. NSCC2009
8. Panchal NB, Patel VR, Pandya II (2014) Optimum angle of diagrid structural system. Int J Eng Tech Res 2(6):150–157
9. Panchal NB, Patel VR (2014) Diagrid structural system: strategies to reduce lateral forces on high-rise buildings. Int J Res Eng Technol 3(03):374–378
10. Yadav S, Garg V (2015) Advantage of steel diagrid building over conventional building. Retrieved from https://www.researchpublish.com/upload/book/Advantage%20of%20Steel%20Diagrid-2182.pdf
11. Boake TM (2013) Diagrid structures: innovation and detailing. University of Water bro, Canada, School of Architecture
12. Singh RK, Garg V, Sharma A (2014) Analysis and design of concrete diagrid building and its comparison with conventional frame building. Int J Sci Eng Technol 2(6)
13. Montuori GM, Mele E, Brandonisio G, De Luca A (2014) Geometrical patterns for diagrid buildings: exploring alternative design strategies from the structural point of view. Elseveir Eng Struct 71:112–127
14. Devaraja R, Megadi MR (2014) Analysis of curved perimeter diagrid lateral system. Int J Eng Res Technol 3(6), ISSN: 2278-0181
15. Jani K, Patel PV (2013) Analysis and design of diagrid structural system for high rise steel buildings. Procedia Eng 51:92–100
16. IS 800:2007 (2007) General constructions in steel—Code of practice. Bureau of Indian Standards, New Delhi
17. IS875 (Part 1)-1987 (1987) Code of Practice for Design Loads (other than earthquake loads) for design of buildings and structures: Part 1, Dead Loads. Bureau of Indian Standards, New Delhi
18. IS875 (Part 2)-1987 (1987) Code of Practice for Design Loads (other than earthquake loads) for design of buildings and structures: Part 2, Imposed Loads. Bureau of Indian Standards, New Delhi

Urban Flood Management Strategies for Bangalore City Using Remote Sensing and GIS Techniques

Arati Reddy Nilap and H. N. Rajakumara

Abstract Cities are often becoming disaster centers where flooding has been occurring rather frequently. Due to the unprecedented frequency and proportion of urbanization taking place, many developed cities are witnessing extreme strain during the monsoons. The rains have caused some of the most advanced cities to face substantial flooding due to major planning lapses. Apart from altering catchments, urban habitats that spread across natural terrain like dry lakes, flood plains, and low-lying areas, commonly encroach over drainage networks and water bodies, eventually resulting in flooding. Insufficient and improper maintenance of stormwater drains have also contributed to urban floods. Because of the exponential increase in concretized areas, the incidence of runoff has increased profoundly, surpassing expected measures. Remote Sensing and GIS technology are used to analyze real-time topographical data, thematic topographical changes along with catchment area analysis of rainfall and drainage patterns. Population has witnessed a 4% average growth rate and the city has received an average of 40% increase in yearly precipitation since 2020. Radical topographical transformation took place from 2003 to 2013 with 40% reduction in waterbodies, 36% reduction in vegetative cover. and 89% increase in concretization. Dynamic structural design planning, flood vulnerability mapping, robust stormwater drainage systems, inbuilt rainwater harvesting and runoff recycling structures, green infrastructures, etc., are management strategies and good practices being suggested for implementation, to regulate the flooding issue.

Keywords Urbanization · Flood estimation · Geographical information system · Remote sensing · Urban storm water runoff

---

A. R. Nilap (✉) · H. N. Rajakumara
Department of Civil Engineering, BMS Institute of Technology and Management (Affiliated to Visvesvaraya Technological University, Belagavi-590018), Bangalore, Karnataka 560064, India
e-mail: aratireddy@gmail.com

H. N. Rajakumara
e-mail: drhnrajkumara@bmsit.in

Y. K. Guruprasad et al. (eds.), *Recent Advances in Infrastructure Development and Management—Volume 2*, Lecture Notes in Civil Engineering 566,
https://doi.org/10.1007/978-981-96-1558-2_17

1 Introduction

Major cities across the Indian Sub-continent are fast expanding, to cater to the development changes and advancement in the present decades, but simultaneously is also causing numerous challenges for city administrators, planners, and policymakers [1]. Bangalore is one of the important cities in South India, and it has experienced a good deal of anxiety related to continuous unplanned expansion, leading to accelerating population pressure, worsening traffic conditions, and flooding scenarios after monsoon-related rainfall events. There are sometimes simultaneous water scarcity issues in one region of the city while the other part experiences floods. Water quality issues and environmental degradation are causing further stress to the city known as Garden City. Looking at the city population, its growth has reached its peak during the last few decades with the initial outskirts now central districts within the city confines. There is an explosion in urbanization to meet this rising population demand [1, 2].

2 Study Area

The study area chosen is the metropolitan city of Bangalore and the main administrative boundary of the Bangalore Urban district, in the south-east region of Karnataka (see Fig. 1). It is almost equidistant from the eastern and western coasts of the South Indian peninsula [2] and has an elevation of 920 m above mean sea level. The mean annual rainfall is 846 mm. It is located 200 km from the Kaveri River. The topology of Bangalore is generally flat, though the western parts of the city are a bit undulating. The central part of the city has a slight ridge running north to south, and the land slopes away gently toward the east and west [3]. There are no major river channels within the study area, but numerous lakes exist within and around the city. Three main watersheds are formed in this slightly sloped topography, Vrishabhavathy Valley, Hebbal Valley, and Koramangala-Challaghatta Valley.

3 Problem Statement

The Karnataka capital has seen increasing annual rainfall over the past few years, leading to submergence in many parts of the city, and causing troublesome traffic congestion for long distances [3–5]. Waterlogged residential localities, flooded roads, overflowing drains, and fallen trees are common problems reported after spells of rainfall. The city has around 860 km length of drainage network both primary and secondary, stormwater drains or Rajakaluve as per BBMP reports. The impact of heavy rain in the IT city has led to drinking water issues related to quality concerns and interrupted supply.

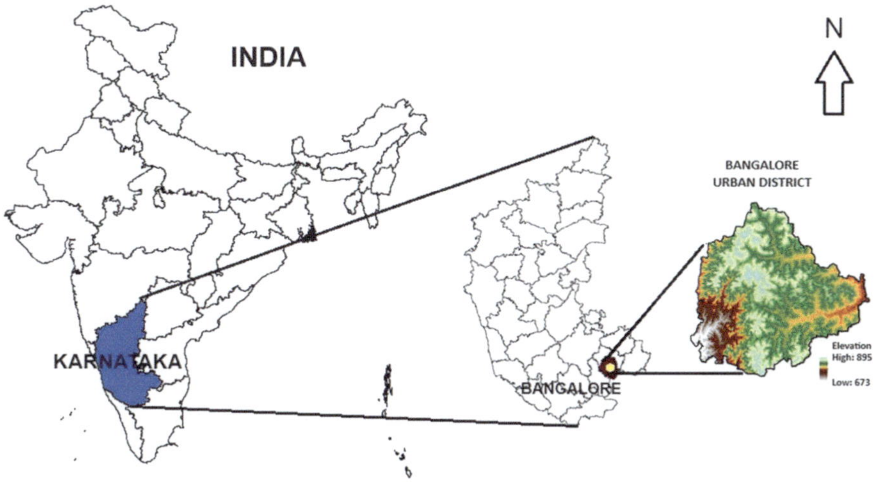

Fig. 1 Study area—Bangalore urban district

The Vrishabhavathy, Hebbal, and Koramangala-Challaghatta watersheds that divide Bangalore city, have many ponds and lakes within them, and during rains, most of the stormwater runoff flows into these water bodies and finally into the Dakshina Pinakini River on the east or the Kaveri River on the west [5]. These lakes are the main source of drinking water for the city along with groundwater. Many lakes are now disappearing, due to unplanned concretization and encroachment. Mismanaged and unchecked dumping of garbage wastes and sewage inflows into these lakes is causing the waters to be highly polluted and unfit for use.

The objectives of this study, therefore, are to analyze various data such as rainfall, population, and thematic maps, for causes leading to flooding in the city and to propose suitable strategies for flood management in the study area.

4 Rainfall and Drainage Analysis

Bangalore population growth studies show the numbers have increased by ten-fold in the past sixty years, and a surge of more than 400% since the last census was taken. To cater to the constant developmental activities and economic expansion, the city census is expanding at a 4% average growth rate [6–8], and most of the open land areas are getting concretized, leading to large volumes of runoff rather than allowing infiltration of rainfall.

For the time period during which the city witnessed this population boost, daily precipitation data was collected and analyzed [9, 10]. Rainfall data analyzed shows a steady annual increase in percentage departure from normal, especially from 2020 to 2022, increasing by 28%, 33%, and 72% each year, with respect to mean average

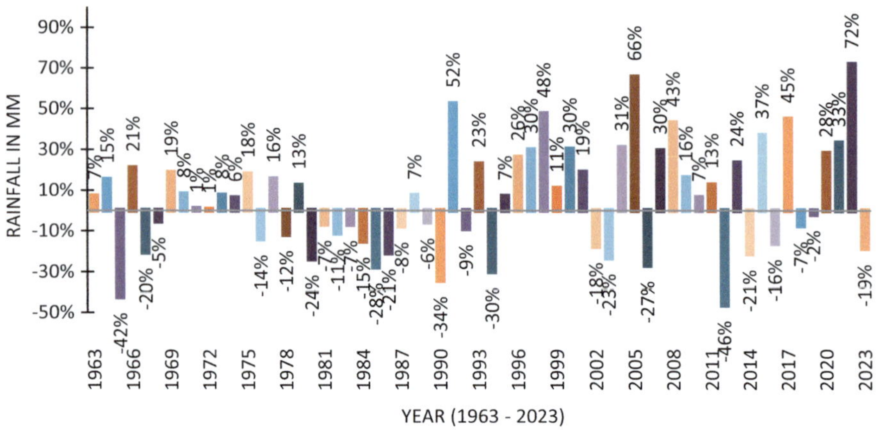

Fig. 2 Annual rainfall % departure from normal

annual rainfall, causing immense stress to the city (see Fig. 2). The annual rainfall in 2020 was 1085 mm and in 2021 was 1128 mm. The highest rainfall the city has ever received was in 2022, which is 1456 mm. The average percentage departure from normal is 6% over the study time period.

Due to its geographical location in the Indian peninsula, being almost midway between both the coasts, and its general topography without much of hilly undulations, Bangalore receives rain from both the southwest monsoon, and the northeast monsoon, and analyzing the average seasonal rainfall variation over the study time period shows the southwest monsoon contributes 55%, and the northeast monsoon contributes 25% of the city's total annual rainfall [6].

Analyzing the month-wise average rainfall variations over the past sixty years, the receding southwest monsoon months of August and September, and the approaching northeast monsoon month of October, together have accounted for 51% of the city's total yearly rainfall (see Fig. 3), whereas the pre-monsoon May month and southwest monsoon months of June and July have accounted for 33% of the yearly rainfall in the study area [6].

Hydrological maps, flow direction, and flow accumulation were generated to extract the drainage network and identify the various watershed basins (see Fig. 4). The drainage maps generated for the study area show a dendritic pattern system, with eighth order streams [6]. The total stream length from the drainage network is 15624 km. The slope map generated shows the terrain has an average slope of 0.7%, causing precipitation to tend to infiltrate rather than runoff. The elevation difference is around 200 m in the study area. Thus, the topography is not rough but very gently sloping.

Fig. 3 Month-wise average rainfall variation over sixty years

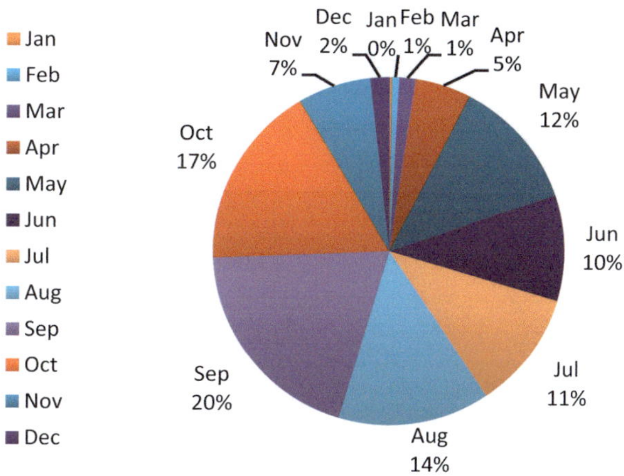

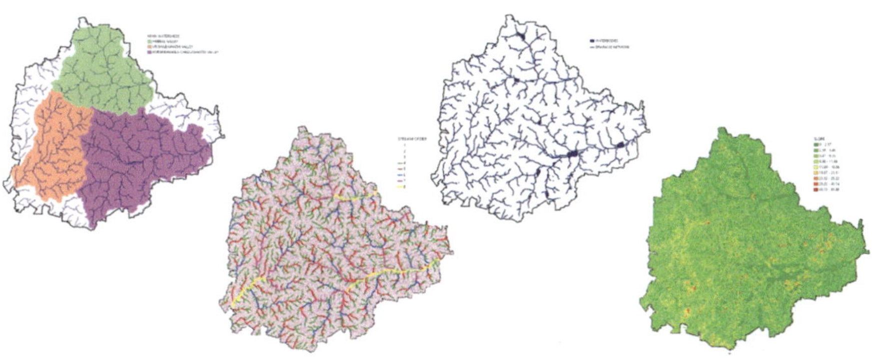

Fig. 4 Drainage analysis—watersheds, drainage network, stream order, and slope maps

5 Thematic Analysis

Bangalore city has shown a drastic transformation during the study period in terms of topographical changes. Landsat satellite data (USGS) and False Color Composite (FCC) images [11] were analyzed and the variation over the study time period of five decades was studied (see Fig. 5).

Land Use Land Cover (LULC) was carried out using the ArcGIS 10.x software, and depicted in ten main categories, Built-up areas such as Built-up Land Dark and Light, Pavements, Non-Built-up areas such as Vegetation (Dark, Medium, and Light), Waterbodies and Others such as Scrub Land, Fallow/Open Land and Bare Land, to evaluate the runoff for urban conditions (see Fig. 6).

The major changes observed these past five decades from 1973 to 2023 was concretization, and most of the open areas have been transformed into built-up lands, with scarcely any bare areas left which would have promoted rainfall infiltration,

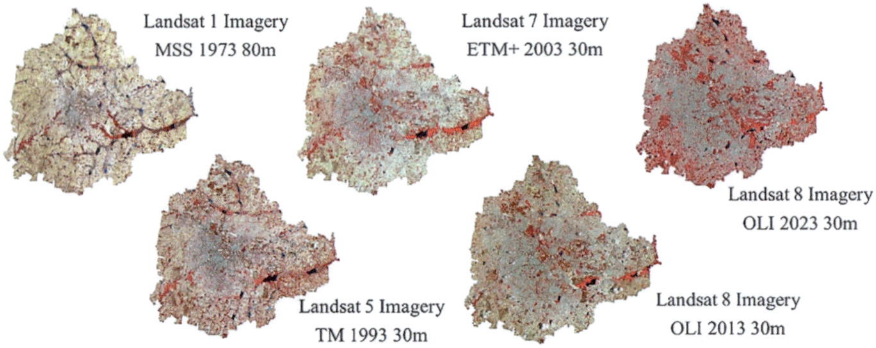

Fig. 5 Landsat FCC satellite imagery from 1973 to 2023 for Bangalore City

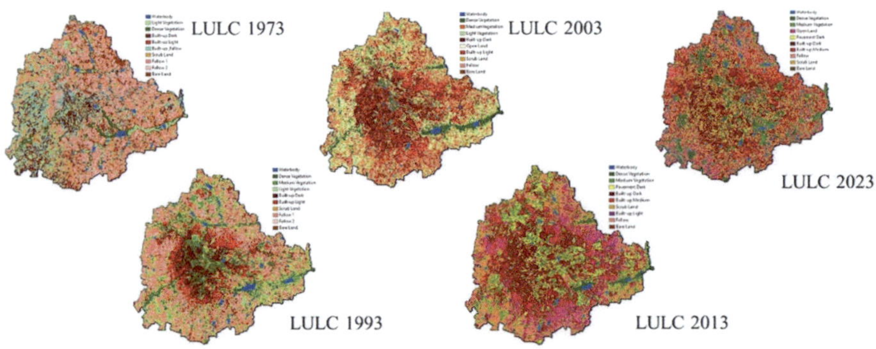

Fig. 6 LULC Comparison from 1973 to 2023 for Bangalore City

resulting in most of the rainfall converting to stormwater runoff. Vegetation areas have reduced thus impeding natural rainfall dispersion. To better grasp the topographical transformations and compare changes in categories of land use, the Built-up Dark, Built-up Light, and Pavements have been collectively catalogued as Built-up areas, Dense, Medium, and Light Vegetation classes are collectively grouped into Vegetation, and areas classified under Open Land, Fallow, Scrub and Bare Land are clubbed as Others. The findings of this analysis show the percentage area of each of these categories occupied (see Fig. 7).

Through two decades from 2003 to 2023, waterbodies have reduced by 14%, open bare land has reduced by 53% and vegetation has reduced by 28%. During the same time period, built-up areas have increased by 118%. Since the past decade alone, built-up area has increased by 15% from 2013 to 2023. Analyzing the land use data over the past five decades from 1973 to 2023, significant transformations show open bare lands have been urbanized by 68% and concretized land under urban built-up areas has increased by 142% during this interval (See Figs. 8 and 9).

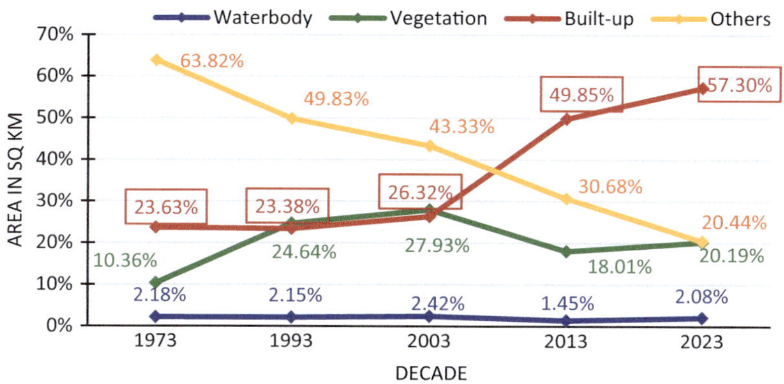

Fig. 7 Land use land cover % Area comparison over five decades

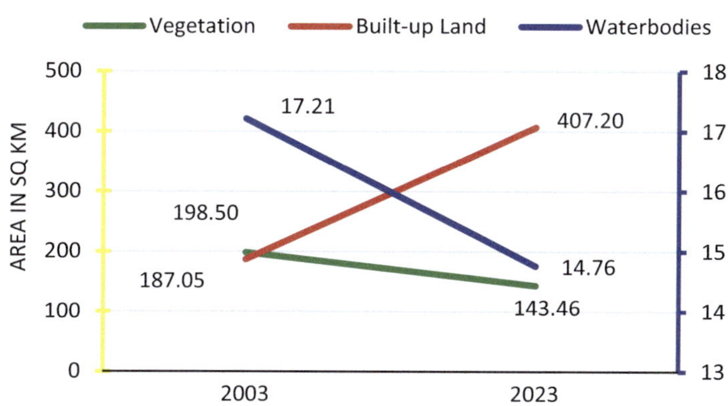

Fig. 8 Land use variations from 2003 to 2023

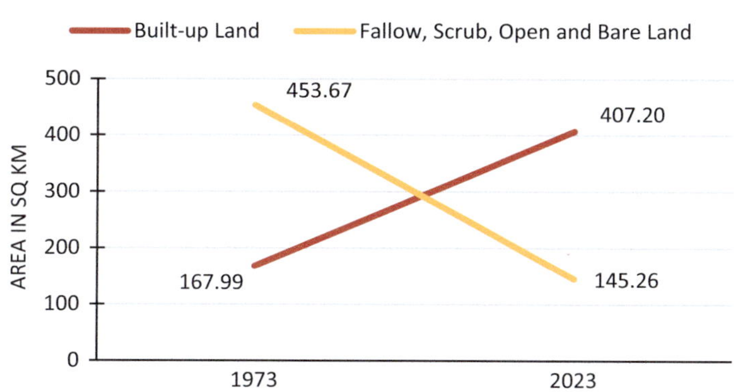

Fig. 9 Land use variations from 1973 to 2023

6 Management Practices and Strategies

To tackle flooding in Bangalore, as rains continue to batter the city every year, bringing daily life activities to a complete standstill, is the need of the hour. Bangalore city has expanded to three times its size in just over a decade while the city inhabitants have increased by more than two-fold since 2001 [3].

Selective urban techniques and management methods that can be implemented for capturing, diverting, and utilization of stormwater runoff are listed below.

- Rooftop rainwater harvesting mitigates flooding in urban surroundings. It can solve severe water deficit conditions by different direct and indirect methods, by reducing wastage of water which would otherwise result in flooding of roads and localities and enhancing surface and groundwater availability for use in various sectors such as farming, household, and commercial uses.
- Urban area rainwater harvesting methods include, recharge pits for individual buildings, recharge trenches for building complexes, recharge tube wells, and trenches with recharge wells.
- Permeable pavements are an effective controlling measure for urban runoff, are a sustainable technology for stormwater management, thereby allowing for percolation into lower strata, and one of the best practices that should be adopted in everyday life, and especially implemented in rapidly urbanizing localities and in dense commercial areas.
- Apart from existing roadside drainage systems, permeable kerb units can be designed to act as a shoulder kerb or a median and can contribute to controlling flooding on the carriageway. The natural camber of the road will aid in directing rainwater to the sides of the carriageway for infiltration through the pervious base of the kerbs and medians. The lower part of the kerb block can be molded using pervious concrete, allowing drainage of precipitation that could not dissipate at the roadsides, through the block and into the underlying soil below. Thus, even during heavy rains, the road pavement is prevented from getting flooded and traffic is not hampered. This will avoid disruption of everyday business life. Groundwater recharge through the underlying soil will be facilitated and will be an added advantage of pervious surfaces.
- Green infrastructure such as rooftop garden areas, sometimes known as green roofs, contribute to urban rooftop rainwater harvesting and prevent rain falling on building roofs from adding to runoff causing flooding in cities.Similarly, vertical gardens along building walls would serve the same purpose.
- Rain gardens built across cities in urban localities, and preferably along a natural slope, would allow for good infiltration of rainwater into the underlying strata and reduce the portion of runoff which would otherwise result in flooding. Rain gardens are water smart landscape features which will catch and filter rainwater runoff and prevent it from overfilling stormwater drains and sewers that effectuate polluting the water supplies.

- City planning methods should include infiltration trenches or rock-filled trenches, along roadsides and between pavements and highways to aid excess runoff to infiltrate into the lower strata and clear off.
- Similarly, swales or infiltration basins should be planned to manage runoff and increase the outflow of excess runoff. An old method of runoff management, they are usually open, shallow, and broad channels along roadsides, filled with a natural cover of vegetation allowing runoff and promoting rainwater infiltration. Swales are built along with Berms, or raised beds, which help direct water to swales and work like the camber in pavements.
- Roadside planting of trees and vegetation also aids in runoff dissipation. It acts as a flood control measure, though slow, but helping to absorb road runoff thereby reducing flooding and waterlogging along pavements and side roads.

The above-mentioned methods and practices would be best suited in the present scenario of urban surrounding and landscaping, and if implemented at characterized low-lying areas would subsequently provide some relief from the flood situation during rainfall that the city so often witnesses.

Regular monitoring for lake encroachment if standardized avoids situations of flood. Lake capacities are being monitored to check for affluents being pumped into lakes from surrounding industrial areas, or drainage from sewage systems, and to prevent flooding of buffer areas. Drainage systems carrying sewage across the city be planned for capacity standards and dimensions to prevent rainfall from causing sewer water to flow into lakes and cause pollution. Stormwater drains be maintained and upgraded as per the growing population and subsequent increasing capacity [3, 6]. Regular desilting of stormwater drains be implemented and clearing garbage dumped in them.

Dynamic flood vulnerability maps generated, assist in monitoring changing land use and subsequent runoff generated. Urban planning to avoid vulnerable areas and high-risk zones [5]. Studies are to be carried out to analyze human and environmental health risks and benefits associated with use of urban stormwater [12].

7 Conclusion

The management of urbanization-induced flooding in Bangalore city has been studied. Rainfall, drainage, and population analysis show that due to increasing urbanization and the subsequent concretization, any amount of rainfall exceeding a certain minimum is resulting in flood situations. In addition, inadequate opportunity for the stormwater runoff to disperse is causing the city to constantly face flood-like situations. Stormwater drains while allowing runoff to flow should be maintained. The steadily increasing annual rainfall year on year has added to the growing problem of Bangalore's flood occurrence.

From the thematic analysis of satellite images, it was seen that rainfall infiltration is being impeded due to concretization, which is a major cause of urban floods.

Significant shrinking of areas available for infiltration, such as empty land or open spaces and vegetation areas has reduced considerably resulting in massive amounts of stormwater surface runoff leading to flooding.

Different stormwater runoff management plans and strategies being adopted across other cities worldwide were studied and then suitable techniques pertaining to the study area were recommended.

References

1. Ahluwalia IJ (2019) Urban governance in India. J Urban Aff 41(1):83–102 https://doi.org/10.1080/07352166.2016.1271614
2. Ramachandra TV, Shivamurthy V, Aithal BH (2017) "Frequent floods in Bangalore: causes and remedial measures", ENVIS Technical Report: 123 https://doi.org/10.13140/RG.2.2.17517.90088
3. Bhandiwad R (2015) Urban flood monitoring and management. Int J Prog Civ Eng (IJPCE), ISSN (Print):2394-4676, (Online):2394-4684, vol-2, issue 1
4. Avinash S (2014) Flood related disasters: concerned to urban flooding in Bangalore, India. IJRET: Int J Res Eng Technol eISSN: 2319-1163 pISSN: 2321-7308
5. Gupta AK, Nair SS (2011) "Urban floods in Bangalore and Chennai: risk management challenges and lessons for sustainable urban ecology", General Articles, Current Science, Vol. 100, No. 11
6. Nilap AR, Rajakumara HN (2024) Analysis of urbanization-induced flooding in the Koramangala-Challaghatta valley of Bangalore city using remote sensing and GIS. In: Menon NVC, Kolathayar S, Rodrigues H, Sreekeshava KS (eds) Recent advances in civil engineering for sustainable communities. IACESD 2023. Lecture Notes in Civil Engineering, vol 459. Springer, Singapore https://doi.org/10.1007/978-981-97-0072-1_43
7. India Census https://censusindia.gov.in/census.website/. Last Accessed 2023/08/31
8. United Nations https://population.un.org/wpp/. Last Accessed 2023/08/31
9. KSNDMC https://www.ksndmc.org/en. Last Accessed 2023/09/30
10. IMD https://mausam.imd.gov.in/. Last Accessed 2023/09/30
11. USGS https://earthexplorer.usgs.gov/. Last Accessed 2023/09/29
12. Jiang SC, Lim KY, Huang X, McCarthy D, Hamilton AJ (2015) Human and environmental health risks and benefits associated with use of urban stormwater, vol-2. Wiley Interdisciplinary Reviews: Water, pp 683–699. https://doi.org/10.1002/wat2.1107

Blending Plastic Waste and M-Sand to Produce Sustainable Building Blocks

Vinayak Bagale and Mangala Keshava

Abstract This research unveils an inventive strategy to combat India's considerable plastic waste predicament, standing at 3.4 million metric tonnes yearly, compounded by inadequate recycling capabilities. India has witnessed a substantial uptick in plastic usage, registering a compounded annual growth rate of 9.7% over five years. The study concentrates on incorporating plastic waste into construction, particularly accentuating masonry bricks—a sector that has received comparatively scant attention. Employing a mix ratio of High-Density Polyethylene (HDPE) and M-Sand in a 1:2 proportion, the research encompasses material collection, preparation of brick moulds, casting, mixing, moulding, and testing. The research scrutinizes the properties of these plastic bricks, covering dimensionality, density, water absorption, compressive strength, and flexural characteristics. Outcomes indicate that these plastic bricks showcase properties akin to various brick types available in the market. Particularly noteworthy is their low water absorption rate, positioning them as a preferred choice for water-retaining structures in comparison to traditional brick variants. This research introduces an inventive resolution to address plastic waste challenges while championing sustainable construction practices.

Keywords Plastic waste · High-density polyethylene (HDPE) · M-Sand · Plastic bricks · Tests on bricks · Low water absorption rate · Sustainable construction practices

1 Introduction

In the twenty-first century, the construction industry has become a major player in both economic growth and environmental impact, particularly evident in cities like Delhi where construction contributes significantly to pollution, with 30% attributed to the sector. Concrete mixing alone accounts for 10% of PM2.5 particulate matter

V. Bagale (✉) · M. Keshava
Department of Civil Engineering, BMS College of Engineering, Bengaluru, India
e-mail: vinayakcbagale@gmail.com

Y. K. Guruprasad et al. (eds.), *Recent Advances in Infrastructure Development and Management—Volume 2*, Lecture Notes in Civil Engineering 566,
https://doi.org/10.1007/978-981-96-1558-2_18

in Delhi's air. To tackle these issues, the adoption of sustainable construction practices, including green building design, recycling, and energy-efficient technologies, is crucial. This can help mitigate environmental impact, reduce resource depletion, and improve air quality in rapidly growing urban areas.

Meanwhile, a recent report sheds light on a pressing problem in India, where approximately 3.4 million metric tonnes of plastic waste are generated, with only 30% undergoing recycling. This highlights a significant gap in waste management, leading to environmental concerns as a substantial amount of plastic waste ends up in landfills or aquatic environments. The report reveals a notable increase in plastic consumption in India, with a compounded annual growth rate of 9.7% over five years, reaching 20 million tonnes in 2019–20 from 14 million tonnes in 2016–17. Collaboratively produced by the Marico Innovation Foundation, the Indian Institute of Science, and Praxis Global Alliance, the study identifies Maharashtra, Gujarat, and Tamil Nadu as collectively contributing 38% to India's total plastic waste generation.

2 Statistics on Plastic Waste

The research conducted by the Central Pollution Control Board (CPCB) in January 2015, focusing on 60 prominent cities in India, reveals that within the overall composition of plastic waste in municipal solid waste, approximately 94% consists of recyclable thermoplastics. The distribution of various types of plastics is as follows: PET (Polyethylene Terephthalate) accounts for 8.66%, HDPE/LDPE (High-density Polyethylene/Low-density Polyethylene) constitutes 66.91%, PVC (Polyvinyl Chloride) makes up 4.14%, PP (Polypropylene) comprises 9.90%, PS (Polystyrene) contributes 4.77%, and other unclassified plastics/Rubbers account for 6.43%. Non-recyclable thermoset plastics make up the remaining 6% (Fig. 1).

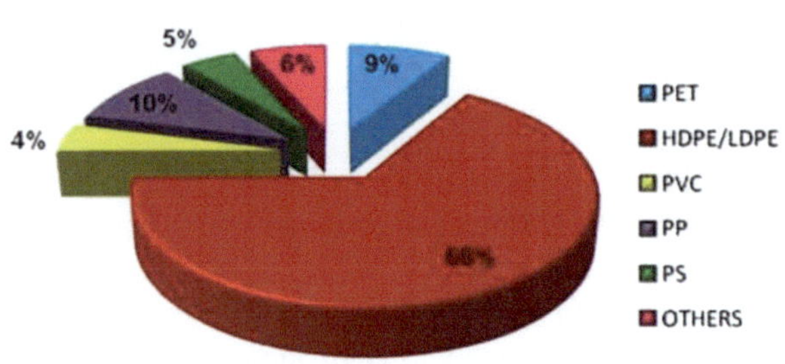

Fig. 1 Percentage of classified plastics

3 Materials and Methodology

3.1 Materials

Four categories of plastic namely HDPE, LDPE, PET, PPE along with M-Sand were taken up for material characterization.

- High-Density Polyethylene (HDPE)
- Low-Density Polyethylene (LDPE)
- Polyethylene Terephthalate (PET)
- Polyphenylene Ether (PPE)
- Crushed Stone Aggregate (M-Sand)

The specific gravity of the plastic waste was examined, and the results have been documented in the Table 1 presented

Testing was conducted on the M-Sand before its application in the experiment, and the results of these tests are detailed in the Table 2.

4 Methodology

An experimental setup with a binder-to-filler ratio of 1:2 was selected. To identify the optimal plastic composition for masonry unit production, a set of trial mixes was formulated. The attained compressive strengths of each mix were documented and organized in a table for further analysis.

1. The necessary plastic shredded pieces underwent a cleaning process and were subsequently dried.
2. Approximately 200 g of plastic were placed in a pan and melted using an electric burner. Observations, such as the odour of gas and the texture of the molten plastic, were recorded.

Table 1 Specific gravity of shredded plastic waste samples

Plastic type	HDPE	LDPE	PET	PPE
Specific gravity	0.9	0.89	1.13	0.83

Table 2 Test results of M-Sand

Tests	Results
Sieve analysis (Fineness modulus)	3.76
Specific gravity	2.29
Water absorption (in %)	2.16

Fig. 2 Process involved in trail mix

3. Once the plastic was melted, 400 g of M-Sand was introduced into the molten plastic and thoroughly mixed.
4. The resulting mixture was then poured into a cube mould measuring 70.6 mm, and the mix was allowed to cool for a duration of 2 h.
5. After cooling, the cube was demoulded, and the cubes were subjected to testing in a compression testing machine (Fig. 2).

The observations are recorded in the Table below

Sl. No.	Plastic Mix (1:2)	Density of Mix (gm/cc)	Compressive Strength (MPa)	Type of Failure	Workability	Time to melt (mins)	Finishing	Remarks
1	HDPE	1.58	12.86	Ductile Spalling	Good	45 min	Good	No odour, Black colour
2	PET	1.84	11.6	Brittle Cracks	Poor	10 min	Poor	Foul Smell, Greyish black colour
3	PP	1.59	10.18	Ductile Spalling	Very good	20 min	Very good	No odour, Black colour
4	50% HDPE + 50% PET**	1.7	14.5	Ductile Spalling	Good	20 min	Good	No odour, Black colour
5	LDPE*		–	–	–	15 min	–	Catches fire at around 200° C

* The absence of results in density, compressive strength, failure type, workability, and finish are attributed to LDPE's combustion at 200 °C during melting.
** Tests focused on a 50% HDPE and 50% PET mix, chosen for their prevalence in India's plastic waste (66.9% HDPE, 8.66% PET).

Due to HDPE constituting the largest proportion at 66.91% of municipal solid waste, a thorough examination was selected to concentrate on the blend incorporating HDPE and M-Sand for the project. Additionally, the mixture utilizing HDPE demonstrated commendable compressive strength, workability, and finish.

5 Production of Bricks

Following steps are adopted to produce plastic bricks.

Material Collection: The HDPE waste necessary for the project was obtained from a plastic scrap vendor, cleaned, and weighed to meet the required quantity. Simultaneously, M-Sand, passing through a 4.75 mm sieve, was dried and weighed to match the specified amount for the mixture (Fig. 3).

Plastic Melting: Using a gas burner setup, the plastic melting process generated the required heat. An open pan was positioned over the flame, creating an optimal environment for the melting procedure. Pre-weighed plastic materials were carefully introduced into the pan, transforming from a solid to a molten form under the flame's heat. This melting phase was closely monitored to ensure the plastic reached an optimal state for further mixing and brick production (Fig. 4).

Mixing Plastic and Sand: After achieving complete plastic liquefaction, M-Sand was introduced and thoroughly blended to create a strong connection between plastic and sand particles. This systematic fusion resulted in a cohesive mixture ideal for brick production, ensuring uniform distribution and a strong bond between plastic and sand (Fig. 5).

Brick Moulding: The subsequent step involved pouring the prepared mixture into steel moulds and compressing it using a tamping rod to ensure even filling. The filled moulds were then left to cool in the open air. Before filling the moulds, a layer of oil was applied to facilitate easy removal of the bricks once solidified, a crucial step to prevent challenges in removing solidified bricks from the moulds (Fig. 6).

Fig. 3 Collection of plastic waste

Fig. 4 Melting of Shredded plastic waste

Fig. 5 Mixing molten plastic and M-Sand

Fig. 6 Moulding of plastic bricks

Table 3 Test results on HDPE

Sl. No.	Test	Results
1	Softening point Flash point	220 °C 284 °C

6 Results and Discussion

6.1 Results

Several tests were carried out in accordance with both IS Codes and ASTM Codes, and the results are presented in the Table 3.

6.2 Discussion

- Blisters form on the bricks during production due to the attempt of hot air to escape from within.
- The reduced adhesion of mortar on the brick's smooth surface can be improved by applying a layer of sand to its surfaces (Tables 4, 5 and 6).

Table 4 Test results on HDPE and M-Sand mix

Sl. No.	Test	Results
1	MOE of cylinder Compressive strength	2292.4 N/mm^2 11.835 MPa
2	Thermal conductivity	0.0986 W/mK

Table 5 Test results on plastic bricks

Sl. No.	Test	Results
1	Dimensionality test	226 × 104 x 71.3 mm
2	Density	1768.46 kg/m^3
3	Initial rate of absorption	0.056 $kg/m^2/min$
4	Water absorption	0.44%
5	Compressive strength	11.13 N/mm^2
6	Flexural test	1.97 N/mm^2

Table 6 Test results on plastic brick prism

Sl. No.	Test	Results
1	MOE of plastic brick prism Compressive strength	3223.95 N/mm^2 4.875 MPa

Table 7 Comparison of plastic brick with other types of bricks

Tests	Table moulded	concrete blocks	Brunt brick	Fly ash brick	Wire cut brick	3 core hollow clay bricks	Plastic bricks
Dimensionality (mm)	226 × 104 x 71	400 × 100 x 200	230.25 × 110.5 x 70.5	220 × 100 x 77	220 × 100 x 77	220 × 100 x 77	226 × 104 x 71.3
Density (Kg/m3)	1841	2146	1849	1692.6	1797.8	1780.9	1768.5
Water absorption (%)	6.62	4.1	11.78	11.96	12.81	13.08	0.44
Wet compressive (MPa)	0.94	8	11.05	11.51	10.59	11.9	11.13

7 Comparison and Conclusions

7.1 Comparison

A comparison between the plastic bricks and the bricks currently available in the market is conducted, and the findings are outlined and recorded in the Table 7.

7.2 Conclusions

- Plastic bricks display promising performance, making them a viable alternative to conventional bricks.
- The water absorption rate of the plastic bricks is measured at 0.44%, rendering them suitable for use in water-retaining structures.
- The thermal conductivity of the plastic bricks is 0.0986 W/mK, which is lower than that of regular softwood at 0.12 W/mK.

Acknowledgements The authors would like to sincerely thank BMS College of Engineering, Bengaluru, India for providing the necessary facilities for carrying out the experimental work needed for this research.

References

1. Kulkarni P, Ravekar V, Rao PR, Waigokar S, Hingankar S (2022) Recycling of waste HDPE and PP plastic in preparation of plastic brick and its mechanical properties. Clean Mater 5:100
2. Shiuly A, Hazra T, Sau D, Maji D (2022) Performance and optimisation study of waste plastic aggregate based sustainable concrete—a machine learning approach. Clean Waste Syst 2
3. Thiam M, Fall M, Diarra MS (2021) Mechanical properties of a mortar with melted plastic waste as the only binder: influence of material composition and curing regime, and application in Bamako. Case Stud Constr Mater 15
4. Agyemana S, Obeng-Ahenkora NK, Assiamah S, Twumasi G (2019) Exploiting recycled plastic waste as an alternative binder for paving blocks production. Case Stud Constr Mater 11:e00246
5. Seghiri M, Boutoutaou D, Kriker A, Hachani MI (2017) The possibility of making a composite material from waste plastic. In: International conference on technologies and materials for renewable energy, environment and sustainability, TMREES17. Beirut Lebanon, pp 163–169
6. Chauhan SS, Kumar B, Singh PS, Khan A, Goyal H, Goya S (2019) Fabrication and testing of plastic sand bricks. In: IOP conference series: materials science and engineering
7. Aneke FI, Awuzie BO, Mostafa MMH, Okorafor C (2021) Durability assessment and microstructure of high-strength performance bricks produced from PET waste and foundry sand. Materials
8. Siddique R, Khatib J, Kaur I (2008) Use of recycled plastic in concrete: a review. Waste Manag 28:1835–1852
9. Babatunde Y, Mwero J, Mutuku R, Jimoh Y, Oguntayo (2022) Influence of material composition on the morphology and engineering properties of waste plastic binder composite for construction purposes. Heliyon 8(10)
10. Biswasa A, Goela A, Potnis S (2020) Performance comparison of waste plastic modified versus conventional bituminous roads in pune city: a case study. Case Stud Constr Mater 13
11. IS: 2386-1963 Methods of test for aggregates for concrete (1963). Bureau of Indian Standards, New Delhi, India
12. IS: 383-1970 Specification for coarse and fine aggregates from natural sources for concrete (1970). Bureau of Indian Standards, New Delhi, India
13. IS: 1077-1992 Common burnt clay building bricks-specification (1992). Bureau of Indian Standards, New Delhi, India
14. IS: 3495-1992 Methods of testing of burnt clay building bricks (1992). Bureau of Indian Standards, New Delhi, India
15. IS: 1205-1978 Indian standard methods for testing tar and bituminous material (1978). Bureau of Indian Standards, New Delhi, India
16. ASTM C67 Standard testing methods for sampling and testing brick and structural clay tile. ASTM International, West Conshohocken, PA
17. ASTM C469 Standard test method for static modulus of elasticity and poisson's ratio of concrete in compression. ASTM International, West Conshohocken, PA
18. ASTM C1317 Standard test method for compressive strength of masonry prisms. ASTM International, West Conshohocken, PA

Finite Element Investigation of the Influence of FRP on RC Beams with Structural Openings

A. N. Thanushree, Neethu Urs, Tahera, and M. Meghashree

Abstract In current times, the mixing of essential services like sewage, communication cables, and water supply often necessitates the use of pipes and ducts, which could significantly lessen a shape's clean height. To maintain aesthetics, incorporating openings in beams turns into a practical option, which could compromise the beam's structural integrity, leading to issues including cracking, excessive deflection, and reduced stiffness. Addressing this venture includes employing fiber-reinforced polymer (FRP) sheets like carbon fiber-strengthened polymer (CFRP), glass fiber-strengthened polymer (GFRP), or basalt fiber-strengthened polymer (BFRP) to retrofit and enhance the region surrounding those openings in concrete structures. This technique gives numerous advantages over conventional strategies due to its lightweight nature, high power, resistance to corrosion, fatigue resistance, ease of installation, and minimal alteration to the structural geometry. This research focuses on analyzing the shear behavior of numerous beam openings regarding their shapes, sizes, and the application of FRP retrofitting, aiming to offer insights into improving beam overall performance while also accommodating necessary service installations.

Keywords Opening in beams · Finite element method · Retrofitting · FRP

A. N. Thanushree (✉)
Shree Designs, Hassan, Karnataka, India
e-mail: thanushreeachar123@gmail.com

N. Urs · Tahera · M. Meghashree
Department of Civil Engineering, Dayananda Sagar of Engineering, Bengaluru, Karnataka, India
e-mail: neethu-cvl@dayanandasagar.edu

Tahera
e-mail: tahera-cvl@dayanandasagar.edu

M. Meghashree
e-mail: meghashree-cvl@dayanandasagar.edu

Y. K. Guruprasad et al. (eds.), *Recent Advances in Infrastructure Development and Management—Volume 2*, Lecture Notes in Civil Engineering 566,
https://doi.org/10.1007/978-981-96-1558-2_19

1 Introduction

The fast evolution of global instances is driving the need to adapt current homes to apply essential services including power cables, AC or cooling ducts, water supply, sewage pipes, gas pipes, and pc lining networks, and many others. Several systems stumble upon constraints in vertical area, hence prompting the want to include openings within reinforced beams. Those openings manifest in numerous shapes rectangular, square, circular, and other normal or irregular form with rectangular and square shapes being predominantly applied. They are categorized into small sizes and large-sized openings based on their dimensions, with massive openings $A = \pi r^2$ exceeding 0.4 multiples the general beam depth, thereby drastically affecting beam conduct by using diminishing stiffness, inducing deflection, and weakening electricity. Considerably, sharp corners in rectangular or square openings precipitate strain concentrations, ensuing in ugly cracks. This investigation employs nonlinear numerical analysis to address the challenges related to introducing openings in strengthened concrete beams. These challenges encompass issues such as intensified localized cracking at edges, discount in stiffness and energy, and an increase in deflection. The objective of these studies is to recommend answers via exploring how carbon fiber-strengthened polymer (CFRP) laminates can efficiently reinforce these openings and mitigate the above-mentioned troubles.

1.1 Significance of Work

Ideally, it is preferred to install service pipes beneath a building's beams. However, clients and service consultants oppose this arrangement due to its disturbance in aesthetical appearance. This type of installation would require bending the pipes, resulting in the floor's height reduction. For instance, a Twenty-story structure approved with a height of 60 m. But, incorporating 150 mm pipes on each level would subtract a total of 3 m from the building's overall height. Consequently, it is a loss of an entire floor, preventing the builder from selling the specific floor.

To enhance the strength of beam with opening by providing carbon reinforced polymer both inside and around the opening

Modeling

This study utilizes a nonlinear analysis approach, creating a simultaneous action and bonding between the reinforce element and the concrete element. FEM approach comprises 3 phases primarily: pre-processing phase, the solution phase, and post-processing phase. The element types chosen are shown in the Table 1.

(a) 5 beam models have been generated for its geometric modeling:

- Beam without opening (solid beam); SB
- Beam with an opening; BWO

Table 1 Type of element [7]

Sl No.	Type of element for Material
1	Solid 187 for concrete
2	Link 180 for steel rebar
3	Shell 181 for CFRP

- Reinforced beam with CFRP within opening; BWCW
- Reinforced beam with CFRP surrounding opening; BWCS
- Reinforced beam with CFRP within and surrounding opening; BWCWS

The beam being studied consists of a length of 1600 mm and a cross section of 230 × 300 mm. It is reinforced with two upper longitudinal bars of 8 mm diameter and two 10 mm diameter bars for tensile reinforcement. Stirrups of 8 mm diameter are placed at 300 mm intervals for structural support. For detailed information on the beam's dimensions and reinforcement layout, please see Figs. 1 and 2 (Table 2).

Beam model provides a rectangular opening with rounded corners of cross-section 150 × 100 mm with a chamfer of fifty millimeters (Fig. 3).

After creating the geometric model, establishing the connection between the rebar and concrete involves using the "share topology" option to ensure node-to-node connectivity. The contact between longitudinal bars and stirrups utilizes a node-to-node interface through MPC, connected by a 4 mm pin ball connection, as depicted in the Figure. The beam meshing employs a global concrete mesh with an edge length of 15 mm, utilizing 10-noded tetrahedral elements. For the contact interface

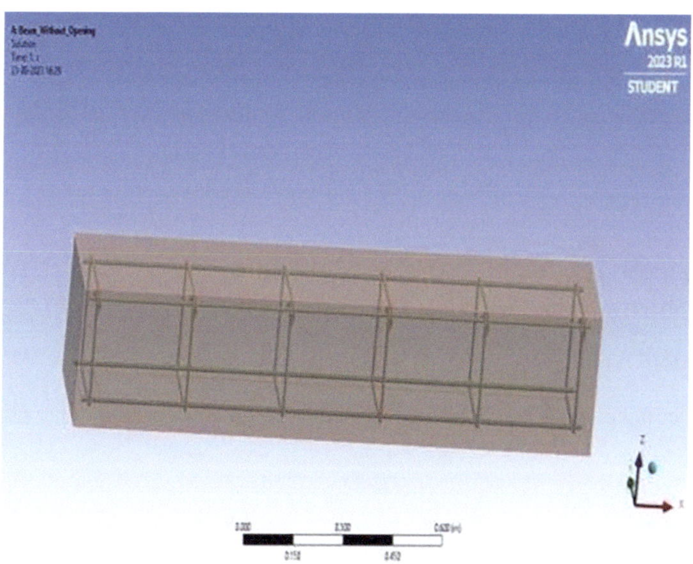

Fig. 1 Beam geometry

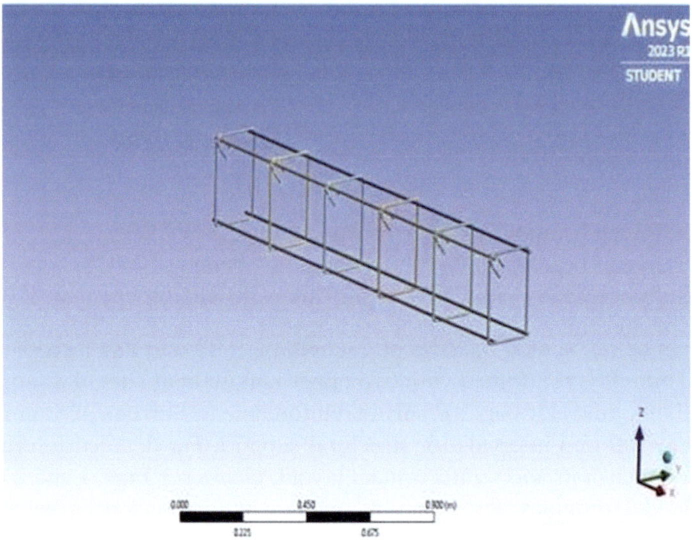

Fig. 2 Reinforcement detail

Table 2 CFRP material properties [7]

Elastic modulus—z direction E_z (MPa)	1.79×10^4
Elastic modulus—y direction E_y (MPa)	1.79×10^4
Elastic modulus—x direction E_x (MPa)	2.3×10^5
Shear modulus—yz direction G_{yz} (MPa)	6880
Shear modulus—xy direction G_{xy} (MPa)	11,790
Shear modulus—xz direction G_{xz} (MPa)	11,790
Poisson's coefficient—yz direction μ_{yz}	0.30
Poisson's coefficient—xy direction μ_{xy}	0.22
Poisson's coefficient—xz direction μ_{xz}	0.22

involving CFRP, a 4 mm edge length is applied, while for the rebar and CFRP, the edge length remains at 4 mm (Figs. 4 and 5).

(b) Boundary condition

Nodes are connected by a master node to provide a beam simply support as shown in Fig. 6.

(c) Load assignment

The beam experiences loading at two points, as depicted in the Fig. 7.

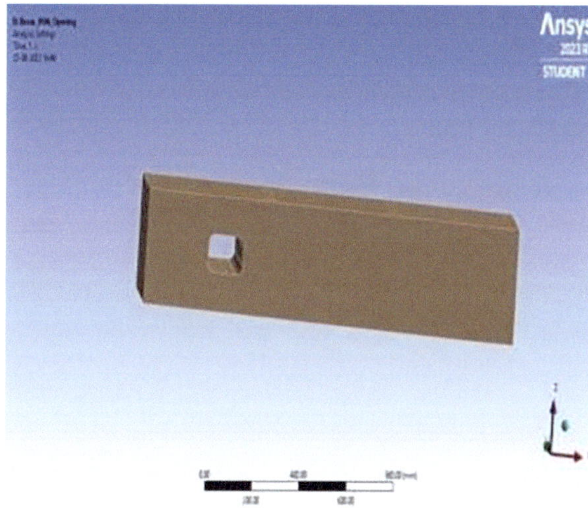

Fig. 3 Beam having rectangular opening with rounded corners

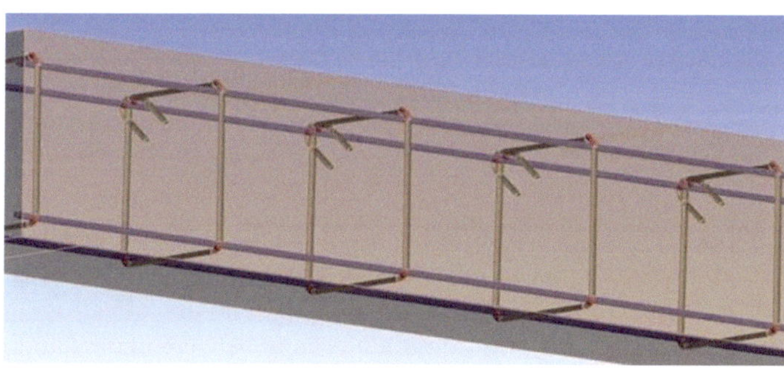

Fig. 4 Beam having rectangular opening with connection between rebars

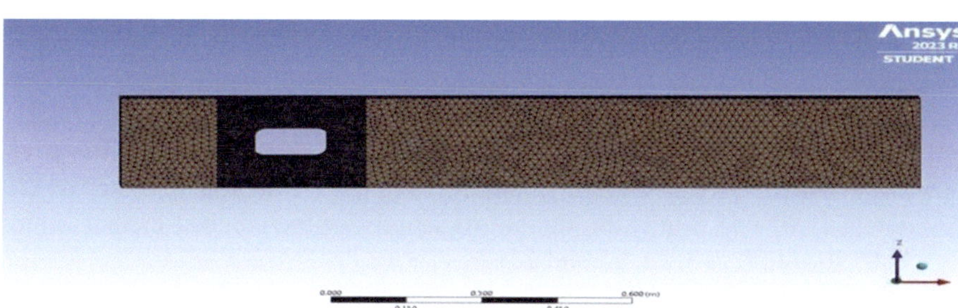

Fig. 5 Meshing of a beam

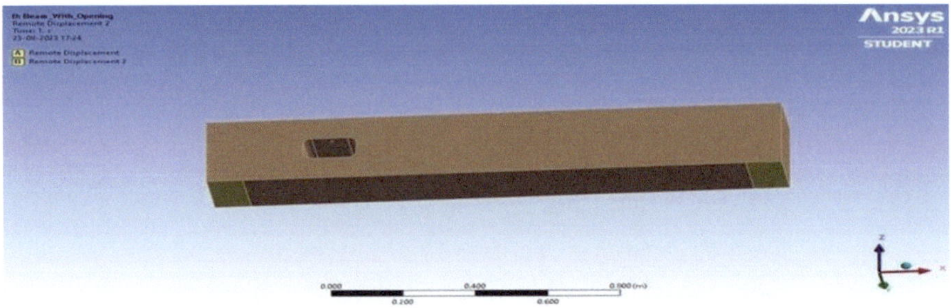

Fig. 6 Support condition of a beam

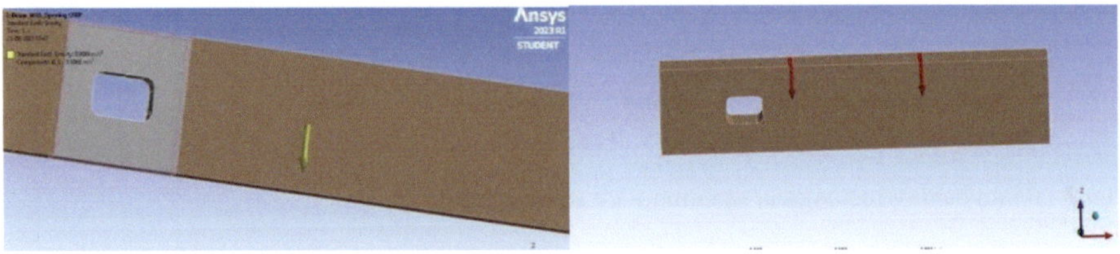

Fig. 7 Beam subjected to self-weight and two point loading

2 Analysis in ANSYS

Examining openings situated in the shear part of beams is pivotal because it reveals intricate structural behaviors. The introduction of an opening in this region significantly alters how shear forces are distributed, potentially affecting the overall load-bearing capacity of the beam. To gain a comprehensive understanding of this phenomenon, an exhaustive examination is necessary, on variations in distribution of shear stress, shear flow, and potential shear-related failure modes. Crucial factors such as the beam material properties, size and position of the opening and loads applied collectively influence the response of the structure. This analysis necessitates combination of experimental tests and computational simulations, such as FEA approach, to precisely note how the modification geometry interacts with shear forces. Understanding behavior of beams having opening is vital in constructing buildings that can withstand diverse loads while maintaining long-term structural uprightness. The stress intensity of five beams throughout its length for incremental loads is noted (Figs. 8, 9, 10, 11, 12, 13, 14, 15 and 16).

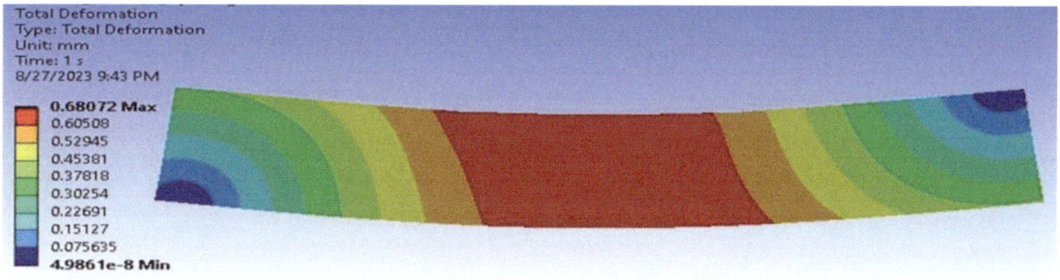

Fig. 8 Deformity of SB

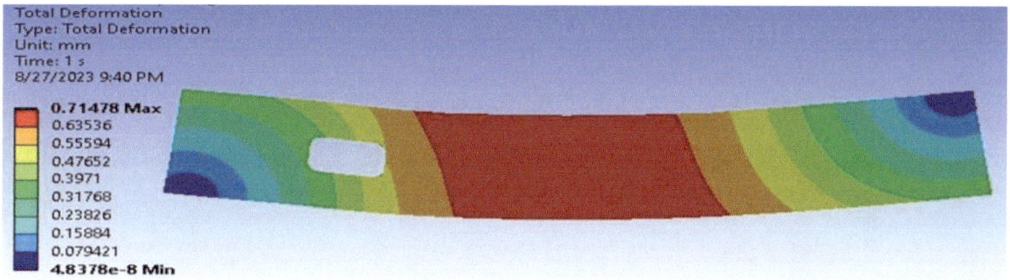

Fig. 9 Deformity of BWO

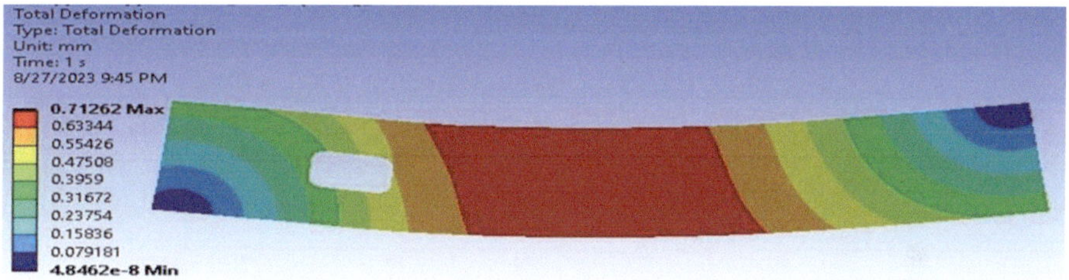

Fig. 10 Deformity of BWCW

Fig. 11 Deformity of BWCWS

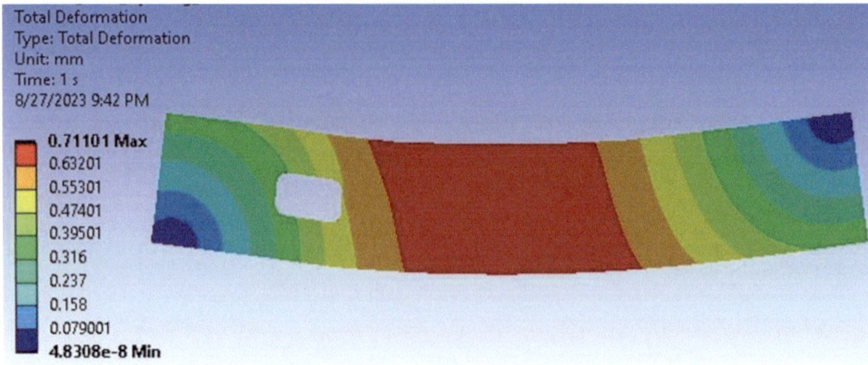

Fig. 12 Deformity of BWCWS

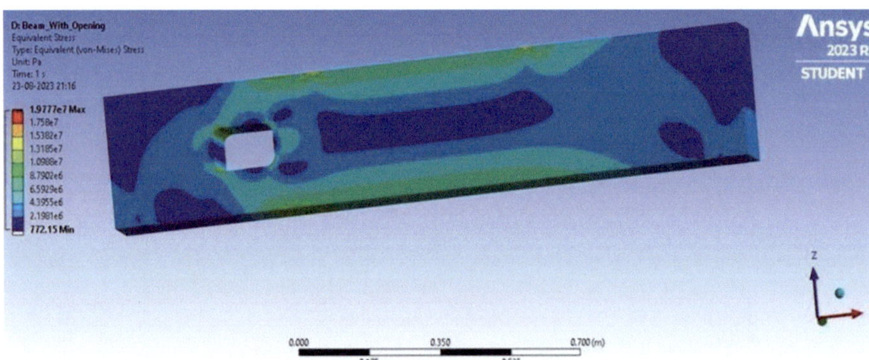

Fig. 13 Equivalent stress

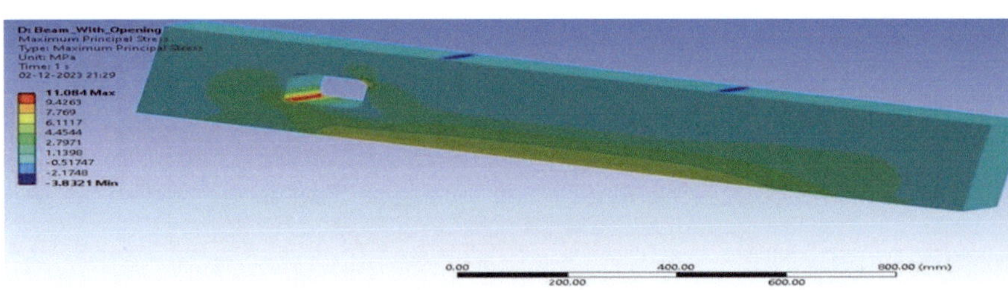

Fig. 14 Maximum principal stress of BWO

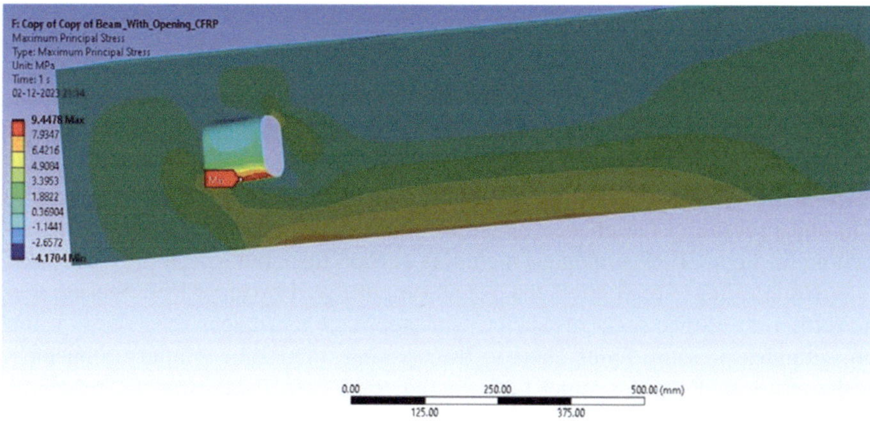

Fig. 15 Maximum principal stress of BWCS

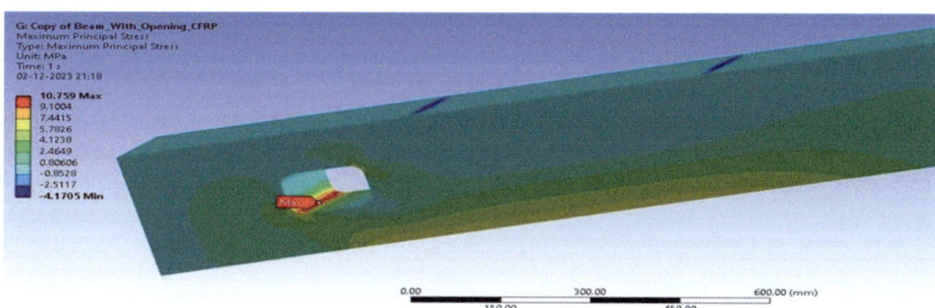

Fig. 16 Maximum principal stress of BWCWS

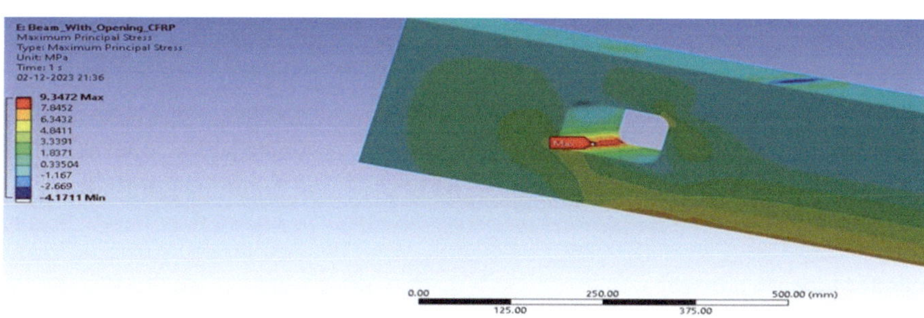

Fig. 17 Maximum principal stress of BWCWS

3 Result and Discussion

The examination results concerning with openings in, jointly with the application of CFRP support strategies, give important experiences into underlying performance and possible enhancements. Presenting openings in this zone is seen to actuate adjustments in stress focus and shear force dissemination, affecting the bearing limit and presenting the chance of failure modes due to shear. Strengthening the opening through CFRP strategies, including carbon fiber-reinforced polymer inside the opening, brings about upgraded execution portrayed by more load-bearing limit and further developed shear resistance. This approach rearranges shear stress values and mitigates pressure focus close to the opening, upgrading primary robustness. Furthermore, CFRP use around the opening reinforces shear strength, restricting disfigurements, stopping of crack expansion, and efficiently sending shear strength, prompting a stronger primary reaction. In addition, utilizing carbon fiber-reinforced polymer encompassing its edges and inside opening gives a significant upgrade shear obstruction, synergistically addressing stress fixation and shear reallocation.

This approach altogether demonstrates load capacity contrasted with individual CFRP support techniques. The conversation accentuates the possible adequacy of CFRP support in moderating unfavorable impacts coming about because of openings inside the shear zone of bars, featuring the significance of cautiously choosing and fitting CFRP fortifying procedures to meet explicit primary prerequisites (Tables 3 and 4).

Table 3 Load and deflection of beams

Load (kN)	Deflection (mm)				
	SB	BWO	BWCW	BWCS	BWCWS
20	0.13614	0.14296	0.14253	0.14263	0.1422
40	0.27229	0.28591	0.28506	0.28527	0.28441
60	0.40843	0.42887	0.42758	0.4279	0.42661
80	0.54457	0.57183	0.57011	0.57053	0.56882
100	0.68072	0.71478	0.71262	0.71317	0.71101

Table 4 Maximum principal stress of concrete in beams

Types of beams	Maximum principal stress
SB	9.3839
BWO	11.084
BWCW	9.4478
BWCS	10.759
BWCWS	9.3472

From above, observation of maximum principal stress in various types of beams wrapping within the opening is a better approach than incorporating CFRP surrounding the opening.

4 Conclusion

In conclusion, the study demonstrates the effectiveness of Carbon Fiber Reinforced Polymer (CFRP) in enhancing the structural performance of RC beams with openings. Openings within the shear span adversely affect stress distribution and reduce the beam's ultimate load-carrying capacity. However, the integration of CFRP within these openings significantly improves the beam's thrust-carrying capacity and post-failure stability by effectively redistributing stress and curtailing crack propagation. Additionally, reinforcing the perimeter of the openings with CFRP acts as a crack detention mechanism, further enhancing the load-bearing capacity and overall structural integrity. The load-deflection analysis reveals improved ductility and a more robust response to varying loads, highlighting the potential of CFRP reinforcement. A combined approach of integrating CFRP within the openings and along their perimeters effectively mitigates stress concentrations and demonstrates the efficacy of a Comprehensive strengthening strategy, offering a reliable solution for addressing structural challenges in RC beams with openings.

References

1. Diggikar R, Mangalagi S, Harsoor R (2013) Strengthening R.C.C. beams with rectangular openings using CFRP and GFRP sheets
2. Mansur MA, Tan KH, Lee S-L (1984) Collapse loads of R/C beams with significant openings. J Struct Eng 110(11):2602–2618
3. Mansur MA (1998) Influence of openings on the behavior and strength of R/C beams in shear. Cement Concr Compos 20(6):477–486
4. Almusallam T, Al-Salloum Y, Elsanadedy HM, Alshenawy AO, Iqbal RA (2018) Behavior of FRP-strengthened RC beams with large rectangular web openings in flexure zones: experimental and numerical study. Int J Concr Struct Mater 12(1)
5. Mareesha Susan B, Mathew A (2016) Analytical Study on the effects of circular openings in RC Beams and strengthening with CFRP and BFRP sheets. Int J Sci Technol Eng 3(2):248–255
6. Nie X, Zhang S, Chen G, Yu T (2020) Strengthening of RC beams with rectangular web openings using externally bonded FRP: numerical simulation. Compos Struct 248:112552
7. Neethu Urs, Tahera, Meghashree, M. (2024). Experimental Study on Effect of CFRP Sheathing on the Behaviour of RCC Beams with Opening. In: Sreekeshava KS, Kolathayar S, Vinod Chandra Menon N (eds) Recent advances in structural engineering. IACESD 2023. Lecture notes in civil engineering, vol 455. Springer, Singapore https://doi.org/10.1007/978-981-99-9502-8_48

Assessing the Performance of Chloride Ion Ingress in Concrete Infused with Copper Slag

K. Asha and M. L. Tejaswini

Abstract Significant issues arise when the chloride ions are on higher side in concrete mixture. These chloride ions break down the submissive layer due to their electrochemical nature. This protects the reinforcing bars without varying the pH. Hydrochloric acid is produced, which is a chemical process when these chloride ions come in contact with steel and the surrounding passive materials. This acid corrodes the steel reinforcement, leading to concrete cracking, spalling, and eventual failure. Therefore, chloride-induced corrosion is a major issue affecting the durability of reinforced concrete (RC) structures. This study investigates the ingress of chloride ions in copper slag concrete. Corrosion tests were conducted using thermo-mechanical treatment (TMT) steel. A study on the Rapid Chloride Ion Penetration Test, conducted according to ASTM C1202, found that the values for all grades of concrete, both with and without incorporated copper slag, fell into the "very low" category. An important observation was that the addition of slag significantly reduces the pores in the concrete, making it impermeable.

Keywords Copper slag · Chloride penetration · Durability · Concrete

1 Introduction

The deterioration of reinforced concrete (RC) structures is significantly influenced by the corrosion of rebar, representing a key factor in their degradation. Precise forecasts of corrosion-related damages serve as the foundation for assessing the safety of existing RC structures and improving the design of new constructions. Elevated levels of chloride ions in concrete pose a significant concern. These chloride ions break down the submissive layer due to their electrochemical nature. This protects

K. Asha (✉)
Department of Civil Engineering, B.M.S. College of Engineering, Bangalore, India
e-mail: asha.civ@bmsce.ac.in

M. L. Tejaswini
Department of Civil Engineering, Alliance University, Bangalore, India

Y. K. Guruprasad et al. (eds.), *Recent Advances in Infrastructure Development and Management—Volume 2*, Lecture Notes in Civil Engineering 566,
https://doi.org/10.1007/978-981-96-1558-2_20

the reinforcing bars without varying the pH. Hydrochloric acid is produced, which is a chemical process when these chloride ions come in contact with steel and the surrounding passive materials. This corrosive hydrochloric acid then erodes the steel reinforcement, ultimately causing concrete to develop cracks, undergo spalling, and eventually lead to structural failure. Thus, the salinity of the atmosphere increases Corrodent damage rates in a very distinctive way, and chlorides in atmosphere have a direct participation in electrochemical corrosion. Hence chloride-induced corrosion is indeed a significant issue affecting the resilience of the concrete with reinforcement. Sand is a fundamental component of traditional concrete, playing a crucial role in meeting the specified fineness modulus and particle graduation requirements. Historically, natural sand, primarily extracted from river beds has been the prevalent source for infrastructure construction. Nevertheless, escalating environmental apprehensions have prompted a shift in depletion of natural sand resources, prompting the concrete production industry to seek alternative substitutes for sand. Conversely in the process of copper smelting, Copper slag a variety of industrial byproduct is produced [1, 2]. The manufacturing process involved in producing one ton of copper results in the creation of approximately 2.2–3.0 T of copper slag, contributing to an annual global output of around 51.7 MT of residue from copper production [3, 4]. Also, the major material in concrete production is the aggregates that constitute 70% of its volume and globally consume 8–12 MT of natural aggregate annually [5]. Insufficient natural resources are unable to meet the growing demand for aggregates in the building industry. Hence the waste residue generated during the production of minerals serves as better alternative substances in construction sector. Studies of significance have shown that utilizing recycled copper slag (CS) as a fine aggregate for concrete production proves to be an efficient solution for addressing the pollution issues associated with copper slag [6, 7]. Based on previous studies [8, 9], it was noted that the presence of copper slag in fresh concrete leads to an elevation in bleeding water, primarily attributed to its exceptionally low absorption characteristics. The introduction of slag induces pozzolanic reactions, resulting in a reduction of calcium hydroxide, a decrease in the volume of capillary pores, and a subsequent improvement in compactness and durability, thereby enhancing the overall quality of the concrete. As a result, this fortifies the strength of the concrete and imparts an anti-corrosion effect [10]. Electrochemical measurements serve as a valuable tool for evaluating corrosion properties, given that the corrosion of reinforcement involves an electrochemical process through reactions with water and oxygen. Consequently, these measurements are commonly utilized as a non-destructive testing (NDT) method [11]. This study primarily focuses on investigating the penetration of eroding materials such as chloride ions and water containing dissolved oxygen, and their correlation with the corrosion behavior when copper slag (CS) is integrated with fine aggregates in the concrete. The research is conducted to generate precise experimental data pertaining to the corrosion and durability attributes of concrete incorporating copper slag in its fine aggregates.

Table 1 Physico-chemical parameters of CS

Physical parameters	Values
Sp. gravity	3.54
Water absorptions	0.32%
FM	5.55
Chemical properties	(%)
Al_2O_3	3.470
SiO_2	29.19
CaO	2.650
Cl	0.001
Cu	0.810
Fe_2O_3	60.070
Pb	0.080
Mg	1.370
K_2O	0.330

2 Investigation on the Experiments

2.1 Resources

Portland cement from Birla Company was utilized in the present work. This type of cement is extensively utilized in the construction sector in India. Locally sourced coarse aggregates with a size of 20 mm and fine aggregates classified as Zone II were obtained. Utilization of CS sourced from M/s Sterlite Industries India Limited (SIL) in Thoothukudi, Tamil Nadu, India was undertaken. Physico-Chemical Parameters of CS are tabulated in Table 1. The physical parameters of CS, such as specific gravity, particle size, and fineness modulus, were determined in accordance with the Bureau of Indian Standards (BIS) specifications outlined in BIS: 2386 Part III-1963 [12]. Figure 1 depicts the sieve analysis curve of CS. The graph reveals that CS comprises 18% of particles in the gravel size range, 78% in the sand size range, and 4% in the silt size range.

2.2 Mix Proportioning

Mix proportioning in this study adheres to the guidelines outlined in IS 10262: 2009 [13] and IS 456:2000. The study includes mix proportioning for various concrete grades of M20 to M40 in steps 5. To achieve the desired workability of 75–100 mm, various trials were conducted initially, fixing the superplasticizer dosage at 0.75%. The proportion of water and cement for the considered grades of concrete was determined through number of trials and found exploring ratios of 0.4, 0.42, and

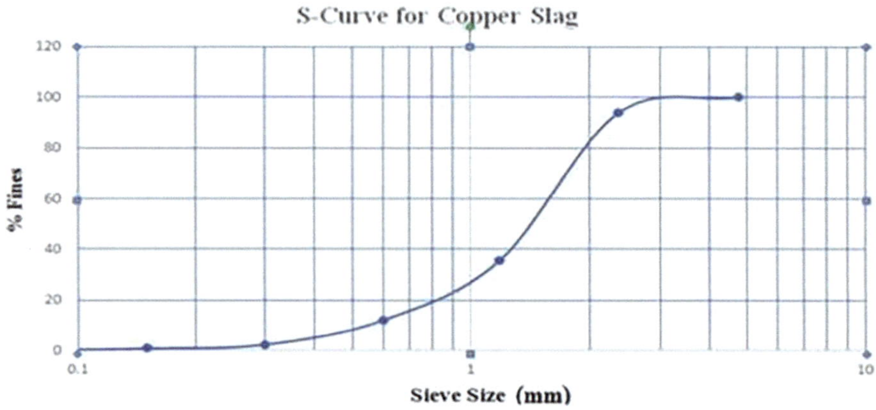

Fig. 1 Sieve analysis curve of CS

Table 2 Concrete-mix proportion

Mix	Cement	Fine aggregates	Copper slag (CS)	Coarse aggregates	W/c ratio
M20	1.0	0.93	0.79	3.02	0.44
M25	1.0	0.93	0.69	3.02	0.42
M30	1.0	0.93	0.86	3.44	0.42
M35	1.0	0.81	0.68	2.73	0.40
M40	1.0	0.96	0.81	3.25	0.40

0.44. Natural fine aggregates were partly substituted with 40% copper slag in all the concrete mixes (Table 2). The test on Coarse and Fine aggregates are done using IS 383:2016 [14].

2.3 Stray Electric Current Corrosion

Stray electric current corrosion within concrete pertains to the corrosion of reinforcing steel within a concrete structure caused by unintended electrical currents. This occurrence commonly happens when stray electric currents, frequently, from external sources like direct current (DC) systems or electrical faults, flow through the concrete and come into contact with the reinforcing steel. The presence of these stray currents can lead to accelerated erosion of the reinforced steel, compromising the integrity of structure of the concrete. In present work, bars of steel with dimensions of 150 mm length and 8 mm diameter were centrally inserted into concrete cubes of dimensions 150 × 150 × 150 mm, as illustrated in Fig. 2. The cubes were then surrounded by pouring concrete. Initial weight Measurements are documented for each cube. The cubes are submerged alternately in a 3.5% sodium chloride solution

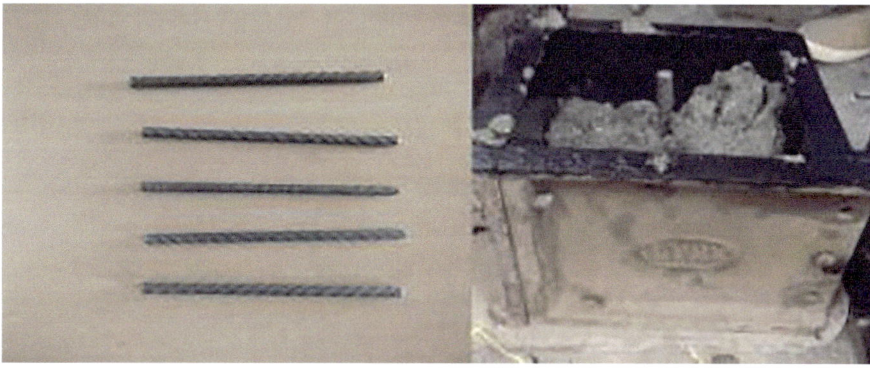

Fig. 2 Steel rods inserted in concrete cubes

and consistently subjected to a continuous DC electric flow of 12 V for a duration of 15 days as depicted in Fig. 3. The voltmeter's positive and negative terminals were connected to the soldered wires and the copper plate. Subsequently, the cubes were taken out from the tank, and the connections were disconnected. The steel rod, initially positioned at the midpoint of the concrete, is extracted, and its mass was determined. Following the duration of corrosion, the rod was retrieved and weighed to calculate the loss of materials. The rate of corrosion was then calculated from the weight loss values using the relationship specified in ASTM G-1. Continuous DC electric flow of 12 V for a duration of 15 days as depicted in Fig. 3. The voltmeter's positive and negative terminals were connected to the soldered wires and the copper plate. Subsequently, the cubes were taken out from the tank, and the connections were disconnected. The steel rod, initially positioned at the midpoint of the concrete, is extracted, and its mass was determined. Following the duration of corrosion, the rod was retrieved and weighed to calculate the loss of materials. The rate of corrosion was then calculated using the relationship specified in ASTM G-1.

$$\text{Corrosion Rate}\left(\frac{\text{mm}}{\text{yr}}\right) = \frac{KW}{ATD}$$

2.4 *Rapid Chloride Permeability Test (RCPT)*

The Rapid Chloride Permeability Test (RCPT) assesses the resistance of concrete blocks to the penetration of chloride ions. Failure to address this can lead to concrete deterioration. RCPT serves as an electrical indicator which showcases the ability of concrete to counterattack the ingress of chloride ions that helps to predict the service life of concrete structures.

Fig. 3 Concrete cubes containing a reinforcing bar (rebar) exposed to a DC power supply

In this study, the Rapid Chloride Permeability Test (RCPT) was conducted in accordance with the ASTM C 1202 [15] standard, as depicted in Fig. 4. This method was chosen for its simplicity and speed in assessing the concrete's resistance to chloride ion ingress. The test apparatus consists of two reservoirs—one containing a 3.0% NaCl solution and the other a 0.3M NaOH solution. The concrete test specimen, measuring 50 mm in thickness and 100 mm in diameter, is subjected to a DC voltage of 60 V applied across the reservoirs for 6 h. The current passing through the concrete is measured at various time intervals, as shown in the experimental setup in Fig. 4. These measurements are recorded over a 6-h duration. The circuit is deactivated and the experiment is repeated for each sample. The current readings are then tabulated, and the total charge passed is computed using the appropriate formula.

$$I = \frac{Q}{t}$$

where Q signifies the charge in coulombs, and t denotes the time period in seconds.

Fig. 4 RCPT-set ups. **a** Mounting concrete specimen in apparatus **b** sealing the gap between the bends. **c** Poring of solutions in the mould **d** connecting the apparactus to DC supply **e** final set up

3 Results and Discussions

3.1 Stray Current Corrosion Analysis

Figure 5 visually represents the variations in Stray Current corrosion analysis for 0 and 40% substitution of copper slag in different concrete grades. This graphical representation provides a clear understanding of the behavioral pattern regarding the reduction. The corrosion rate is slightly elevated compared to that observed in normal concrete specimens without the addition of copper slag. The corrosion rate, as outlined in Sects. 2.3 and 2.4, was calculated to vary between 0.17 and 0.5 mm/

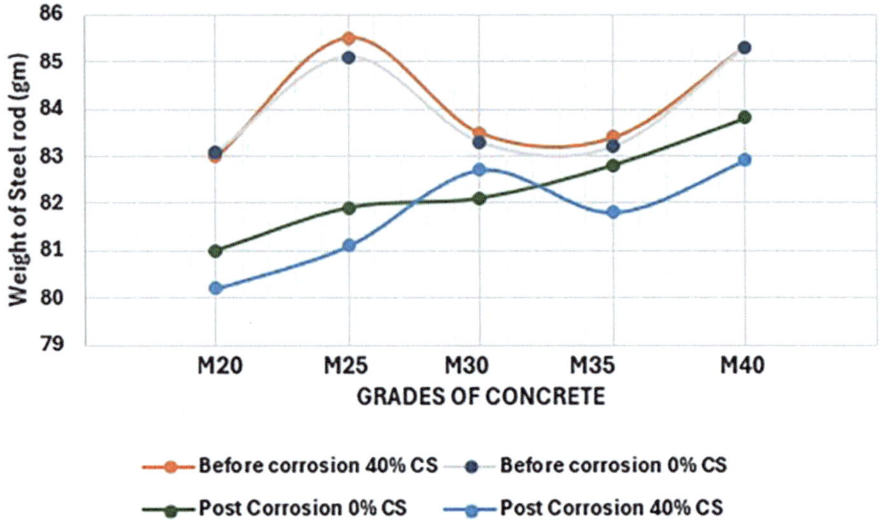

Fig. 5 Graph illustrating stray current corrosion analysis for 0 and 40% substitution of copper slag in different concrete grades

yr. The percentage reduction in the weight of the rod varies from 2.04 to 5.75% in the 15-day immersion test. This indicates that the addition of 40% copper slag does not significantly increase the corrosion rate. However, in view of the durability of concrete, applying a protective coating around the rods is advisable, ensuring the long-term integrity of reinforced concrete.

3.2 Results of the Rapid Chloride Permeability Test (RCPT)

The results of three trials were conducted to determine the average current passed in amperes and, subsequently, the total charge in coulombs over a 6-h duration. This investigation focused on M40 grade concrete, examining scenarios with 0 and 40% replacement of CS. Additionally, the Rapid Chloride Permeability Test (RCPT) was performed for all the previously mentioned concrete grades.

Table 3, provides a comprehensive overview of the Rapid Chloride Permeability Test (RCPT) results at both 0-h and 6-h time durations for diverse concrete grades, encompassing 0 and 40% replacement of CS. The corresponding graphical representation in Fig. 6 unmistakably indicates that minimal charges are transmitted through the concrete under different conditions, aligning with the criteria outlined in ASTM C1202, thus affirming excellent electrical resistance [16, 17].

Table 3 Results of RCPT conducted at both 0 and 6-h time durations for various concrete grades, considering 0 and 40% replacement of CS

Time (*h*)	M20 grade				M25 grade				M30 grade				M35 grade				M40 grade			
	With 0%		With 40%		With 0%		With 40%		With 0%		With 40%		With 0%		With 40%		With 0%		With 40%	
	CS		CS		CS		CS		CS		CS		CS		CS		CS		CS	
	Avg	Avg	Avg	Avg	Avg	Avg	Avg	Avg	Avg	Avg	Avg	Avg	Avg	Avg	Avg	Avg	Avg	Avg	Avg	Avg
	I	Q	I	Q	I	Q	I	Q	I	Q	I	Q	I	Q	I	Q	I	Q	I	Q
0	0.01	0	0	0	0.007	0	0.003	0	0.01	0	0.01	0	0.1	0	0.1	0	0.02	0	0.014	0
6	0.03	684	0.04	936	0.033	702	0.038	829.8	0.03	720	0.03	723.6	0.08	608	0.03	724	0.02	496.8	0.029	626.4

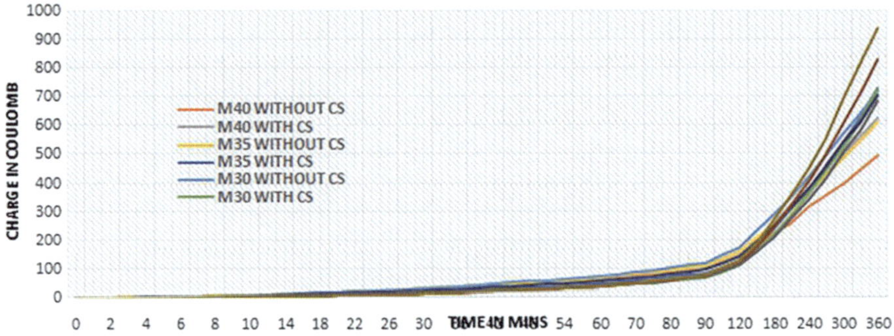

Fig. 6 Graph illustrates the variation in charge passed, measured in coulombs, over a 6-h period for different concrete grades including CS

4 Conclusion

The integration of CS as a partial substitute for fine aggregates consistently enhances the corrosion resistance of steel reinforcements in diverse-grade concrete specimens subjected to stray current tests and RCPT. The current investigation undeniably shows that conducting stray current tests on various concrete grades with a 40% replacement of CS results in an exceedingly low corrosion rate, highlighting its prospective appropriateness for utilization in reinforced concrete structures [18–21]. Furthermore, RCPT tests conducted on different concrete grades with a 40% replacement of CS suggest that, in accordance with ASTM-C-1202, values resulted for CS admixed concrete fall under the very low category, signifying significantly reduced permeability of concrete with slag admixture. An essential examination shows that the inclusion of slag effectively diminishes pores of the concrete, rendering it highly impermeable.

References

1. Dash MK, Patro SK, Rath AK (2016) Sustainable use of industrial-waste as partial replacement of fine aggregate for preparation of concrete—a review. Int J Sustain Built Environ 5:484–516
2. Filipovi S, Doki O, Radevi A, Zaki D (2021) Copper slag of pyroxene composition as a partial replacement of natural aggregate for concrete production. Minerals 11:439. https://doi.org/10.3390/min11050439
3. Tam C, Taylor M, Gielen D, Twigg C, Klee H, Rocha P, Meer R (2009) Cement technology roadmap 2009—carbon emissions reductions up to 2050. World Business Council for Sustainable Development, Geneva, Switzerland
4. Sankh AC, Biradar PM, Naghathan SJ, Ishwargol MB (2002) Recent trends in replacement of natural sand with different alternatives akshay. IOSRJ Mech Civ Eng (IOSR-JMCE) 12:180–183
5. Aguilar RA, Diaz OB, Garcia JIE (2010) Lightweight concretes of activated metakaolin–fly ash binders, with blast furnace slag aggregates. Constr Build Mater 24(7):1166–1175

6. Hong Y, Mohan KK, Froass PM et al (2010) Expandable microspheres and methods of—making and using the same. US Patent,US20100051220A1
7. Kuntikana G, Singh DN (2017) Contemporary Issues Related to Utilization of Industrial-byproducts. Adv Civ Eng Mater 20160050
8. Muttashar HL, Mohd Ariffin MA, Hussin MW, Ishaq SB (2018) Realisation of enhanced self-compacting geopolymer concrete using spent garnet as sand replacement. Mag Concr Res 70(11):558–569
9. Maharishi A, Singh SP, Gupta LK, Shehnazdeep (2020) Strength and durability studies on slag cement concrete made with copper slag as fine aggregates. Mater Today Proc
10. Prasad PS, Ramana GV (2016) Feasibility study of copper slag as a structural fill in reinforced soil structures. Geotext Geomembr 44:623–640
11. Chithra S, Senthil Kumar SRR, Chinnaraju K (2016) The effect of colloidal nano-silica on workability, mechanical and durability properties of high performance concrete with copper slag as partial fine aggregate. Constr Build Mater 113:794–804
12. IS2386:1963. Methods of test for aggregates for concrete
13. IS-10262:2019. Mix design of concrete
14. IS383 : 2016. Specification for coarse and fine aggregates
15. ASTMC 1202. Standard test method for Rapid Chloride Permeability Test
16. Sandra N, Kawaai K, Ujike I (2020) Influence of copper slag on corrosion behavior of horizontal steel bars in reinforced concrete column specimen due to chloride-induced corrosion. Constr Build Mater 255:119265
17. Geetha S, Madhavan S (2017) High performance concrete with copper slag for marine environment. Mater Today Proc 4(2):3525–3533
18. Edwin RS, Schepper MD, Gruyaert E, Belie ND (2016) Effect of secondary copper slag as cementitious material in ultra-high-performance mortar. Constr Build Mater 119:31–44
19. Al-Jabri KS, Hisada M, Al-Saidy AH, Al-Oraimi SK (2009) Performance of high strength concrete made with copper slag as a fine aggregate. Constr Build Mater 23:2132–2140
20. Chithra S, Senthil Kumar SRR, Chinnaraju K, Alfin Ashmita F (2016) A comparative study on the compressive strength prediction models for high performance concrete containing nano silica and copper slag using regression analysis and artificial neural networks. Constr Build Mater 114:528–535
21. Raju S, Dharmar B (2016) Mechanical properties of concrete with copper slag and fly ash by DT and NDT. Periodica Polytech Civ Eng 60(3):313–322. https://doi.org/10.3311/PPci.7904CreativeCommonsAttribution
22. Lee SC, Cunnington GR (2000) Conduction and radiation heat transfer in high porosity fiber thermal insulation. J Thermophys Heat Transfer 14(2):121–136
23. IS13311(Part1):1992. Non-destructive testing of concrete-methods of test

Hemp Fiber-Reinforced Concrete with FGPS: Mechanical Properties Under High-Temperature and Microstructure Analysis

Arjun Kasi and K. Asha

Abstract This experiment explores incorporation of hemp fiber and FlyAsh Geopolymer Sand (FGPS) as composite materials in reinforced concrete, aiming to enhance its mechanical properties and high-temperature resistance. The research focuses on the synergistic effects of these natural and mineral additives on the structural performance, sustainability, and thermal stability of concrete. Hemp fibers, recognized for their tensile strength and flexibility, are incorporated to enhance the ductility and fracture resistance of the concrete matrix. Simultaneously, FGPS is utilized to replace a portion of traditional Fine aggregates, capitalizing on its lightweight properties to improve the concrete's resistance to high temperatures. The study provides insights into how these innovative materials can contribute to the development of more durable and sustainable concrete solutions. Comprehensive experimental analysis, including compressive strength, flexural strength, and high-temperature tests, was conducted to assess the performance of the modified concrete. The results demonstrate significant improvements in both mechanical properties and high-temperature performance in comparison to conventional concrete. The incorporation of hemp fiber and FGPS not only enhances the sustainability of the concrete but also provides a viable solution for construction applications requiring enhanced mechanical robustness and thermal resilience. This research contributes in development of greener and more resilient construction material, paving the way for innovations in sustainable infrastructure.

Keywords Hemp fiber · Flyash geopolymer sand (FGPS) · Mechanical characteristics

---

A. Kasi (✉) · K. Asha
B.M.S College of Engineering, Bengaluru, India
e-mail: arjunk.civ@bmsce.ac.in

Y. K. Guruprasad et al. (eds.), *Recent Advances in Infrastructure Development and Management—Volume 2*, Lecture Notes in Civil Engineering 566,
https://doi.org/10.1007/978-981-96-1558-2_21

1 Introduction

The construction sector is continuously seeking innovative materials that increase the performance and sustainability in concrete, one of the most extensively utilized construction materials globally. In recent years, there has been an increasing focus on incorporating natural fibers and supplementary cementitious materials to enhance the mechanical properties of concrete while minimizing its environmental impact. Among these materials, hemp fiber and Flyash Geopolymer Sand (FGPS) have emerged as promising additives due to their unique properties.

Hemp fiber is derived from the stems of the hemp plant (Cannabis sativa) and is known for its high tensile strength, flexibility, and biodegradability. These characteristics make hemp fiber an attractive reinforcement material for concrete, as it can potentially improve the ductility and crack resistance in the concrete matrix [1]. Furthermore, use of hemp fiber aligns with the principles of sustainable development, as it is a renewable resource with a low environmental impact [2].

Flyash Geopolymer Sand (FGPS), Fly ash, is a derivative of coal combustion from power plants, which has been widely studied like a precursor for geopolymer synthesis because of its abundant reactive silica and alumina. The use of fly ash not only addresses the environmental concerns related to its disposal but also reduces the carbon footprint of construction materials. When activated with alkaline solutions, fly ash can form a geopolymer matrix that exhibits excellent mechanical properties, chemical resistance, and thermal stability [3–5]. Geopolymer sand, produced from fly ash, presents a novel approach to creating sustainable construction aggregates. The process involves the alkaline activation of fly ash to form a geopolymeric binder, which is then processed into sand-sized particles. This FGPS can serve as a substitute for natural sand in concrete and various other construction applications [6, 7]. The use of FGPS not only addresses the environmental challenges associated with sand mining but also enhances the performance characteristics of the resulting composites [8].

The integration of hemp fiber and FGPS in concrete aims to combine the mechanical benefits of natural fibers with the thermal advantages of lightweight aggregates. Previous studies have indicated that natural fibers can significantly improve the tensile and flexural strength of concrete, while FGPS can enhance its thermal performance [9, 10]. However, there is limited research on the combined effects of hemp fiber and FGPS on the overall performance of concrete, particularly in terms of high-temperature resistance.

This paper seeks to address this gap by investigating the mechanical characteristics and high-temperature resistance of hemp fiber-reinforced concrete incorporating FGPS. The basic goals of this research are: evaluate compressive and flexural strength of concrete reinforced with varying proportions of hemp fiber and FGPS. Assess the thermal conductivity and high-temperature resistance of the modified concrete. Analyze the synergistic effects of hemp fiber and FGPS on the overall performance of concrete. Through comprehensive experimental analysis, this study aims to offer valuable insights into the potential of hemp fiber and FGPS as sustainable and effective additives in concrete. The findings are expected to aid in the creation of eco-friendly construction materials that provide superior mechanical strength and thermal durability, leading to more sustainable infrastructure solutions.

2 Materials and Methods

2.1 Materials

Cement

Type: Ordinary Portland Cement (OPC) conforming to ASTM C150.
Purpose: Used as the primary binder in the concrete mix.

Flyash Geopolymer Sand (FGPS)

Source: Obtained from coal combustion by-products processed into sand-sized particles.
Purpose: Partial replacement of natural sand to enhance thermal resistance and sustainability.

Hemp Fiber

Type: Processed hemp fibers with a length of 10–20 mm and diameter of approximately 0.1–0.2 mm.
Purpose: Reinforcement to improve tensile strength, ductility, and fracture resistance of concrete.

Aggregates

Coarse Aggregate: Crushed stone with a particle size of 20 mm was used. Fine aggregate consisted of natural river sand, partially replaced with FGPS.
Purpose: Provide bulk and strength to the concrete mix.

Alkaline Activator

Components: A solution of sodium hydroxide (NaOH) and a solution of sodium silicate (Na_2SiO_3).
Objective: Trigger the activation on fly ash in FGPS in the formation of geopolymer matrix.

Water

Type: Potable water.
Purpose: Hydration of cement and activation of the alkaline activator.

Fig. 1 Materials used in mix design (as shown in Fig. 1)

2.2 *Methodology*

Mix Design

Objective: Develop a mix design that incorporates varying proportions of hemp fiber and FGPS while maintaining workability and desired strength.

Procedure:

- Design multiple concrete mixes with different ratios of FGPS (0, 10, 20, 30%) as a replacement for natural sand.
- Incorporate hemp fibers at varying percentages (0.5, 1, 1.5, 2%) by volume of the concrete.
- Maintain a constant water-to-cement ratio for all mixes.
- Use the alkaline activator solution in appropriate proportions to ensure effective geopolymerization of FGPS.

M30 Grade

Cement used = opc 53 grade
Nominal size of aggregates = 20 mm
Exposure condition = severe
Workability = 120 mm (slump)
Specific Gravity of the obtained FGPS = 2.65
Specific Gravity of crushed Coarse Aggregate = 1.5
Specific Gravity of the Cement used = 3.15.

Mix Proportion:

Cement = 360 kg/m^3
Flyash Geopolymer Sand (FGPS) = 661 kg/m^3
Coarse aggregate = 702.74 kg/m^3
Hemp = 5.385 × 10^−3 kg/m^3
Water = 162 kg/m^3

Mixing Process

Steps:

Dry mix cement and FGPS in a mechanical mixer for 2 min.
Gradually add coarse aggregates and mix for another 2 min.
Introduce hemp fibers slowly to ensure even distribution.
Add the alkaline activator solution to the mix, followed by water.
Mix thoroughly for 3–5 min until a uniform consistency is achieved.

Casting and Curing

Molds: Utilize standard steel molds for the fabrication of specimens (cubes for compressive strength, beams for flexural strength).

Casting: Fill the molds in 3 layers, with every layer compacted using a vibrating table.

Curing:

- Initial curing in molds about 24 h at ambient temperature.
- After demolding, cure the specimens in a water tank maintained at 20 °C for duration of 28 days.
- For high-temperature tests, additional curing at elevated temperatures (e.g., 200, 400, 600 °C) for 24 h.

Testing Procedures

Compressive Strength Test:

- Standard: ASTM C39.
- Specimens: 150 mm cubic samples.
- Procedure: Test at intervals of 7, 28, and 56 days, and conduct compression tests with the help of compression testing machine.

Flexural Strength Test:

- Standard: ASTM C78.
- Specimens: Beam dimensions of 100 × 100 × 500 mm.
- Procedure: Test at 7, 28, and 56 days using a three-point loading setup.

High-Temperature Resistance Test:

- Specimens: 150 mm cubic samples.
- Procedure: Expose specimens to Temperatures of 200, 400, 600, 750, and 900 °C for a duration of 2 h, followed by cooling to room temperature.
- Assess residual compressive and flexural strengths.

3 Results Analysis

Microstructural Analysis

X-ray diffraction (XRD) analysis is a robust method employed for investigating the crystalline structure of materials. For hemp fibers, XRD can provide detailed information about the crystallinity, crystal structure, and orientation of cellulose, which is the primary component of the fibers (Figs. 2 and 3).

Figure 3 displaying an Energy Dispersive Spectroscopy (EDS) spectrum for Area 463, Spot 1. The x-axis represents energy levels in keV, ranging from 0 to 6.03, while the y-axis shows counts up to 25,000. Prominent peaks are labeled with elements: Oxygen (O) at 0.67 keV, Silicon (Si) at 1.34 keV, Aluminum (Al) at 1.34 keV, Carbon (C), Sodium (Na), Potassium (K), and Calcium (Ca). The Oxygen peak is the highest, indicating its abundance. Additional details include a takeoff angle of 35.2, live time of 50, and resolution of 128 eV. A greater crystallinity Index suggests a more extensive presence of crystalline cellulose, which correlates with better mechanical properties and thermal stability of the fibers. Larger crystallite sizes indicate more ordered structures, while smaller sizes suggest more amorphous regions. Understanding the phase composition helps in assessing the quality and potential applications of the hemp fibers.

Results on Fresh Properties of Concrete

Slump Cone Test

The concrete's workability, crucial for ensuring its quality in both fresh and hardened conditions [11], is determined through handling, placing, and compacting processes during the period between mixing and setting. The slump test serves as a crucial assessment tool for evaluating concrete's fresh-state properties (Fig. 4).

Results of Compression Strength of Specimen

To examine the hardened property, the compression strength for specimens is prepared utilizing fly ash geopolymer sand are tested according to IS-456. Compression strength for specimen with complete sand replacement (Fig. 5; Table 1).

Flexural Test

Flexural strength of concrete was done according to IS: 516 (clause 8: Flexural strength of concrete) (Table 2).

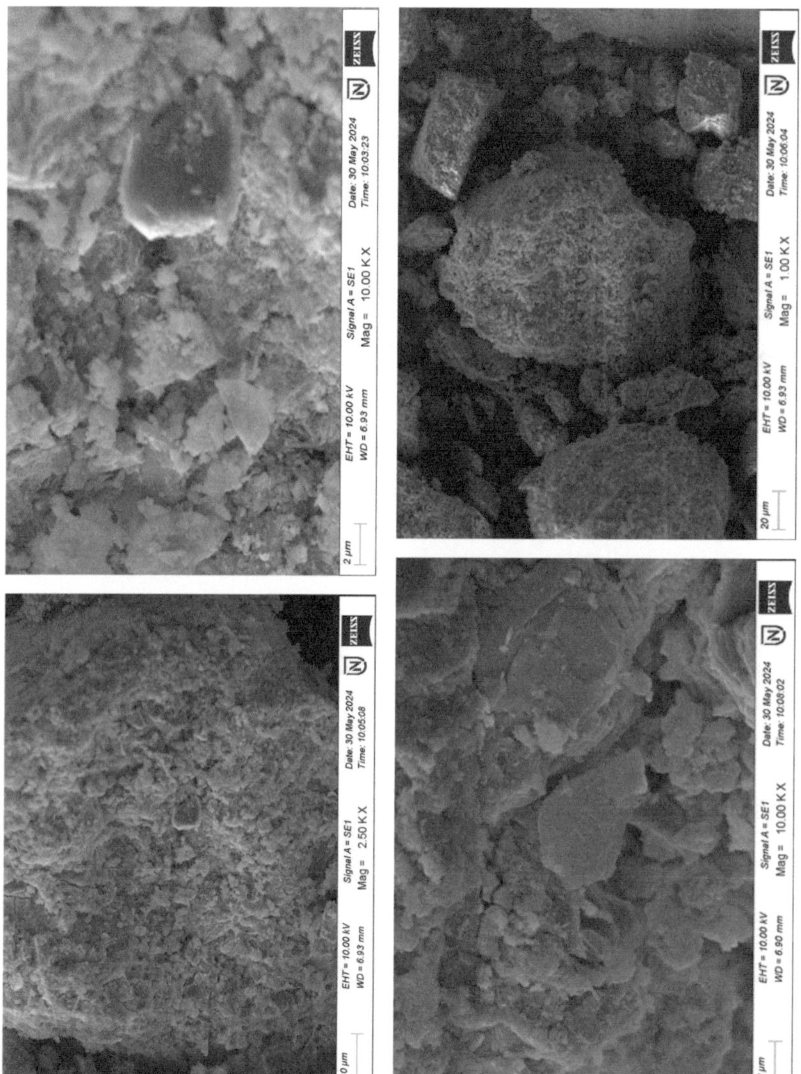

Fig. 2 M30—scanning electron microscopy images

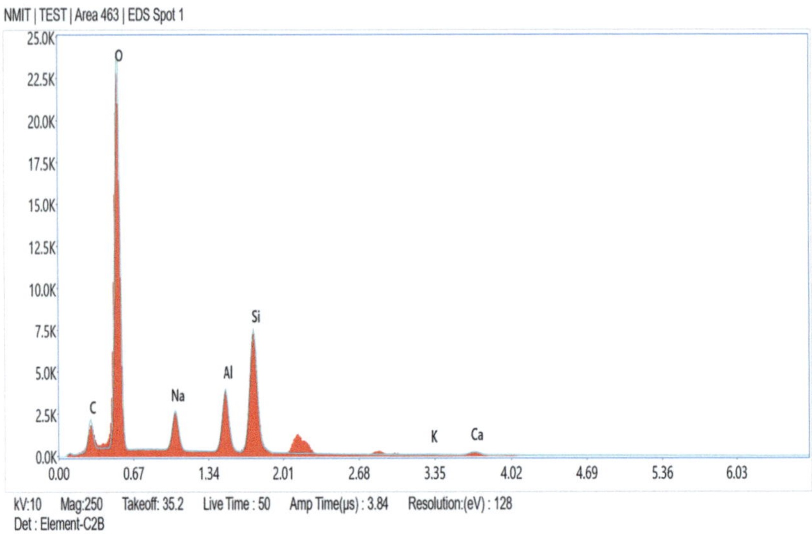

Fig. 3 XRD of FGPS

Fig. 4 Slump cone test

$$\text{Flexural strength} = \frac{PL}{bd^2}$$

High-Temperature Resistance

The Compressive strength of concrete remains constant till 300 °C. The residual strength reduces from 93.59 to 69.23% when the temperature rises from 300 to 450 °C. Further, the strength reduces to 46.15% when the temperature rises from

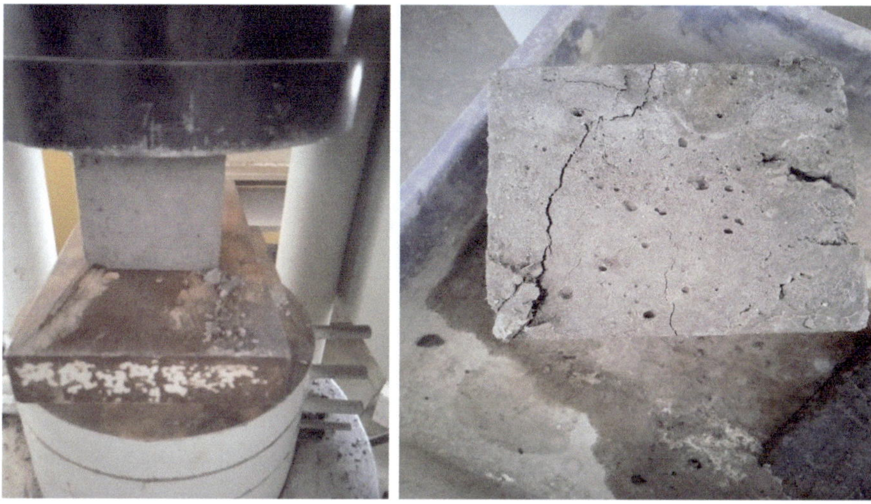

Fig. 5 Testing for compression strength of concrete

Table 1 Compression strength of concrete casted using manufactured fine aggregates

Grade	Mean compression strength		
	7 days (Mpa)	14 days (Mpa)	28 days (Mpa)
M30	12.30	13.93	17.08
M40	12.71	14.12	18.52
M50	13.20	15.64	23.61

Table 2 Flexural strength of concrete

Sl. No.	Days	Load applied (KN)	Flexural strength (N/mm^2)
1	3	4.50	2.25
2	7	5.60	2.8
3	14	6.30	3.15
4	21	6.50	3.25
5	28	6.65	3.325

450 to 600 °C. When the temperature rises to 750 and 900 °C, a significant decline in compressive strength was observed, amounting to a reduction of 11.54% (Fig. 6).

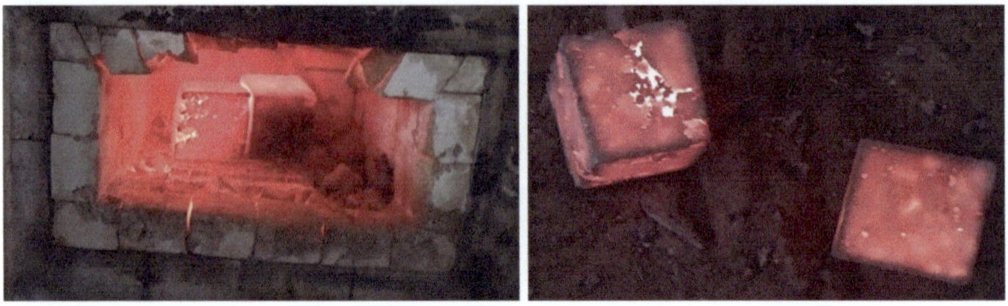

Fig. 6 Concrete specimen kept in furnace

4 Conclusion

In this investigation, the influence of incorporating industrial hemp fibers and Flyash Geopolymer Sand (FGPS) into plain concrete, building upon prior research efforts. The objective was to develop a sustainable and environmentally friendly concrete mixture by integrating natural fibers to compensate for reduced coarse aggregate quantities. Various volumetric percentages of hemp fibers (0.5, 0.75, and 1%) were introduced into the concrete mix, accompanied by corresponding decreases in coarse aggregate content (10, 20, and 30%). A comprehensive study of tests, including evaluations of compressive strength, tensile strength, and thermal conductivity, was conducted on all formulations.

Findings from standard cube tests revealed that incorporation of hemp fiber typically led to a decrease in compression strength compared to conventional concrete. Nonetheless, flexural tests illustrated that the incorporation of hemp fibers, alongside a reduction in coarse aggregates, conferred a ductile nature to the concrete. Furthermore, thermal conductivity assessments indicated that hemp fibers effectively mitigated heat transfer.

The study concluded that the incorporation of 0.75–1.0% fibers, along with a decrease in coarse aggregate by 20–30%, facilitates the development of a novel concrete mixture. While this new formulation demonstrates a reduction in compressive strength by approximately 25%, it demonstrates enhanced ductility, a decrease in the modulus of elasticity by 20–30%, unchanged splitting tensile strength, and a decrease in compressive strength by 25–35% under high temperatures attributed to the lightweight aggregate.

In terms of workability, the slump measurements for mixes containing 0.75% hemp fibers were found to be satisfactory, ranging from 100 to 150 mm. However, mixes with 1.0% hemp fibers showed a minimum slump of around 70 mm, indicating that higher volumetric ratios may not be advisable. The findings related to compression, tension, thermal resistance, and slump suggest that similar hemp fiber concrete mixes could be suitable for non-structural applications like masonry, tiles, and paving blocks, as well as for structural reinforced concrete elements. This study lays the groundwork for future experimental and analytical investigations to further explore and understand the behavior of this innovative material.

References

1. Li Z, Wang L, Wang X (2019) Mechanical properties and microstructure of hemp fiber reinforced concrete. Constr Build Mater 203:65–73
2. Shahzad A (2012) Hemp fiber and its composites—a review. J Compos Mater 46(8):973–986
3. Xu H, Van Deventer JSJ (2000) The geopolymerisation of alumino-silicate minerals. Int J Miner Process 59(3):247–266
4. Carabba L, Santandrea M, Carloni C, Manzi S, Bignozzi MC (2017) Steel fiber reinforced geopolymer matrix (S-FRGM) composites applied to reinforced concrete structures for strengthening applications: a preliminary study. Compos B Eng 128:83–90
5. Nithambigai G, Rameshwaran PM (2021) Effect of binders and hemp fibre in concrete. Mater Today Proc 46:3791–3794
6. Awwad E, Mabsout M, Hamad B, Farran MT, Khatib H (2012) Studies on fiber-reinforced concrete using industrial hemp fibers. Constr Build Mater 35:710–717
7. da Costa Gonçalves LF, Balestra CET, Gil MAR (2023) Evaluation of mechanical, physical and chemical properties of ecological modular soil-alkali activated bricks without Portland cement. Environ Dev 48:100932
8. Provis JL, van Deventer JSJ (2009) Geopolymers: structures, processing, properties and industrial applications. Woodhead Publishing
9. Coutts RSP (2005) A review of Australian research into natural fibre cement composites. Cement Concr Compos 27(5):518–526
10. Rashad AM (2018) A comprehensive overview about the influence of different additives on the properties of alkali-activated slag—a guide for civil engineer. Constr Build Mater 187:897–924
11. Rangan BV (2008) Fly ash-based geopolymer concrete. In: Concrete construction engineering handbook, pp 26-1–26-20). CRC Press
12. Nadi M, Allaoui D, Majdoubi H, Hamdane H, Haddaji Y, Mansouri S, Tamraoui Y, Manoun B, Hannache H, Oumam M (2023) Effect of natural graphite additions on the microstructural and mechanical behavior of metakaolin based geopolymer materials. J Build Eng 78:107562
13. Adesina A (2021) Performance and sustainability overview of sodium carbonate activated slag materials cured at ambient temperature. Resour Environ Sustain 3:100016
14. Das SK, Mustakim SM, Adesina A, Mishra J, Alomayri TS, Assaedi HS, Kaze CR (2020) Fresh, strength and microstructure properties of geopolymer concrete incorporating lime and silica fume as replacement of fly ash. J Build Eng 32:101780
15. Okoye FN, Prakash S, Singh NB (2017) Durability of fly ash based geopolymer concrete in the presence of silica fume. J Clean Prod 149:1062–1067
16. Okoye FN, Durgaprasad J, Singh NB (2016) Effect of silica fume on the mechanical properties of fly ash based-geopolymer concrete. Ceram Int 42(2):3000–3006
17. Gu YM, Fang YH, You D, Gong YF, Zhu CH (2015) Properties and microstructure of alkali-activated slag cement cured at below-and about-normal temperature. Constr Build Mater 79:1–8
18. Kaplan G, Bayraktar OY (2021) The effect of hemp fiber usage on the mechanical and physical properties of cement based mortars. Res Eng Struct Mater 7:245–258
19. Ghafoori N, Diawara H (2010) Use of pumice and scoria aggregates in concrete. J Mater Civ Eng 22(1):52–60
20. Mendes A, Sanjayan J, Collins F (2019) Enhancing high-temperature performance of lightweight concretes with fine ground pumice stone. Constr Build Mater 227:117054
21. Hardjito D, Wallah SE, Sumajouw DMJ, Rangan BV (2005) On the development of fly ash-based geopolymer concrete. ACI Mater J 101(6):467–472

Hazard Mapping Due to Dam Overtopping Failure

N. Soham Gowda, A. S. Suhas, Sudeep M. Haveri, T. V. Reshmidevi, and Suresh Ramaswwamyreddy

Abstract The study of dam breach analysis helps predict the flow in case of dam breaches and helps in disaster preparedness. By using the Hydrologic Engineering Center's River Analysis System (HEC-RAS), an overtopping failure scenario of the Krishnarajasagara dam in the Cauvery River in Karnataka has been simulated. Flood wave propagation is simulated till the Mettur Dam, located about 212 km downstream, in the Salem District in Tamil Nadu. Two-dimensional flow simulation using HEC-RAS is used to estimate the plausible inundation extent and flood wave arrival time at various points downstream of the dam. Probable maximum precipitation for the area is used to compute the flood inflow to the dam at the time of breach. The peak flow is estimated at 25,000 m^3/s. The simulation results show a time interval of 49 h between the breach of the Krishnarajasagara Dam and the flood wave arrival at the Mettur Dam reservoir.

Keywords Dam breach · Overtopping failure · Flood wave · HEC-RAS · Krishnarajasagara dam

1 Introduction

Dams are the mega structures catering to multiple purposes such as flood control, electricity production, irrigation, and drinking water supply. Dams contain huge potential energy which is used for beneficial purposes like hydro-power generation, nevertheless adds to the risk of sudden breach leading to heavy damages on the downstream side. Changes in land use and interrupting activities of human beings in the upstream catchments have increased a lot day by day. These activities cause a larger volume of soil erosion and deposits in reservoirs resulting in a reduction of

N. S. Gowda · A. S. Suhas (✉) · S. M. Haveri
Department of Civil Engineering, B.M.S. College of Engineering, Bengaluru, India
e-mail: suhasa.cv20@bmsce.ac.in

T. V. Reshmidevi · S. Ramaswwamyreddy
Faculty of Civil Engineering, B.M.S. College of Engineering, Bengaluru, India

Y. K. Guruprasad et al. (eds.), *Recent Advances in Infrastructure Development and Management—Volume 2*, Lecture Notes in Civil Engineering 566,
https://doi.org/10.1007/978-981-96-1558-2_22

the useful life of the reservoir and a decrease in storage capacity. When a high flood occurs, there is not enough space to attenuate the approaching flood which causes excess flood overflow over the dam section. This kind of overtopping failure of dams forms a massive flood wave, inundating vast areas on the downstream reaches of the dam.

In the past 50 years, more than 200 major incidents of dam breach have been reported worldwide. Failure of Banqiao Dam in China in the year 1975 [1], Brumadinho Tailings Dam in Brazil in 2019 [2], and Teesta Dam in India in 2023 [3] are just a few examples. Major causes of dam failure include overtopping, piping or seepage, structural failure, and spillway gate malfunctioning, with overtopping failure [4] being the predominant reason. A recent report by the Central Water Commission shows that India has 6281 large dams with a gross storage capacity of 343.6 BCM [5]. Failure of these structures will lead to catastrophic disaster. A dam break may result in flood waves up to tens of meters deep, traveling along the mother valley at quite high speeds. The impact of such a wave on developed areas can be very devastating. Such destructive force can cause devastating damage, including loss of life, if advance warning and evacuation are not possible. Dam breach analysis and hazard maps are the essential requisites for an effective disaster management plan. The analysis can be done by manual calculations, through experimental methods (physical model construction), or by simulations. The most feasible method is to do it by simulation as it consumes less time and the result is comparable with the actual values. There are numerous software available to analyze the dam breach such as MIKE 11 [6], BOSS DAMBRK [7], and HEC-RAS [8]. Using conceptual models and mathematical equations representing various hydrologic and hydraulic processes, simulation provides a powerful tool to project the potential impact of any such hazardous dam breaches.

The present study focuses on the two-dimensional flood hazard analysis of the Krishnarajasagara Dam in Karnataka using HEC-RAS.

2 Study Area

Krishnarajasagara Dam (KRS Dam) is known as Kannambadi Katte, and it is a barrier constructed across the river Kaveri near Srirangapatna town in the Mandya District, Karnataka. The dam is located at Latitude 12° 25′ 30″ North and Longitude 76° 34′ 30″ East. It was commissioned in the year 1931 with a gross storage capacity of 1400 MCM (49.452 TMC) mainly designed to meet the hydro-electric power generation and to supply irrigation to about 505.85 km^2 command area [9]. The catchment area of the reservoir is 10880.63 km^2, mostly in the Western Ghat region, receiving heavy rainfall. The structure of the dam is made of random rubble masonry with surkhi mortar. KRS Dam has a length of 2621 m and has a maximum height of 44.66 m. KRS Dam was the first dam in India to install automated crest gates. A series of 173 sluices at different levels are provided on the dam for surplus discharge. The maximum discharge capacity of these spillways is 9910.8 m^3/s [9].

Fig.1 Location map of the study area from Krishnarajasagara dam to Mettur dam

Outflow from the dam through the River Cauvery reaches the Mettur Dam in the Salem District in Karnataka, after traversing about 212 km. The downstream area for which the analysis is done is from the KRS Dam to the Mettur Dam. Prominent landmarks on this route include Srirangapatna (an ancient township and tourist destination), Mahadevapura, T. Narasipura, Shivanasamudra (hydro-power station), Mekedattu (tourist destination) and Hogenakkal (tourist destination). The location map of the study area is shown in Fig. 1.

3 Input Data and the Methodology

The Hydrologic Engineering Center's River Analysis System (HEC-RAS), developed by the US Army Corps of Engineers, is used in this study to perform the analysis. The version used in this study is HEC-RAS 6.2. Two-dimensional analysis of the dam overtopping scenario is simulated using the software. Numerous input data required for the analysis are obtained from different reliable sources, which are listed below.

The Digital Elevation Model (DEM) of the study area is downloaded from ALOS JAXA. The 30 m resolution data and the projection files are imported into the HEC-RAS software. RAS Mapper is used to create the geometry of the dam, reservoir area, and the downstream area selected for the simulation of flood inundation.

Probable Maximum Precipitation (PMP) is the theoretical maximum precipitation for a given duration that is likely to happen over a design watershed. In this study, rainfall only is considered as the precipitation as other forms of precipitation are

Table 1 Breach parameters defined for the study

Parameter	Value
Final bottom elevation	715 m
Slope	1:1
Breach weir coefficient	1.44
Starting water surface	752.48 m
Failure mode	overtopping
Breach formulation time	1-h

very less and can be neglected. Though the PMP is subjected to change with long-term climatic variations, this progress is slow or their influence on PMP is small compared to other uncertainties. Climatic trends are therefore not considered in this study. The 1-day and 2-day PMP values obtained from the Central Water Commission (CWC), Govt. India, are 20.24 cm and 35.82 cm, respectively [10]. The PMP value is further used in the HEC-HMS tool to simulate the corresponding flood inflow to the reservoir. The flood inflow hydrograph thus estimated is given as the upstream boundary condition in HEC-RAS to simulate the reservoir. Downstream boundary condition is defined by the precipitation.

The elevation-capacity curve of the dam is adapted from the report of the Chief Engineer [9] and is given as input. Breach parameters of the dam are defined as given in Table 1.

HEC-RAS uses the Froelich's [11] equation to compute the breach hydrograph using the above breach parameters. The breach hydrograph is further routed through the channel downstream. HEC-RAS uses the Saint–Venant Equation to model the flood wave propagation under the unsteady flow scenario created due to the breach. Numerical simulation is adopted by discretizing the area into finer meshes. 2D mesh is generated for the entire study area for numerical solution. The cell size of the mesh is 100 m $\times$ 100 m and the total number of cells created is 335562.

During the flood wave propagation through the channel, as the depth of flow increases above the channel depth, bank overflow is detected. The maximum depth of flow observed at any location is compared with the topographic elevation from DEM to identify inundation depth at various points. Also, the velocity profile is generated during the simulation to estimate the flow velocity at the inundated areas. The depth of flow and the flow velocity are further used to calculate the hazard index of the inundated areas.

4 Results and Discussion

With the 1-day PMP of 20.24 cm for the entire catchment area, the corresponding flood hydrograph is generated using the hydrological modeling tool HEC-HMS. The flood hydrograph obtained is shown in Fig. 2. The peak inflow value obtained is 17519.2 m^3/s. Whereas, the maximum spillway capacity is limited to 9910.8 m^3/s.

Using the inflow hydrograph and the breach parameters, the breach scenario is developed and the corresponding breach hydrograph is shown in Fig. 3. Breach formation time is set to 1 h, and the corresponding peak flow in the breach hydrograph is calculated as 25,000 m^3/sec. The flood flow is further routed through the channel and the flood hydrograph at various landmark locations is estimated. The flood wave at T. Narasipura, which is 49 km downstream of the KRS Dam is shown in Fig. 4. It is observed that the peak of the flood wave attenuates from 25,000 m^3/s just downstream of the dam to 18,854.67 m^3/s, as it reaches T. Narasipura.

The flood peak further attenuates to 4299.97 m^3/s as it enters the Mettur Dam. The peak of the flood wave and the time lapse between the breach and the peak flow arrival at various landmarks are presented in Table 2. It has been observed that the average flood wave propagates at the rate of 4.326 km/hr. Thus, with an early warning system in place, most of these places will get more than 8 h to evacuate the people and to take appropriate mitigation measures.

HEC-RAS also provides a graphical representation of the depth of flood inundation and the velocity of flow, as given in Fig. 5a, b, respectively. These maps are used to develop the hazard map. The hazard index is calculated as the product of depth and

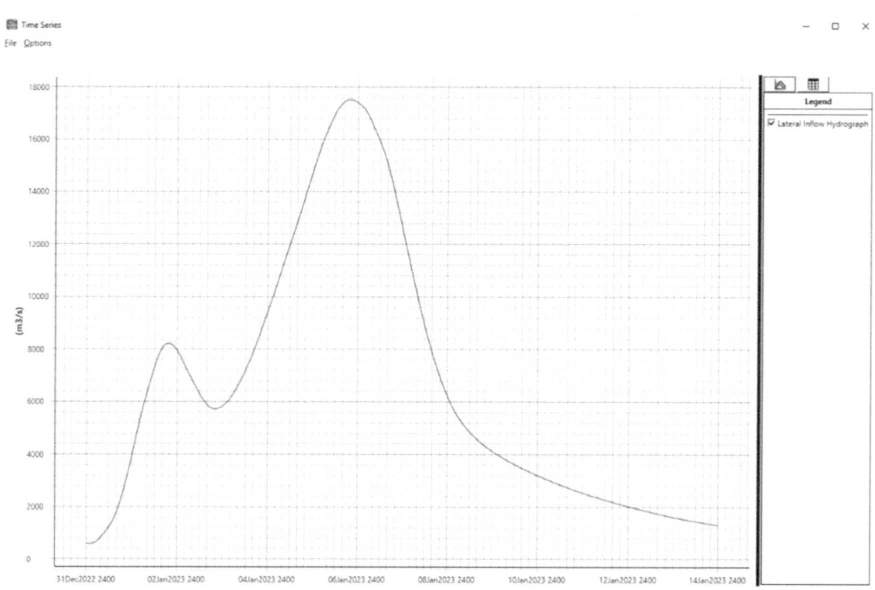

Fig. 2 Inflow flood hydrograph estimated from the PMP

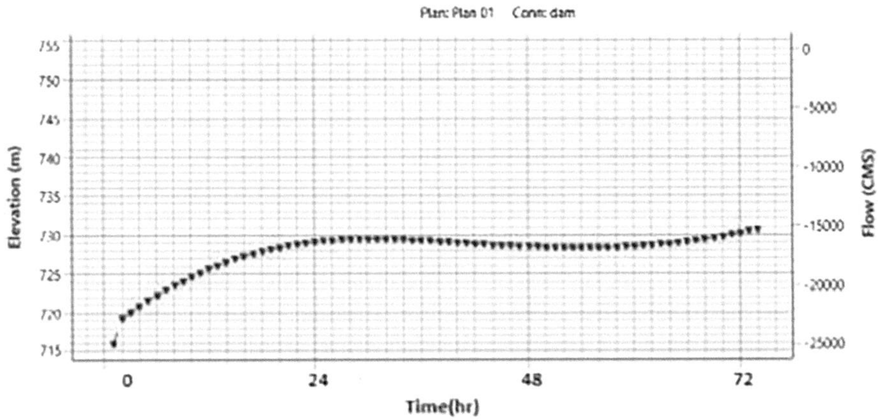

Fig. 3 Breach hydrograph for 1 h breach formation time

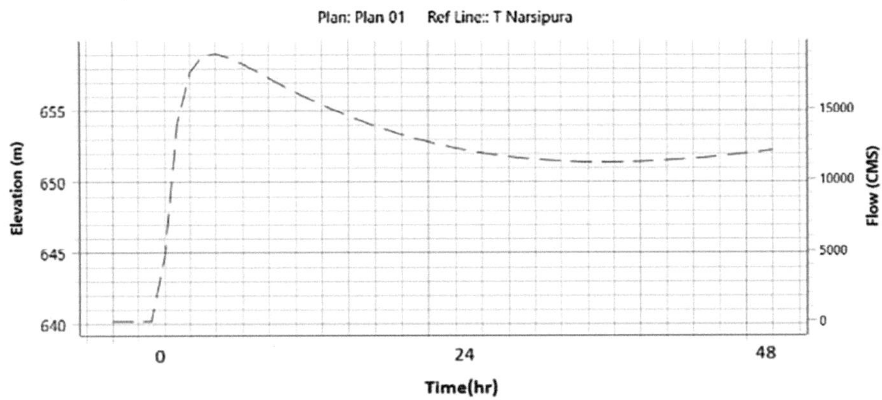

Fig. 4 Flood hydrograph at T. Narasipura

Table 2 Flood peak and the arrival time at various landmarks

Landmark	Flood peak (m^3/s)	Flood peak arrival time (hr)
Srirangapatna	19,800.64	1.5
T. Narasipura	18,854.67	8
Shivanasamudra	846.82	29
Mekedatu	14,009.45	31
Hogenakkal	9969.31	33
Mettur dam	4299.97	49

velocity [4]. As per the CWC classification, hazard index values are classified into 6 classes: 0 (safe), 0–0.6 m^2/s (low risk), 0.6–0.8 m^2/s (moderate risk), 0.8–1.2 m^2/s (high risk), and > 1.2 m^2/s (extremely high risk). Classified flood hazard map of the area is given in Fig. 6.

Based on the hazard map, critical areas identified are Srirangapatna, Somanathapura, Talakadu, Hampapura, Sosale, and Hale Megadahalli. A total of 289.83 km^2 area has been identified to be under extremely high risk in case of the KRS Dam breach as simulated in this study. Areas classified as moderate and high-risk locations are estimated as 11.54 km^2 and 16.87 km^2, respectively.

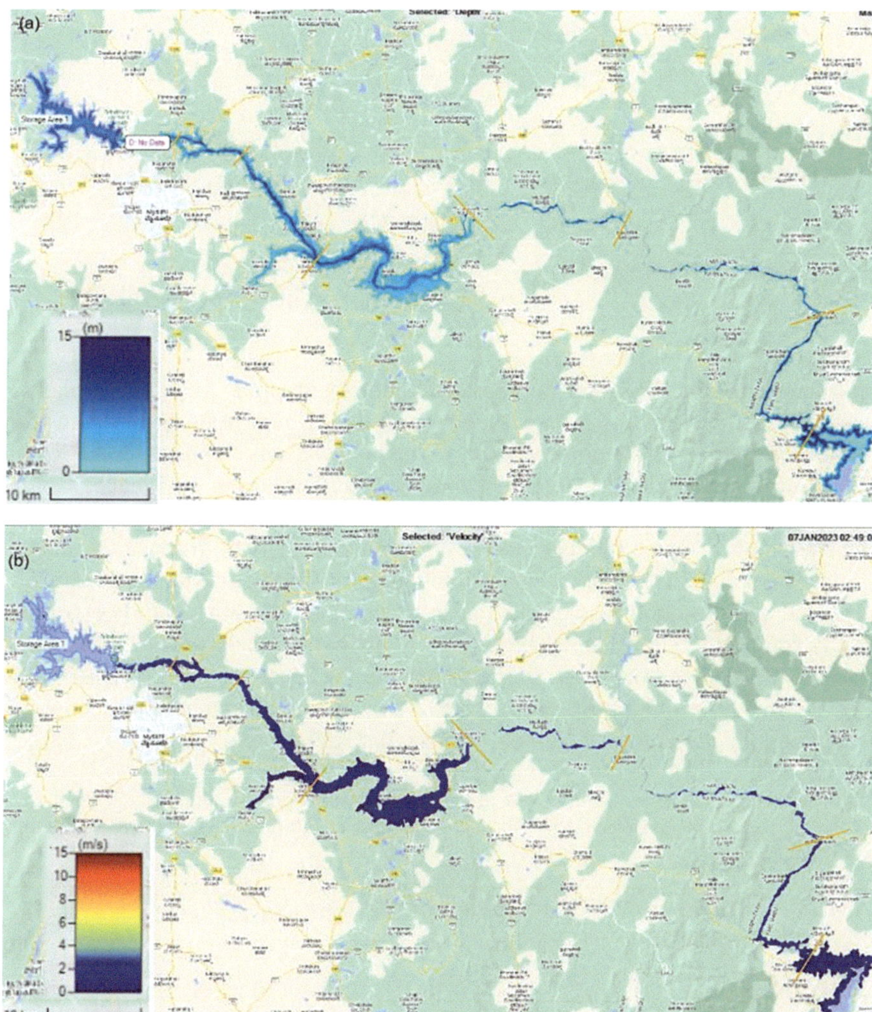

Fig. 5 **a** Depth of inundation and **b** velocity of the flood wave

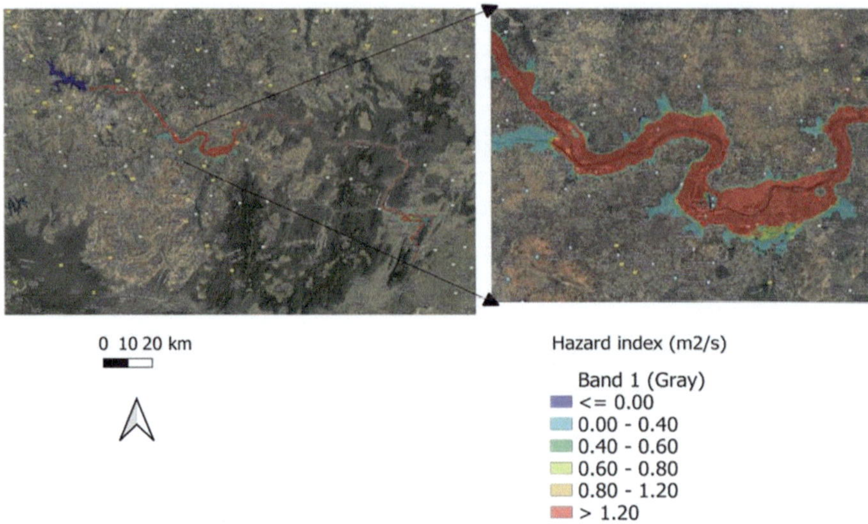

Fig. 6 Classified flood hazard map from KRS dam to Mettur dam

5 Conclusions

A Hazard map of the KRS Dam is generated for the overtopping failure of the dam. It is observed that overtopping failure is likely to produce a huge flood wave with a peak flow of 25,000 m^3/s. The average velocity of the flood wave is estimated as 4.326 km/hr. The huge volume of water released from the reservoir during the breach is likely to cause inundation along the banks of the river. Around 16.87 and 289.83 km^2 area downstream of the dam, are classified as high and extremely high-risk areas in case of a dam breach as simulated in this study under the PMP. The time lapse between the breach and the flood wave arrival at the downstream Mettur Dam is estimated to be 49 h. The study thus shows the extent of inundation likely to be caused by the breach of the dam and the reaction time available for rescue and evacuation in case of a disaster. The inundation depth and the flood hazard map can be of immense use in preparing the disaster management plan for the KRS Dam.

Acknowledgements The work presented in this paper is supported by the research grant from La Fondation, Dassault Systèmes, India through DSF Project ID: IN-2023-2-02 and is gratefully acknowledged.

References

1. Fish E (2013) The forgotten legacy of the Banqiao dam collapse. Economic Observer, Dt 02 Aug 2013
2. Grebby S, Sowter A, Gluyas J et al (2021) Advanced analysis of satellite data reveals ground deformation precursors to the Brumadinho Tailings aam collapse. Commun Earth Environ 2. https://doi.org/10.1038/s43247-020-00079-2

3. International Rivers (2023) International rivers statement on Teesta dam disaster in Sikkim, 5th Oct
4. CWC (2018) Guidelines for mapping flood risks associated with dams. Central Water Commission, Govt. India. Doc. No. CDSO_GUD_DS_05_v1.0
5. CWC and NDSA (2023) National register of large (specified) dams. Central Water Commission (CWC) and National Dam Safety Authority (NDSA), Government of India
6. Jibhakate S, Timbadiya PV, Patel PL (2021) Dam break flood wave analysis from Ukai dam in lower Tapi river, India. In: ICOLD symposium on sustainable development of dams and river basins, 24–27th Feb 2021, New Delhi
7. George AC, Nair BT (2015) dam break analysis using BOSS DAMBRK. Aquat Procedia 4:853–860. https://doi.org/10.1016/j.aqpro.2015.02.107
8. Balaji B, Kumar S (2018) Dam break analysis of Kalyani dam using HEC-RAS. Int J Civ Eng Technol 9(5)
9. CNNL (2019) Operation and maintenance manual for Krishnarajasagara dam. Prepared by the Chief Engineer, CNNL Irrigation South Zone, Mysore, India
10. Hydrology Directorate (1986) Flood estimation report for Kaveri basin subzone. Design office report, Hydrology (small catchments) Directorate, Central Water Commission, New Delhi
11. Froelich D (2016) Predicting peak discharge from gradually breached embankment dam. J Hydrolo Eng 21(11):04016041

Adapting Existing Spillways for Higher Flow Rates: Performance Analysis Through CFD Modelling

D. M. Prajwal, B. N. Darshan, T. V. Reshmidevi, Geetha Kuntoji, and Alok Pandey

Abstract Spillways are hydraulic structures designed to safely discharge excess water from reservoirs and prevent potential damage caused by uncontrolled overflow. With changing climate patterns and associated more frequent and intense rainfall events, many spillway structures are being exceeded in terms of their original design capacities. Modifying existing spillways to safely handle increased discharge is of foremost importance. This study focuses on analyzing hydraulic characteristics and flow patterns of a modified spillway of the Dhanikhari da, using Computational Fluid Dynamics (CFD) simulations. Flow characteristics are investigated using the SST K-omega turbulence model. The model is validated by comparing the simulated flow depth and velocity at the toe of the spillway with that reported in the literature. The model is subsequently employed to examine the pressure and velocity variations over the spillway. With the appropriate design of the spillway profile, it has been observed that cavitation does not form in the case of the modified spillway while passing the augmented design discharge. The insights gained from the CFD simulations can help in retrofitting and upgrading existing spillway structures, ensuring the functionality in the face of change in hydrological conditions.

Keywords Spillways · CFD simulations · Computational modelling · SST K-omega model · Volume of flow

1 Introduction

Spillways serve as controlled outlets for excess water during high-flow events, ensuring the structural integrity of the dam. Spillway capacity is decided by the storage requirements, highest design flood, highest flood level reached this inflow,

D. M. Prajwal (✉) · B. N. Darshan
Department of Civil Engineering, B.M.S. College of Engineering, Bengaluru, India
e-mail: dmprajwal20@gmail.com

T. V. Reshmidevi · G. Kuntoji · A. Pandey
Faculty of Civil Engineering Department, B.M.S. College of Engineering, Bengaluru, India

Y. K. Guruprasad et al. (eds.), *Recent Advances in Infrastructure Development and Management—Volume 2*, Lecture Notes in Civil Engineering 566,
https://doi.org/10.1007/978-981-96-1558-2_23

free board available, geology, and structural safety [1]. Current scenarios of rising water demand increased inflow to the reservoirs due to the variations in the precipitation pattern and reduction in the reservoir storage capacity due to sedimentation, etc. often raise a demand to augment the reservoir storage capacity. However, this engineering endeavour brings its own set of challenges, one of which is the need to modify spillway systems to accommodate the augmented flow. Spillway modification is a common engineering approach used to increase a dam's capacity to handle larger water volumes while ensuring the safety and stability of the structure.

Any such augmentation in the discharge capacity is to be carefully analyzed for the corresponding changes in the pressure, flow profiles, and the downstream energy dissipation capacity. Hydraulic models were used to assess the performance of modified spillway and energy dissipator since model studies have demonstrated high effectiveness in the performance assessment of these structures [2]. Physical modelling is highly expensive, time-consuming, and requires a high level of expertise while conducting and interpreting the results. Simulations using Computational Fluid Dynamics (CFD) are largely supplementing the physical model studies. Availability of the high-performing computational facility, efficient and ready to use tools for modelling, and simplicity in making changes in the conceptual model makes CFD a popular tool in spillway analysis. Several studies are available where CFD simulations have been employed to analyze the hydraulic characteristics of spillways [3–6].

In this study, the hydraulic performance of the Dhanikhari dam spillway is analyzed for the augmented discharge using CFD modelling. Results of the physical modelling studies reported by Ramarao and Bhajantri [7] are used for the model validation.

2 Dhanikhari Dam Spillway

Dhanikhari dam is located across the river Dhanikhari River in Port Blair, India was originally constructed with a length of 132 m, height of 32.25 m with 4.61 Mm^3 of storage capacity at a full reservoir level (FRL) of 61 m. The original dam spillway consisted of two spans, each of 7.9 m width and 2.2 m height. Design capacity of the spillway was 145 m^3/s when the crest level was at 58.5 m. With increase in the water demand, the storage capacity of the reservoir was increased by 3.23 Mm^3 by raising the FRL to 66 m. The design discharge of the modified spillway was also recalculated as 225 m^3/s, at 10% of gates at inoperative condition. To ensure the safe passage of the increased discharge through the spillway, rather than reducing the crest level, the original open spillways were modified into breast wall type spillways with 3 spans each of 3.6 m wide and 3.6 m height. The crest level was raised to 59.0 m. Appropriate quadrant of ellipse approach was provided on the spillway glacis as per IS 6934:1998 [8]. Downstream quadrant and the downstream profile were also modified as per the standards. Specifications of the altered spillway can be found in the literature [7].

3 CFD Modelling of the Dhanikhari Dam Spillway

In this study, a CFD model of the Dhanikhari dam spillway is developed. Main objective of the study is to evaluate the CFD model using the results from physical modelling reported by Ramarao and Bhajantri [7]. The validated model is further used to analyze the hydraulic characteristics of the flow over the spillway, like pressure and velocity. Fluid Scenario Creation tool available in the SIMULIA software is used for the simulations. Capability to perform CFD simulations with free water surface, multiple options for selecting the governing equations and solvers, makes SIMULIA a potential tool for modelling the hydraulic characteristics of the flow over the spillways. A recent study by Sharma et al. [6] presented the application of SIMULIA to study the flow over spillway and the jump characteristics in the stilling basin at the downstream.

Spillway modelling is initiated by defining the three-dimensional structure of the dam using the Generative Shape Design tool in the SIMULIA. Dimensions of the structure used in this study are adopted from the report by Ramarao and Bhajantri [7] and are described under Sect. 2. The fluid domain is defined using the Fluid Assistant option, by defining four walls around the spillway, leaving the inlet and outlet openings. From fluid physics, the gravity flow is enabled. The selected fluid domain is discretised using hexagonal mesh of minimum and maximum sizes 0.14 m and 1 m, respectively. Top view, front elevation, and the mesh generated for the simulation of the original spillway are shown in Fig. 1.

In a similar way, the 3D model of the modified spillway is developed. Side view and top view of the modified spillway model are shown in Fig. 2. Discharge is set

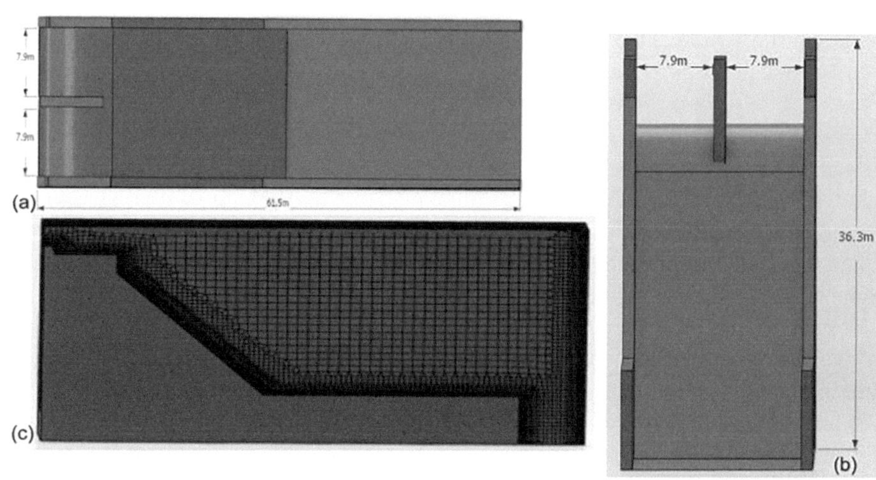

Fig. 1 **a** Plan **b** front elevation and **c** mesh of the model of the original spillway of the Dhanikhari dam spillway

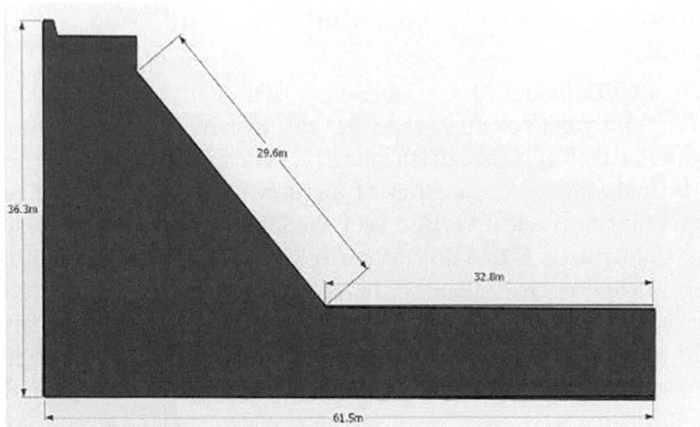

Fig. 2 Top and side view of the model of the modified breast wall type spillway of the Dhanikhari dam spillway

as the inlet boundary condition. Local atmospheric pressure 101,325 Pa is set at the top and outlet boundaries.

4 Results and Discussions

Flow over the original and the modified spillways are simulated using the SST k-omega model in the SIMULIA tool. Discharge over the old spillway is set as 145 m^3/s, whereas the flow over the modified spillway is simulated for a discharge of 225 m^3/s at 10% of the gates inoperative condition. The corresponding VOF profiles are shown in Fig. 3. VOF values close to 1 indicate water, whereas smaller values close to zero indicate the air. In this case, a cutoff value of 0.1 is selected to define the free water surface. Depth of flow and the velocity at the toe of the spillway are estimated. Table. 1 shows the comparison of the model simulated depth and velocity with that reported by Ramarao and Bhajantri [7]. Simulation results are in close agreement with the values reported in the literature and thereby show the efficiency of the SIMULIA model in simulating the flow over the spillways.

Froude's number at the toe of the spillway is calculated for the old and the new spillway using the depth and velocity values obtained from the simulation. The values were found to be almost comparable with Froude's number 10.09 for the old spillway 9.74 for the modified spillway.

Advantage of a computational model is that it can provide details of the minute variations of the hydraulic variables in the flow domain. Validated CFD model of the spillway is used further to analyze the pressure and velocity variations over the spillway. One of the commonly observed problems in spillways is the excessive negative pressure development and the cavitational damage. The absolute pressure

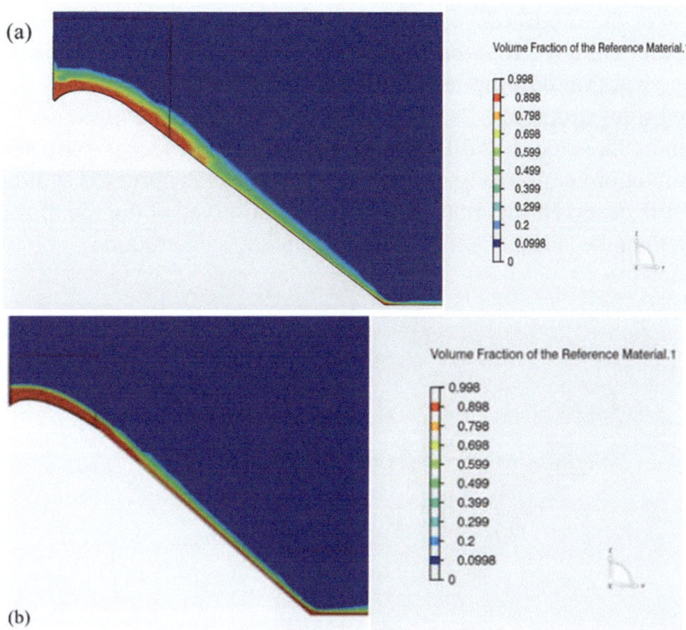

Fig. 3 VOF profile indicating the depth of flow at various sections along the **a** old spillway and **b** new spillway

Table 1 Validation of the hydraulic parameter of the flow over the spillway

Parameter		Old spillway	Modified spillway
Discharge intensity q (m^3/s/m)		8.5	13.2
Reservoir water level RWL (m)		61	66
Crest elevation (m)		58.5	59.5
Stilling basin invert El (m)		31.2	31.2
Hydraulic variables at the toe of the spillway, as reported in the literature	Depth of flow (m)	0.4	0.6
	Velocity (m/s)	21.8	23.5
Hydraulic variables at the toe of the spillway, obtained from the simulation	Depth of flow (m)	0.454	0.614
	Velocity (m/s)	21.3	23.9

variations over the modified Dhanikhari dam spillway are shown in Fig. 4. From the results, it is observed that the right selection of the downstream profile helps to prevent the cavitation development on the spillway and hence ensures the effective functioning of the structure.

Variation in the velocity at different depths from the surface is extracted from the velocity simulation outputs (Fig. 5). Maximum velocity is observed at almost half of the depth. It is observed that maximum velocity observed at the toe of the spillway is 8% more than the surface velocity observed at the same section.

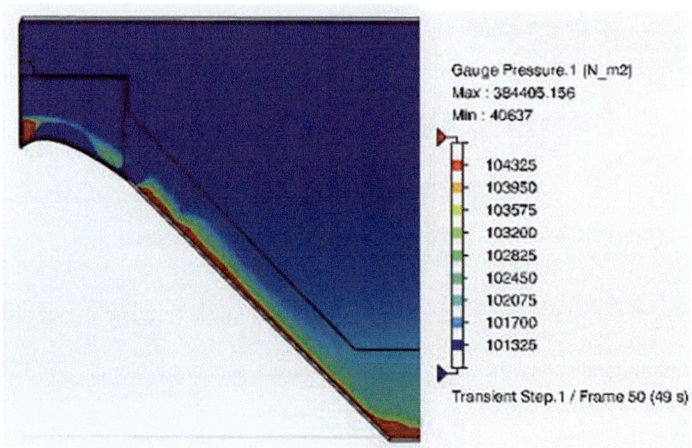

Fig. 4 Pressure variation along the modified Dhanikhari dam spillway with a discharge of 225 m^3/s at 10% gates inoperative condition

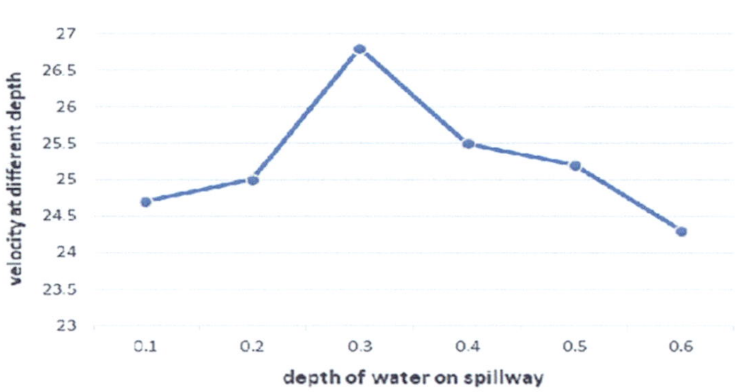

Fig. 5 Velocity profile over the depth of flow at the toe of the spillway

5 Conclusions

Hydraulic analysis of spillways under the design discharge is essential to ensure the efficient functioning. Further, during retrofitting or redesigning, they need to be tested for the augmented discharge. It is essential to make sure that the spillways are capable of safely discharging the augmented flow, without developing any cavitation on the downstream face. Computational modelling complements the widely adopted physical modelling. In this study, CFD model of the Dhanikhari dam spillway is developed using the SIMULIA tool to analyze the performance of the spillway under the original discharge and the augmented discharge conditions. The CFD model results are found to be in close agreement with the actual values reported in the literature, which shows the efficiency of the model to simulate the spillway flow conditions. The model outputs are further used to study the pressure and velocity variations over the spillway. The pressure profile shows that no negative pressure is formed along the spillway surface, thereby eliminating the cavitation issues. In addition to the overall performance, the computation model helps to simulate minute variations in the hydraulic variables along the spillway. Such models are of immense benefit for the engineers to study the performance of the spillways under various hydrologic and hydraulic conditions.

Acknowledgements The work presented in this paper is supported by the research grant from La Fondation, Dassault Systèmes, India through DSF Project ID: IN-2023-2-02 and is gratefully acknowledged.

References

1. Bhajantri M, Eldho TI, Deolalikar P (2006) Hydrodynamic modelling of flow over a spillway using a two-dimensional finite volume-based numerical model. Sadhana Acad Proc Eng Sci 31:743–754. https://doi.org/10.1007/BF02716893
2. Bhosekar VV, Sridevi MI, Deolalikar PB (2005) Hydraulic model studies for a dual purpose spillway with 21 M depth of overflow—a case study. ISH J Hydraul Eng 11(3):32–46
3. Ho D, Boyes K, Donohoo S, Cooper B (2003) Numerical flow analysis for spillways. In: Proceedings of 43rd ANCOLD Conference, Hobart, Tasmania, 24–29 Oct 2003, pp 227–237
4. Kumcu SY (2017) Investigation of flow over spillway modeling and comparison between experimental data and CFD analysis. KSCE J Civ Eng 21:994–1003. https://doi.org/10.1007/s12205-016-1257-z
5. Bhajantri M, Eldho TI, Deolalikar P (2007) Numerical modelling of turbulent flow through spillway with gated operation. Int J Numer Methods Eng 72:221–243. https://doi.org/10.1002/nme.2016
6. Sharma U, Prajwal DM, Pandey A, Reshmidevi TV (2023) Numerical modeling investigations of hydraulic jump characteristics over a chute spillway. In: International conference on interdisciplinary approaches in civil engineering for sustainable development IACESD-2023. Organised by Jyothi Institute of Technology, Bengaluru, 7–8 July 2023
7. Ramarao VS, Bhajantri MR (2020) Performance of modified spillway for heightened gravity dam. ISH J Hydraul Eng 26(3):325–333. https://doi.org/10.1080/09715010.2018.1491809
8. IS 6934 (1998) Indian standard: hydraulic design of high ogee overflow spillways—recommendations (first revision). Bureau of Indian Standards, New Delhi

Nanomaterials and Their Impact on Cement Paste Characteristics: A Review

T. S. Harish and B. S. Santhosh

Abstract This review examines the transformative influence of nanomaterials on the rheological characteristics of cement paste, unraveling their profound impact on the construction industry's technological landscape. Nanoparticles, such as silica fume, carbon nanotubes, and graphene oxide, are explored for their unique ability to modulate the rheological behavior of cementitious materials, ushering in a new era of advancements in construction practices. Nanomaterials play a crucial role in tailoring the flow and viscosity of cement paste, offering unprecedented control over workability, and setting time. Silica nanoparticles, for instance, contribute to improved dispersion, resulting in enhanced fluidity without compromising the material's overall structural integrity. Similarly, the addition of carbon nanotubes and graphene oxide influences the paste's rheological properties, imparting stability and preventing segregation during transport and placement. The implications of these rheological enhancements extend to the construction industry's efficiency and productivity. The improved workability facilitated by nanomaterials allows for more intricate and streamlined construction processes, reducing the need for excessive water content, and optimizing the overall material utilization. Additionally, the controlled rheology paves the way for innovations in formwork design and facilitates the construction of complex structures with greater precision. In conclusion, the integration of nanomaterials into cement paste presents a paradigm shift in the rheological control of construction materials. This review highlights the potential of nanotechnology to revolutionize the construction industry by offering a nuanced understanding of how nanomaterials influence the rheological characteristics of

T. S. Harish (✉)
Research Scholar, Assistant Professor, Department of Civil Engineering, Vivekananda Institute of Technology, Bengaluru 560074, Karnataka, India
e-mail: harishts.civil@gmail.com

T. S. Harish · B. S. Santhosh
Visvesvaraya Technological University, Belagavi 590018, Karnataka, India

B. S. Santhosh
Research Supervisor, Professor & Head, Department of Civil Engineering, JSS Academy of Technical Education, Bengaluru 560060, Karnataka, India

Y. K. Guruprasad et al. (eds.), *Recent Advances in Infrastructure Development and Management—Volume 2*, Lecture Notes in Civil Engineering 566,
https://doi.org/10.1007/978-981-96-1558-2_24

cement paste, ultimately fostering advancements in construction practices, efficiency, and the realization of complex and sustainable structures.

Keywords Cement paste · Innovative materials · Nanomaterials · Rheology · Workability

1 Introduction

1.1 *Brief Overview of Nanomaterials and Their Applications in Cement-Based Materials*

Nanomaterials have been extensively researched for their applications in cement-based materials. They have been found to improve the mechanical properties and durability of cementitious materials [1]. Nanoparticles, such as cellulose and chitin nanomaterials, have been investigated as additives in cementitious systems, and they have shown significant improvements in the macroscopic properties of cement paste, mortar, and concrete [2]. Additionally, nanomaterials have been used to modify protective coatings for cement-based materials, enhancing their functional properties and providing better durability and functionality [3]. Graphene, in particular, has shown promise as an additive for cement composites, improving flexural and compressive strength and increasing durability [4]. Carbon-based nanomaterials, including graphene oxide, have also been explored as reinforcement for cement and concrete, resulting in increased strength and durability at low concentrations [5]. Overall, nanomaterials offer great potential for enhancing the properties and performance of cement-based materials.

1.2 *Importance and Motivation Behind Exploring the Impact of Nanomaterials on Cement Paste Characteristics*

The exploration of the impact of nanomaterials on cement paste characteristics is important for several reasons. Firstly, nanomaterials such as graphene oxide (GO), carbon nanotubes (CNTs), and nano-silica (NS) have been introduced to improve the microstructures and performances of cement-based materials (CBM) [6]. Secondly, the addition of nanomaterials like nano-SiO_2 (NS) and nano-Al_2O_3 (NA) has been found to enhance the mechanical and durability properties of cement-based materials [7]. Thirdly, the specific surface area of nano-silica (NS) has been shown to influence the pore structure, hydration reaction, and overall properties of cement paste [8]. Additionally, the use of nanoparticles (NPs) of different types, such as $nSiO_2$, nAl_2O_3, and $nCaCO_3$, has been found to accelerate the hydration kinetics and affect the apparent activation energy of Portland cement [9]. Lastly, the effects of contrasting types of cellulose nanomaterials (CNMs) on cement properties have been studied, highlighting the need to appropriately specify the type and dosage of CNMs for concrete applications [10].

2 Types of Nanomaterials Utilized

2.1 *Description of Various Nanomaterials Used, Such as Nanoparticles, Nanofibers, and Nano-Additives*

Nanomaterials, such as nanoparticles, nanofibers, and nano-additives, have been extensively studied for their use in cement paste. These materials have shown the ability to improve the mechanical properties and durability of cement-based materials. Nanoparticles, due to their high surface energy and specific surface areas, can affect the packing characteristics of cementitious particles, leading to agglomeration and entrapped water [1]. Polysaccharide nanomaterials, including cellulose and chitin, have gained research interest as multifunctional additives in cementitious systems [2]. Cellulose nanomaterials have been extensively investigated, while chitin nanomaterials and starch derivatives are emerging research areas [10]. Different types of cellulose nanomaterials, such as cellulose nanocrystals (CNCs) and mechanically fibrillated cellulose nanofibers (CNFs), have distinct effects on the rheology, setting times, and compressive strength of cement paste [11]. Other nanomaterials commonly used in cementitious composites include carbon nanotubes, nanocellulose, nanographene, graphene oxide, nano-silica, and nano-TiO_2, which can improve various properties of cementitious composites [12]. Additionally, the incorporation of nanoparticles, such as nano-SiO_2, nano-TiO_2, and nano-Fe_2O_3, in cement mortar can significantly improve its fresh properties, mechanical properties, and microstructure.

2.2 *Specific Properties and Advantages of Different Types of Nanomaterials in Cement Paste*

Nanomaterials, such as cellulose nanocrystals (CNCs), mechanically fibrillated cellulose nanofibers (CNFs), carbon nanotubes, nanocellulose, nanographene, graphene oxide, nano-silica, nano-TiO_2, nano-SiO_2, nano-Al_2O_3, and nano-clay, have been studied for their effects on cement paste properties. CNCs reduce viscosity and provide longer slump retention, while CNFs increase viscosity [10]. Nanomaterials can improve the physical, mechanical, thermal, and electrical properties of cementitious composites, including compressive and tensile strength, hydration acceleration, porosity reduction, and fire resistance enhancement [11]. In geopolymer composites, nanomaterials like nano-SiO_2, nano-Al_2O_3, nano-TiO_2, carbon nanotubes, and nano-clay enhance fresh and hardened state properties, microstructure, and durability [13].

Graphene-based nanomaterials, including graphene oxide, reduced graphene oxide, and pristine graphene nanoplatelets, improve the workability, hydration, mechanical, microstructure, and transport properties of cement paste composites [14]. Multi-walled carbon nanotubes with specific properties, such as surface area, length, functional group content, and outer diameter, can significantly affect cement paste properties, including electrical resistivity, splitting tensile strength, CH content, water absorption, and porosity [15].

3 Influence on Cement Paste Properties

3.1 Mechanical Properties Enhancement: Discussion on How Nanomaterials Affect Compressive Strength, Flexural Strength, Etc.

Nanomaterials, such as nanoparticles and carbon-based nanoreinforcers, have been shown to enhance the mechanical properties of cement paste. These materials improve the flexural and compressive strength of cement-based materials, with some studies reporting increases of over 100% [1]. The addition of silica nanoparticles to cement mortars has been found to significantly increase both compressive and flexural strengths, with the highest improvement observed at a dosage of 4% by weight [16]. Similarly, the incorporation of cellulose nanocrystals (CNCs) and mechanically fibrillated cellulose nanofibers (CNFs) in cement paste has been shown to increase compressive strength by 17–18% [17]. The effects of nanomaterials on other mechanical properties, such as flexural strength and elastic modulus, have also been investigated [18]. Overall, nanomaterials have the potential to enhance the mechanical properties of cement paste, making it more durable and suitable for practical engineering applications.

3.2 Durability Improvements: Exploration of Enhanced Resistance to Chemical Attacks, Permeability Reduction, and Durability in Harsh Conditions

Nanomaterials have shown promise in improving the durability of cement paste in various ways. They enhance resistance to chemical attacks, reduce permeability, and improve durability in harsh conditions [19]. Cellulose nanomaterials, such as cellulose nanocrystals (CNCs) and cellulose nanofibers (CNFs), have contrasting effects on the properties of cement paste, with CNCs reducing viscosity and providing longer slump retention, while CNFs increase viscosity and act as a set retarder [20]. Graphene oxide (GO) dispersion properties significantly affect the properties of cement pastes, leading to improvements in mechanical properties, hydration rate, and thermal conductivity [21]. COOH-functionalized multi-walled carbon nanotubes (MWCNTs) have been found to increase the flexural and compressive strength of cement mortars, improving their durability in sulfate and water environments [22]. Various nanomaterials, including nano-silica, nano-magnesium oxide, and nano-alumina, have been shown to enhance the fresh-state and hard-state properties of cement mortar, including mechanical properties and durability [23].

3.3 Microstructural Changes: Insights into Alterations in the Microstructure of Cement Paste Due to Nanomaterial Integration

Microstructural changes in cement paste due to nanomaterial integration have been investigated in several studies. Ying et al. found that the addition of 3D graphene materials improved the micromechanical properties and chloride diffusion resistance of hardened cement paste [24]. Echeverry-Cardona et al. studied the influence of changing the molarity of dispersed multiple wall carbon nanotubes (MWCNTs) on the mechanical properties of cement paste and found that a molarity of 10 mM exhibited the best results [25]. Singh and Saini highlighted the wide range of applications of nanomaterials in the cement industry and their effects on key properties of cementitious products [26]. Jiao and De Schutter investigated the viscoelastic properties of cement paste and found that factors such as water-to-cement ratio, fly ash content, and the addition of nanoparticles and superplasticizers influenced the storage modulus, critical strain, and viscoelastic yield stress of cement paste [27].

4 Dispersion and Compatibility Challenges

4.1 Addressing Issues Related to Achieving Proper Dispersion and Ensuring Compatibility of Nanomaterials Within the Cement Matrix

Proper dispersion and compatibility of nanomaterials within the cement matrix are crucial for maximizing their benefits and effectiveness. Different surfactants, both anionic and nonionic, have been used to disperse nano-silica in cement pore solution, leading to improved dispersibility and increased compressive strength of cement-NS pastes [28]. Carbon nanomaterials, such as carbon nanotubes and graphene oxide, have been widely used in cementitious composites to enhance their physical and mechanical properties. The dispersion methods used during mixing significantly influence the mechanical properties and the percentages of nanomaterials used in the mixes [17]. Nanomaterials, including carbon nanotubes, nanocellulose, and nano-silica, can reinforce the nanoscale of cementitious composites, improving their performance by increasing compressive and tensile strength, accelerating hydration, reducing porosity, and enhancing fire resistance [11]. The combined use of carbon nanotubes and metakaolin in cementitious matrices has shown mixed results, with the dispersion method and the presence of supplementary cementitious materials affecting the mechanical properties [29]. Different methods and mechanisms have been explored to disperse nanoparticles in cementitious matrices, with a combination of methods being effective in achieving the desired dispersion [30].

5 Future Perspectives and Challenges

5.1 Discussion on Potential Advancements, Unresolved Challenges, and Opportunities for Further Research

Nanomaterials have shown potential advancements in cement paste. Cellulose nanomaterials (CNMs) such as cellulose nanocrystals (CNCs) and mechanically fibrillated cellulose nanofibers (CNFs) have contrasting effects on cement properties, including rheology, setting times, and compressive strength [10]. Nano-modified composite coatings have also been explored for protecting cement-based materials, with nanomaterials affecting the physical and chemical properties of the coatings [4]. Dispersion properties of nanomaterials, such as graphene oxide (GO), have been found to significantly impact the properties of cement pastes, including splitting tensile strength, ultrasonic pulse velocity, hydration rate, thermal conductivity, electrical resistance, porosity, and water absorption [31]. Different types of nanoparticles (NPs), such as $nSiO_2$, nAl_2O_3, and $nCaCO_3$, have been studied for their effects on cement hydration kinetics and apparent activation energy, with the specific surface area (SSA) of the NPs playing a role in their performance [9]. Carbon nanomaterials have also been investigated for their impact on the mechanical properties of cementitious composites, with dispersion methods influencing both the mechanical properties and the percentages of nanomaterials used [17]. Further research is needed to address unresolved challenges and explore opportunities for the use of nanomaterials in cement paste.

5.2 Consideration of Economic, Environmental, and Scalability Aspects for Practical Implementation

The use of nanomaterials in cement paste has economic, environmental, and scalability aspects. Economically, the addition of nanomaterials such as graphene oxide

(GO) and carbon nanotubes (CNTs) improves the mechanical properties of cement-based materials, leading to greater durability and higher strength [31, 32]. This enhancement in performance can contribute to cost savings in terms of maintenance and repair of concrete structures [33]. From an environmental perspective, the use of nanomaterials can enhance the ecological sustainability of concrete by reducing its carbon footprint and improving its durability, thus extending the lifespan of structures [34]. Additionally, the incorporation of nanomaterials can optimize the use of supplementary cementitious materials, such as ground granulated blast furnace slag, leading to a synergistic effect on the hydration kinetics and overall properties of cement [10]. In terms of scalability, the development of field-applicable methods for growing nanoparticles within cement slurries using inexpensive precursors enables large-scale application of nanomaterials in the construction industry.

6 Conclusion

6.1 Summary of Key Findings and Implications of Employing Nanomaterials in Cement Paste

Nanomaterials, such as polyurethane-decorated MgO nanohybrids, carbon-based nanoreinforcers, and cellulose nanomaterials, have been found to have significant effects on the properties of cement paste. These nanomaterials can improve the mechanical performance of cement paste, including increasing compressive and flexural strength [35]. Carbon-based nanoreinforcers, particularly graphene oxide (GO), show promise due to their low cost, ease of use, and high modification potential [16]. The incorporation of nanomaterials into cement composites accelerates hydration, optimizes microstructure, and enhances strength [36]. Cellulose nanocrystals (CNCs) and mechanically fibrillated cellulose nanofibers (CNFs) have contrasting effects on cement paste properties, with CNCs reducing viscosity and providing longer slump retention, while CNFs increase viscosity and retard setting times [10]. Different types of nanoparticles, such as $nSiO_2$, nAl_2O_3, and $nCaCO_3$, can accelerate cement hydration, with the effect depending on the replacement level and specific surface area of the nanoparticles [9]. These findings have implications for tailoring the properties of cementitious systems and can aid in the selection and dosage of nanomaterials for concrete applications.

6.2 Highlighting the Importance of Continued Research and Development in This Field

Continued research and development in employing nanomaterials in cement paste is important for several reasons. Nanoparticles have been shown to improve the mechanical properties and durability of cement-based materials [1]. Nanomaterials retrieved from biomass sources, such as cellulose and chitin, have gained research interest due to their remarkable properties and abundance [2]. Carbon-based nanomaterials, like graphene oxide, have also shown significant improvements in the strength and durability of cement [16]. Additionally, cellulose nanomaterials have been studied in cementitious systems, but the effects of different types of cellulose nanomaterials on cement properties have not been fully understood [10]. Furthermore, the effects of different types of nanoparticles, such as $nSiO_2$, nAl_2O_3, and $nCaCO_3$, on the hydration kinetics and strength development of cement have been analyzed [9]. Continued research will help in understanding the mechanisms and optimizing the use of nanomaterials in cement paste for enhanced performance and application in practical engineering.

References

1. Li Q, Fan Y, Shah S (2022) Rheological properties and structural build-up of cement based materials with addition of nanoparticles: a review. Buildings 12(12)
2. Zhong T, Jian G, Chen Z et al (2022) Interfacial interactions and reinforcing mechanisms of cellulose and chitin nanomaterials and starch derivatives for cement and concrete strength and durability enhancement: a review. Nanotechnol Rev 11(1):2673–2713
3. Yuan J, Lu D, Wu H et al (2023) Carbon nanotubes-coated cement particles for cement-based sensors with excellent piezoresistivity. Smart Mater Struct 32(6)
4. Yin B et al (2021) Research and application progress of nano-modified coating in improving the durability of cement-based materials. Prog Org Coat 161:106529
5. Polverino S, Castillo A, Brencich A et al (2022) Few-layers graphene-based cement mortars: production process and mechanical properties. Sustainability (Switzerland) 14(2)
6. Yang H, Monasterio M, Zheng D et al (2021) Effects of nano silica on the properties of cement-based materials: a comprehensive review. Constr Build Mater 282
7. Zhang A, Yang W, Ge Y, Du Y (2021) Effects of nano-SiO_2 and nano-Al_2O_3 on mechanical and durability properties of cement-based materials: a comparative study. J Build Eng 34
8. Bai S, Guan X, Li H et al (2021) Effect of the specific surface area of nano-silica particle on the properties of cement paste. Powder Technol 680–689:392
9. Sargam Y, Wang K (2021) Hydration kinetics and activation energy of cement pastes containing various nanoparticles. Compos Part B Eng 216
10. Nassiri S, Chen Z, Jian G et al (2021) Comparison of unique effects of two contrasting types of cellulose nanomaterials on setting time, rheology, and compressive strength of cement paste. Cem Concr Compos 123
11. Metaxa Z, Tolkou A, Efstathiou S et al (2021) Nanomaterials in cementitious composites: an update. Molecules 26(5)
12. Ng S, Paul D, Anggraini S et al (2020) Influence of SiO_2, TiO_2 and Fe_2O_3 nanoparticles on the properties of fly ash blended cement mortars. Constr Build Mater 258
13. Jindal B, Sharma R (2020) The effect of nanomaterials on properties of geopolymers derived from industrial by-products: a state-of-the-art review. Constr Build Mater 252
14. Qureshi T, Panesar D (2020) Nano reinforced cement paste composite with functionalized graphene and pristine graphene nanoplatelets. Compos Part B Eng 197
15. Şimşek B (2020) Multi-walled carbon nanotubes with different features reinforced cement pastes: a compressive and systematic approach using principal component analysis. J Build Eng 32

16. Sheikh T, Anwar M, Muthoosamy K et al (2021) The mechanics of carbon-based nanomaterials as cement reinforcement—a critical review. Constr Build Mater 303
17. Yaghobian M, Whittleston G (2022) A critical review of carbon nanomaterials applied in cementitious composites—a focus on mechanical properties and dispersion techniques. Alexandria Eng J 61(5):3417–3433
18. Behnia B, Safardoust-Hojaghan H, Amiri O et al (2021) High-performance cement mortars-based composites with colloidal nano-silica: synthesis, characterization and mechanical properties. Arab J Chem 14(9)
19. Du Y, Korjakins A (2023) Synergic effects of nano additives on mechanical performance and microstructure of lightweight cement mortar. Appl Sci (Switzerland) 13(8)
20. Gebre Y, Lahmer T, Müller M et al (2023) Properties of concrete with coated aggregates under different loading. J Res Eng Appl Sci 8(1):449–459
21. Li Q, Li J, Zhang S et al (2023) Research progress of low-carbon cementitious materials based on synergistic industrial wastes. Energies 16(5)
22. Mansouri Sarvandani M, Mahdikhani M, Aghabarati H et al (2021) Effect of functionalized multi-walled carbon nanotubes on mechanical properties and durability of cement mortars. J Build Eng 41
23. Singh NB, Middendorf B, Tomar R (2023) Effect of nanomaterials on the properties of binding materials in the construction industry. Mater Res Found 135:226–264
24. Ying J, Jiang Z, Deng C (2023) Microstructure, mechanics, and chloride diffusion of three-dimensional graphene reinforced cement paste. J Mater Civ Eng 35(4)
25. Laura E, Rafael C, Jorge Q et al (2022) Effects of molarity and storage time of MWCNTs on the properties of cement paste. Materials 15(24)
26. Singh G, Saini B (2022) Nanomaterial in cement industry: a brief review. Innov Infrastruct Solutions 7(1)
27. Jiao D, De Schutter G (2022) Insights into the viscoelastic properties of cement paste based on SAOS technique. Constr Build Mater 357
28. Sargam Y, Wang K, Tsyrenova A et al (2021) Effects of anionic and nonionic surfactants on the dispersion and stability of nano SiO_2 in aqueous and cement pore solutions. Cem Concr Res 144
29. Neto A, Santos J, Pinto T et al (2021) Effect of the combined use of carbon nanotubes (CNT) and metakaolin on the properties of cementitious matrices. Constr Build Mater 271:121903
30. Korayem A, Tourani N, Zakertabrizi M et al (2017) A review of dispersion of nanoparticles in cementitious matrices: nanoparticle geometry perspective. Constr Build Mater 346–357:153
31. Şimşek B, Doruk S, Ceran Ö et al (2021) Principal component analysis approach to dispersed graphene oxide decorated with sodium dodecyl sulfate cement pastes. J Build Eng 38
32. Makul N (2020) Advanced smart concrete—a review of current progress, benefits and challenges. J Cleaner Prod 274
33. Bhojaraju C, Di Mare M, Ouellet-Plamondon C (2021) The impact of carbon-based nanomaterial additions on the hydration reactions and kinetics of GGBS-modified cements. Constr Build Mater 303
34. Mehairi A, Husein M (2020) Enhancement of cement properties by means of in situ grown nanoparticles. Constr Build Mater 261
35. Fang Y, Ning W, Li Y et al (2022) The effects of the addition of polyurethane–MgO nanohybrids on the mechanical properties of ordinary Portland cement paste. Nanomaterials 12(22)
36. Luan C, Zhou Y, Liu Y et al (2022) Effects of nano-SiO_2, nano-$CaCO_3$, nano-TiO_2 on properties and microstructure of the high content calcium silicate phase cement (HCSC). Constr Build Mater Part A 314:125377

Short Column of Fiber-Reinforced Concrete Under Axial Loading: Experimental and Theoretical Modeling

A. Meenachi, R. Malathy, V. Barath Kumar, and S. Dhanush

Graphical Abstract

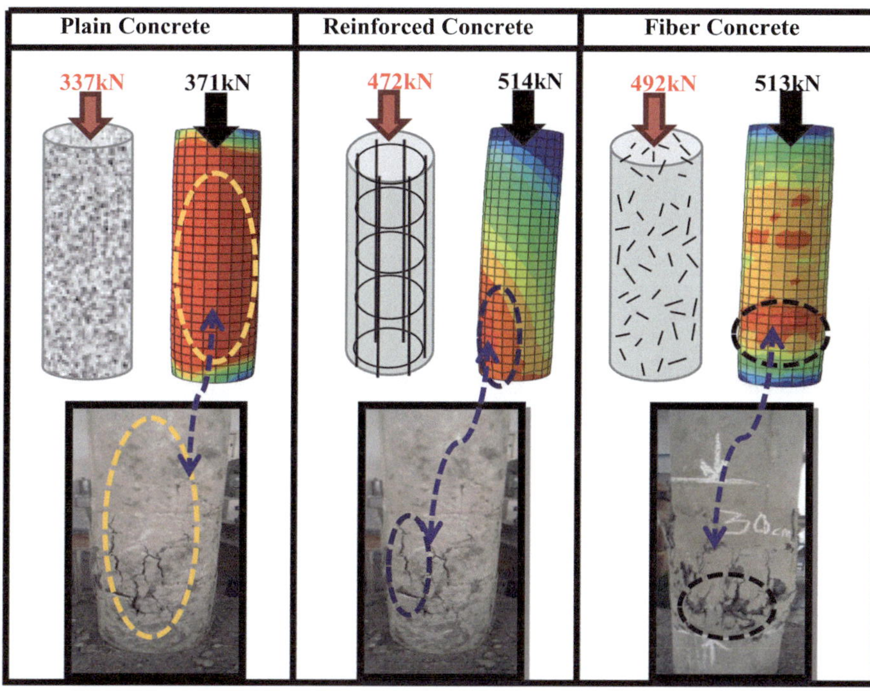

Abstract Extensive studies on short discrete steel fibers show better performance in diverse parameters like strength in compression, tension, flexure, and increase in modulus of elasticity and more currently researched. The reinforced concrete short column takes predominantly axial load than the flexure but still, minimum reinforcements are provided even when concrete alone can take the compression force due

A. Meenachi (✉) · R. Malathy · V. Barath Kumar · S. Dhanush
Department of Civil Engineering, Sona College of Technology, Salem, India
e-mail: meenachia@sonatech.ac.in

Y. K. Guruprasad et al. (eds.), *Recent Advances in Infrastructure Development and Management—Volume 2*, Lecture Notes in Civil Engineering 566,
https://doi.org/10.1007/978-981-96-1558-2_25

to the axial load. The minimum reinforcement is required to resist the temperature stresses, avoid brittle failure, to take tension developed due to the minimum eccentricity, and to increase the ductility of the column. This paper investigates the effect of short discrete fibers experimentally and analytically on short concrete column behavior. The experimental program consists of ten 600 mm-long, 150 mm diameter specimens with circular cross-section specimens. It includes six concrete specimens with fiber in different percentages (0.5, 1, and 1.5% of the volume of the fraction), two concrete specimens with reinforcement, and two-plain cement concrete specimens. The specimens were tested under compression and loaded up to failure. The analytical program consists of model creation and to validate the experimental test results. The finite element model was created using the FEM code, ABAQUS. FE model. All of the specimens were compared after an experimental and analytical analysis of the failure pattern and load-carrying ability was conducted. It can be found that plain cement short column demands minimum reinforcement from the failure pattern and that can be satisfied with the randomly dispersed steel fibers as an alternate to reinforcement cost-effectively.

Keywords Discrete steel fiber · Short concrete column · Axial loading · Finite element model · Cost-effective

1 Introduction

The column is a vertical load-carrying member predominantly a member in compression entirely if the load coincides with the axis of the column. Moreover, in the short column, the failure mode is by crushing of concrete irrespective of reinforcement provided. The short columns are subjected to predominantly to axial compressions and a minimum moment due to external loads. When a column's effective length to its smallest lateral dimension is less than or equal to 12, that column is said to be a short column [1]. Even when concrete alone is good in carrying the compressive forces, still reinforcements are provided. The provision of reinforcements is when the axis of the load does not coincide with the axis of the column there will be an eccentric loading, it will greatly reduce the size of the columns, and also it will impart a ductile behavior of the column. Because concrete is very weak in tension, various national codes specify the minimum reinforcement in concrete members in order to prevent cracking due to thermal shrinkage and creep. They also specify that water must not enter the concrete element through cracks in order to prevent corrosion of the reinforcement and weathering of the concrete, both of which have an impact on durability in the short- and long-term. Also during an earthquake force due to the lateral excitation longitudinal bending of the column may occur which results in the generation of longitudinal will tension along with one phase in such cases also it is essential to have steel in the column to prevent the brittle cracking of concrete. The vertical reinforcements help in taking up loads axially and the lateral ties do not take up the load acting on the column but help in keeping in position the vertical ties and are spaced in the equal distance along with the full height of the column.

Under axial and shear load developed due to earthquake, it is very dangerous because the concrete has shown very [2] brittle failure, so under loads whether axial compression or shear, the brittle characteristics of ultra-high-strength concrete should be altered to more ductile characteristics. The addition of steel fiber will reduce the brittleness of concrete and leads to delay in concrete cover spalling [3–5], improve the ductility. In terms of mechanical qualities [8], energy absorption capacity [9], fatigue performance [10], and durability [5, 11–15], concrete containing steel fibers [3, 6, 7] (often 2% by volume) [5] performs exceptionally well. Steel fibers [7] are having the property to control cracking due to the more closely spaced fibers. The crack propagation starts in concrete and the randomly distributed fibers start taking up the load, thus arresting the cracks and improve strength and ductility. Apart from normal steel reinforcement, various types of fibers are also used to improve the concrete properties, [16, 17] mainly for enhancing the tensile strength and thus satisfy the requirements of reinforcements in a short column.

2 Experimental Program

The main ideology represented in this paper is to use short discrete steel fibers as an alternate to reinforcements in reinforced concrete short columns. Circular column of height 600 and 15 mm diameter is chosen to satisfy short column criteria. Totally, 10 specimens were casted, 2 with plain cement concrete (PCC), two standard reinforced concrete (RCC) and six columns with three different volume fractions (0.5% (FC 0.5), 1% (FC 1.0) and 1.5% (FC 1.5)) of fibers in concrete without reinforcements. The plain cement concrete and reinforced cement concrete specimen were used to compare the behavior with fiber in a short column. The analysis comprises experimentally and analytical and compared with standard codes and empirical formula proposed earlier. Mix design of grade M20 was used in this process. Short discrete crimped steel fibers of aspect ratio 50 were used for good dispersion of fibers in concrete.

2.1 Test Matrix

Ten concrete columns with a circular cross-section were prepared and tested under concentric axial loadings. Six out of ten specimens were concrete with different percentages of crimped steel fiber (0.5, 1, 1.5% of the volume of the fraction) without reinforcement. Four specimens consisting of two plain concrete and two reinforced concrete specimens. In the reinforced column, 6 numbers of 8 mm diameter rods were used as vertical reinforcement in circular patterns. Lateral ties of 6 mm diameter with 150 mm spacing were used throughout the length of the column. In preparation of

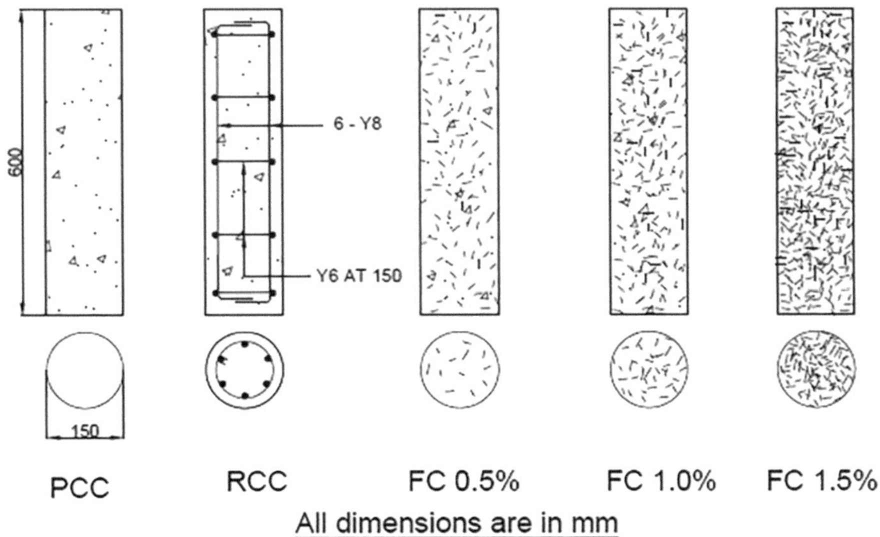

Fig. 1 Experimental specimen

fiber concrete, the fiber was added and mixed thoroughly, to endure even distribution. A total of six mixes were prepared. Two mixes, with conventional design method and the other three with 0.5, 1, and 1.5% percentage of fiber, by volume of crimped steel fiber. The experimental matrix of five typical specimens is shown in Fig. 1.

2.2 *Material Properties*

The materials used in this study are natural aggregate as coarse aggregate, manufactured sand as fine aggregate, cement, crimped steel fiber, and water. The properties of these materials play a significant role in the required properties of concrete such as strength and workability. The most crucial component of concrete, which serves as a binding agent for both fine and coarse aggregate, is cement. Regular Portland cement, grade 43, compliant with IS:8112-2013, was used. Fine aggregate going through an IS sieve measuring 4.75 mm. For this experiment, manufacturer sand was chosen. Sand that has been manufactured from hard granite stone aggregates that have been crushed, sorted, and washed. Zone II-related IS 383-1970 is confirmed. The use of fine aggregate in concrete mixes improves the consistency and workability by keeping the aggregate particles suspended in the cement paste. Coarse aggregate is hard, dense, clean, durable, and strong. A large part of the concrete volume (60–80%) is occupied by coarse aggregates. The optimal size of the coarse aggregate used is 20 mm natural gravel. It should have a high crushing value and lowest porosity. A coarse aggregate used is roughly cubic and flaky pieces is avoided.

Fig. 2 Crimped steel fiber

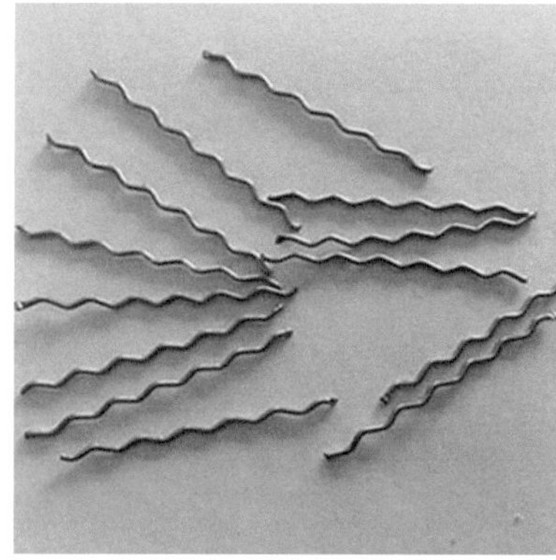

Table 1 Mechanical properties of steel fiber

Type of fiber	d_f[a] (mm)	L_f[a] (mm)	Aspect ratio (L_f/d_f)	Density (g/cm^3)	Tensile strength (MPa)	Elastic modulus (GPa)	Poisson ratio
Crimped steel fiber	0.5	25	50	7.9	2800	200	0.3

[a] d_f Diameter of fiber; L_f Length of fiber (from manufacturer)

Crimped steel fiber is circular, untwisted, and made of steel shown in Fig. 2. The length of each fiber was 25 mm and its diameter was 0.5 mm. The aspect ratio (length/diameter) of the fiber is 50 and used in proportions of 0.5, 1, and 1.5% of the volume of the concrete mass and the properties are given in Table 1. The fiber was mixed well with the concrete matrix and the resulting concrete was cast into the cylinder molds of the short column and cast in molds of diameter 150 mm and length 600 mm to satisfy the short column criteria.

2.3 Axial Load Carrying Capacity

The casting was done after mixing the concrete thoroughly and cast into the greased molds with regular compaction. The concrete short column was demolded after 24 h and was placed carefully in a curing tank. The cylinder specimen was completely immersed in water for 28 days after which it was allowed to dry for 24 h before testing.

The dried concrete columns specimens were loaded along the axis of the short column as shown in Fig. 3 in a universal testing machine (UTM) with a capacity of 100 tones. During the experiment, the column's behavior under load and the pattern of failure were observed. Each short column's final load and deflection were monitored,

Fig. 3 Experimental setup

Table 2 Load deflection values of specimen

Specimen	PCC		RCC		FC 0.5%		FC 1.0%		FC 1.5%	
	Spec1	Spec2	Spec1	Spec2	Spec1	Spec2	Spec1	Spec2	Spec1	Spec2
Ultimate load (kN)	353	389	530	498	431	421	420	401	528	498
Deflection at ultimate load (mm)	2.2	2.4	6.8	6.1	5.2	5.1	3.4	3.3	6.9	6.7

and the findings are shown in Table 2. The ultimate load increased as a result of the steel fiber addition, as can be seen.

3 Analytical Program

The purpose of a finite element analysis is to mathematically restore the behavior of an actual technical system [18]. The structural characterization of the PCC column, RCC column, and the fiber column prototype was discussed here. The details of the finite element analysis and its results will be explained in the following sessions as follows.

In this model of the RCC column, the prototype was designed using the concrete and steel properties for the M20 grade and Fe415, respectively. To model, the column, vertical bars, lateral ties, and steel fiber were formed as the part module as per the dimensions [19]. Properties were assigned to all the elements and the interactions for equal load distribution throughout the column were given. The ABAQUS/CAE components and assemblies can generate meshes using the Mesh module. The modeling aspects at the various level is shown in Fig. 4 and the load values given in Table 3.

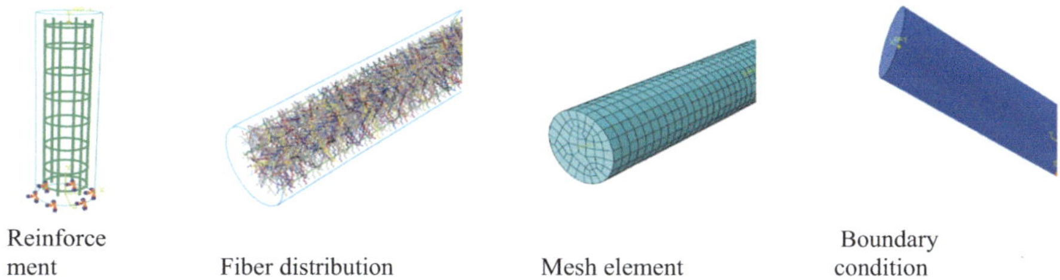

Fig.4 Modeling aspects in ABAQUS/CAE

Table 3 Load values of specimen

Specimen	PCC	RCC	FC 0.5%	FC 1.0%	FC 1.5%
Ultimate load (kN)	337	472	466	473	492

4 Results and Discussion

4.1 Failure Patterns

Experimental The failure patterns of the specimens indicate that the columns except PCC were affected by the shear failure. The lowest point of the crack was always 150 mm above the base of the column. PCC specimens failed with a crushed top and a longitudinal crack perpendicular to the plane of loading showing a brittle failure. The other specimen behaved in the same way starting with a crack at the top edge which continued approximately till the depth of 150 mm above the base resulting in the formation of a shear plane along which the failed. All specimens' failure patterns are displayed in Fig. 5a–e. For the RCC specimen, the first crack appeared at the column's highest point and the fissures at its base appeared when the loading was applied beyond the ultimate load and buckling of reinforcement is also seen. The fiber concrete exhibited a different failure pattern. The failure occurred due to shear along a plain that was present between around 300 mm from the bottom of the column to the top of the column and was inclined at an angle of 45°–50°. These failures in all the concrete specimens were similar and matching with RCC specimens.

Analytical The data acquired from the column are grouped in this section according on the applied axial load. The following figures show that the height of a fully open zone increases with an increase in time interval while loading is constant. Additionally, an explanation for the potential rise in the height of the totally open zone in conjunction with the growing axial loading may be found. The tensile face is the face where this zone is taken into account. On this face of the column, direct stresses resulting from bending that are tensile in character develop as the column is pushed laterally. Over the column's height, the axial load—which is imposed as a pressure load—has a compressive quality. As a result, the value of the compressive

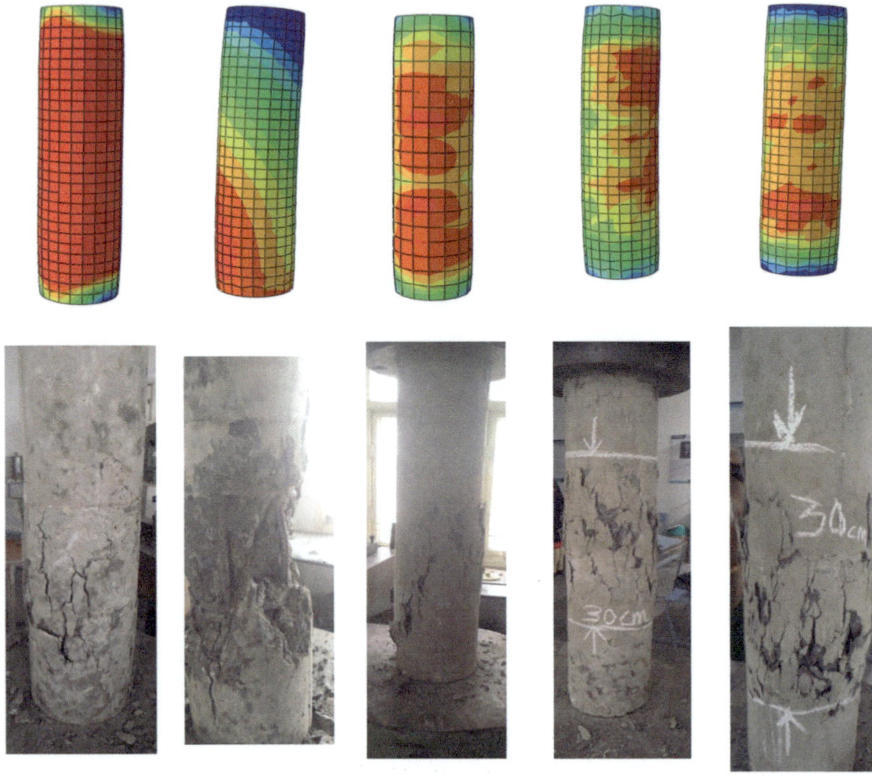

`(a) PCC (b) RCC (c) FC 0.5% (d) FC 1.0% (e) FC 1.5%

Fig. 5 a–e Failure pattern

stress increases together with the axial load ratio. Therefore, the total stress—which is the net sum of the direct compressive stress from the axial load, which is opposite in sign to the tensile stress, and the direct tensile stress from bending—is lowered at a given cross-section. Consequently, cracks only appear in the area of the concrete where the lowered overall stress exceeds the stress needed to cause the concrete to rupture. The images obtained from the analysis are displayed first in each section, followed by a graphical representation of the results based on the variation in time in the images obtained from the analysis. The middle zone is represented by the green color zone, and the total failure zone is shown by the red color zone. The bottom zone of the model is represented by the color blue**.** Figure 5a–e shows the deformed Shape of the specimens both experimentally and analytically.

A graphical representation of the ultimate loads from experiments and finite element analysis for all the specimens are shown Fig. 6.

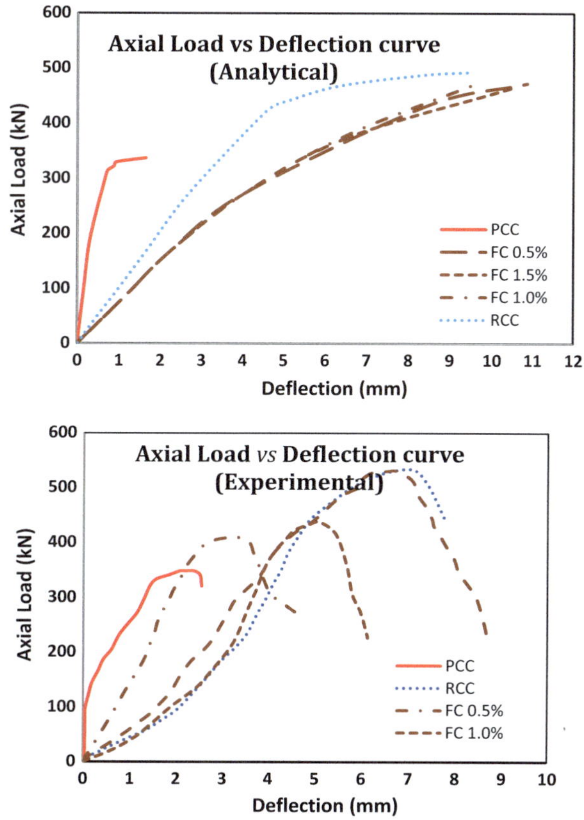

Fig. 6 Axial load versus deflection for all specimens (analytical and experimental)

4.2 Empirical Validation

The theoretical ultimate loads that can be carried by specimens were calculated using codal provisions and theoretical formulas proposed by researchers. For plain cement concrete specimen, it is calculated by the formula proposed by Rankine and for the reinforced concrete column, it is calculated from clause 39.3 of IS 456:2000 [1]. The empirical formula formulated by Bae et al. [20] for calculation of mechanical compressive strength based on different variables such as percentage of steel fiber, water-cement ratio, aspect ratio, percentage of cement and aggregate has been used in this calculation of ultimate load in specimen adding fibers and has been formulated in Table 4.

Table 4 Theoretical load calculations

Specimen	PCC	RCC	FC 0.5%	FC 1.0%	FC 1.5%
Ultimate load (kN)	141	225	168	174	181

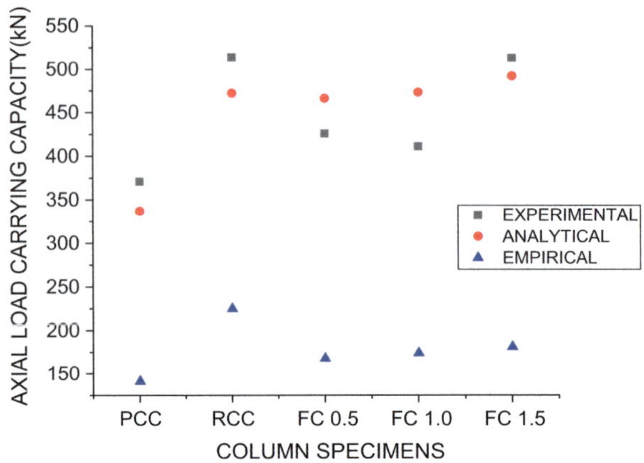

Fig. 7 Comparison of axial load-carrying capacity

5 Conclusions

In this study, the performances of short concrete columns reinforced with crimped steel fiber were investigated experimentally and analytically and compared theoretically is shown in Fig. 7.

- The failure pattern was analyzed and it can be found that steel fiber reinforced the concrete short column replacing the minimum reinforcement.
- Steel fiber incorporation into concrete significantly imparts tensile strength and makes concrete more ductile as observed from failure modes of specimens. The increase in ductility prevents sudden failure and hence can be used effectively in earthquake-resistant structures.
- All the fiber reinforced concrete exhibited shear failure. The lowest point of crack was at least 150 mm above the base of the column where 150 mm is the diameter of the specimens and defines the clear shear cracks.
- The finite element model and the experimental data showed that steel fiber incorporated in concrete can provide better structural integrity.
- The results obtained from the experimental process indicate that when 1.5% by volume of steel fiber is added to concrete in a short column, the behavior is comparable to that of a reinforced concrete short column.
- The empirical pattern from the comparison graph shows it follows a similar pattern from the experimental and analytical results.
- It can be clearly stated that the cost of reinforcement is 20% higher than that of using steel fiber and prove to be an effective alternative when used in a short column.
- The further parametric study, it is possible to define the equation that suggests the structural element's capacity using the established finite element model.

References

1. IS 456 (2000) Plain and reinforced concrete-code of practice. Bureau of Indian Standards, Dehli, pp 1–114
2. Somma G (2008) Shear strength of fiber reinforced concrete beam-column joints under seismic loading. Concrete 2008
3. Rosidawani, Imran I, Sugiri S, Pane I (2015) Behaviour of macro synthetic fiber reinforced concrete columns under concentric axial compression. Procedia Eng 125:987–94. https://doi.org/10.1016/j.proeng.2015.11.152
4. Yuan F, Chen M, Pan J, Khani J, Bozorg M, Zahrai SM (2018) Investigating seismic behavior of reinforced concrete columns with SFRP using finite element method. Compos Part B Eng 176:107272. https://doi.org/10.3844/ajeassp.2018.996.1004
5. Balanji EK, Sheikh MN, Hadi MN (2016) Behaviour of high strength concrete columns reinforced with steel fibres under different eccentric loads. In: 4th Conference, mining and environmental engineering, University of Wollongong
6. Hadi MNS (2009) Reinforcing concrete columns with steel fibres. Asian J Civ Eng 10:79–95
7. Yoo DY, Yoon YS (2015) Structural performance of ultra-high-performance concrete beams with different steel fibers. Eng Struct 102:409–423. https://doi.org/10.1016/j.engstruct.2015.08.029
8. Luhania EN, Goswami EN, Nagar B (2017) Performance of concrete by using steel fibre an experimental study. Int Res J Eng Technol
9. Pang Y, Cai L, Ouyang H, Zhou X (2019) Seismic performance assessment of different fibers reinforced concrete columns using incremental dynamic analysis. Constr Build Mater 203:241–257. https://doi.org/10.1016/j.conbuildmat.2019.01.087
10. Use of steel fibre-reinforcement for impact resistance and machinery foundation. Composites 1980. https://doi.org/10.1016/0010-4361(80)90385-7
11. Atea RS (2019) A case study on concrete column strength improvement with different steel fibers and polypropylene fibers. J Mater Res Technol 8:6106–6114. https://doi.org/10.1016/j.jmrt.2019.10.005
12. Ganesan N, Indira PV, Abraham R (2007) Steel fibre reinforced high performance concrete beam-column joints subjected to cyclic loading. ISET J Earthq Technol 44
13. Shende AM, Pande AM, Pathan MG (2012) Experimental study on steel fiber reinforced concrete for M-40 grade. Int Ref J Eng Sci 1:43–48
14. Vairagade VS, Kene KS (2012) Introduction to steel fiber reinforced concrete on engineering performance of concrete. Int J Sci Technol Res 1
15. Bharti R, Chidambaram RS, Kwatra N (2017) Influence of fiber reinforced concrete on plastic behavior on exterior beam column joint under cyclic loading. Procedia Eng 173:1122–9. https://doi.org/10.1016/j.proeng.2016.12.077
16. Hannant DJ (2003) Fibre-reinforced concrete. Adv Concr Technol. https://doi.org/10.1016/B978-075065686-3/50292-5
17. Plizzari G, Mindess S (2019) Fiber-reinforced concrete. Dev Formul Reinf Concr. https://doi.org/10.1016/B978-0-08-102616-8.00011-3
18. Bae BI, Choi HK, Choi CS (n.d.) FE analysis of behavior of steel fibre reinforced ultra-high strength concrete columns under reversed cyclic loading
19. Bae B, Choi H, Choi C (2012) FE analysis of behavior of steel fibre reinforced ultra-high strength concrete columns under reversed cyclic loading. In: 15th World conference on earthquake engineering, pp 1–8
20. Yazici Ş, Inan G, Tabak V (2007) Effect of aspect ratio and volume fraction of steel fiber on the mechanical properties of SFRC. Constr Build Mater. https://doi.org/10.1016/j.conbuildmat.2006.05.025

Experimental Study on the Hydraulic Jump in Rectangular Sloping Channel

N. Mythri, N. S. Prasanna Rao, V. Y. Muhammad Zain Hasan, Alok Pandey, and Geetha Kuntoji

Abstract Hydraulic jumps are made in open channels to counteract energy dissipation in turbulent flows, which can cause scouring and erosion downstream. This study examines how obstacle height and slope affect the frequency of these jumps. A hydraulic experiment was carried out to compare hydraulic jumps and investigate hydraulic properties, such as variations in water levels, velocities, and energy. In order to manage water resources sustainably, it is essential to comprehend and maximize energy dissipation since it guards against detrimental effects on ecosystem health and channel stability.

Keywords Hydraulic jump · Open channel · Energy dissipation

1 Introduction

Hydraulic jumps happen when water moves from a supercritical (high velocity) to a subcritical (low velocity) condition, increasing the depth of the water. This phenomenon usually happens when there is a sudden shift in the flow depth over a short distance along the length of the channel due to fast changes in flow conditions. By dissipating energy, the jump shields buildings from the damaging effects of high pressure and velocity. Hydraulic jumps can occur as a result of various conditions, including flow velocity/discharge, bed slope, channel form, and surface roughness. The classification of hydraulic jump types into A, B, C, D, E, and F is the result of research on leaps on moderate, steep, and unfavorable slopes.

The experimental phase of this study was conducted in carefully designed laboratory setups that replicated open rectangular channels with slopes subject to controlled

N. Mythri (✉) · N. S. Prasanna Rao · V. Y. Muhammad Zain Hasan
Department of Civil Engineering, B.M.S. College of Engineering, Bengaluru, India
e-mail: mythrin.cv20@bmsce.ac.in

A. Pandey · G. Kuntoji
Faculty, Department of Civil Engineering, B.M.S. College of Engineering, Bengaluru, India

Y. K. Guruprasad et al. (eds.), *Recent Advances in Infrastructure Development and Management—Volume 2*, Lecture Notes in Civil Engineering 566,
https://doi.org/10.1007/978-981-96-1558-2_26

variations. These experiments were meticulously carried out to capture the characteristics of hydraulic jumps under different slope conditions. Parameters such as flow velocity, Froude number, and energy dissipation were calculated from the observed data, providing a set of empirical data crucial for validating theoretical models.

This research is significant as it aims to bridge the gap in our understanding of hydraulic jumps, with a focus on varying slopes in open channels. The empirical and numerical data generated are expected to contribute not only to the academic knowledge base but also to offer practical insights for engineers and researchers involved in the design and maintenance of water conveyance systems.

2 Governing Equations of Hydraulic Jump

Boussinesq's equations form the basis for the one-dimensional rapid flow equations in rectangular channels. In deriving these equations, several assumptions are made: The channel has a prismatic, rectangular cross-section; the velocity is zero at the bottom of the channel and reaches its maximum at the water's surface, or free surface; there is zero velocity perpendicular to the direction of flow; the depth has a consistent distribution of flow velocity; the fluid is incompressible; and the longitudinal bottom or channel bed slope is small. The sudden turbulent transition of water from a supercritical to subcritical state occurs when water moves from a shallow stream to a relatively larger depth. Here, (y_1) represents the initial depth, or the height of the water before the jump, and (y_2) is the sequent depth, or the height of the water after the jump. Figure 1 illustrates the hydraulic jump, with the horizontal distance between these two depths representing the length of the jump (L_J).

2.1 Relationship Between Initial and Sequent Depths

The momentum equation is formulated by assuming a hydrostatic distribution of pressure, ignoring the influence of roughness, and obtaining the conjugate depth

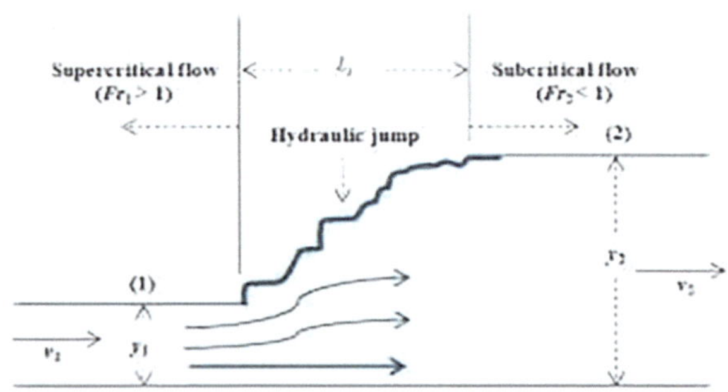

Fig. 1 Physical domain of classical hydraulic jump

ratio between Sections I (beginning of the leap) and II (finish of the jump). This formula is written as.

$$\frac{y_2}{y_1} = 0.5\left(\sqrt{1 + 8G_1^2} - 1\right) \tag{1}$$

where $G1$ is parameter at section I value of which can be calculated using empirical equation [9]:

$$G_1 = (\theta, F_{r,1}) \tag{2}$$

$$G_1^2 = K_1^2 F_1^2 \tag{3}$$

$$Fr, 1 = {V_1}\Big/\sqrt{g y_1 \cos\theta} \tag{4}$$

Here, $F_{r,1}$ is Froude's number of flow and V_1 is average velocity at Section I where the jumps starts and

$$K_1 = 10^{0.027\theta} \tag{5}$$

where θ is angle of channel bed with the horizontal and is in degree [12].

2.2 *Length of the Jump*

The longer distance between the jump's beginning and the point on the level water surface directly downstream of the roller or the point in the subcritical flow area where the streamline separates from the floor is known as the length of the jump (L_J). The water surface continues to increase in an unfavorably sloped channel, therefore the maximum height of the water level cannot be used to indicate the conclusion of the leap. As a result, the jump's end is indicated by the surface rollers' end.

2.3 *Relative Energy Loss*

The difference in total energy before and after the leap, written as ΔE, represents the energy lost in a hydraulic jump. The ratio of energy loss to total energy previous to the leap ($\Delta E/E_1$) is the definition of the relative energy loss, sometimes referred to as jump efficiency. The decrease in flow energy (ΔE) downstream of the structure can be represented as follows when the weight of water in a sloping channel is taken into consideration.

$$\Delta E = (E_1 + L_J \tan\theta) - E_2 \tag{6}$$

$$\Delta E = \left(y_1 \cos\theta + \frac{V_1^2}{2g} + L_J \tan\theta - y_2 \cos\theta - \frac{V_2^2}{2g} \right) \tag{7}$$

3 Experimental Setup and Methodology

The study was carried out at the B.M.S. College of Engineering, Bengaluru, at the Applied Hydraulics Laboratory. A rectangular open channel with dimensions of 8 m in length, 0.397 m in width, and 0.53 m in depth was used in the setup (Fig. 2).

To create a hydraulic leap, the sluice gate was opened to a predetermined depth and the open channel was adjusted to the necessary slope. Measurements were then made of parameters such the starting depth (Y_1), the depth following the leap (Y_2), and the jump's length. With the use of an electromagnetic flow meter, the channel's discharge was documented. By altering the height, sluice gate opening, and channel slope, these procedures were repeated in order to gather data.

Fig. 2 Rectangular open channel 8 m long, 0.397 m wide and 0.53 m depth at Applied Hydraulics Lab

4 Results and Discussion

Table 1 lists the parameters of the hydraulic jump, including discharge, sluice gate opening, obstruction height, slope, initial and sequent depths, and velocities at the start and end of the jump. Table 2 provides the calculated Froude number, specific energy, and relative energy loss percentage at Sections I and II.

The water surface in sloping channels keeps rising even after the leap is over, demonstrating that the jump's termination cannot be precisely marked by the water surface's highest elevation. It was also observed that the subsequent depth ratio (Y_2/

Table 1 Lists the observed values for the following: Length of jump (L_J), initial depth (Y_1), subsequent depth (Y_2), obstruction height (H_1), slope (S_o), sluice gate opening (SG), discharge (Q), and average velocities at the beginning and end of the leap (V_1)

S. No.	Slope	SG (cm)	$H1$ (cm)	$Y1$ (cm)	$Y2$ (cm)	L_J (m)	Q (m^3/s)	$A1$ (m^2)	$A2$ (m^2)	$V1$ (m/s)	$V2$ (m/s)
1	0	6	5	3.3	18	2	0.0276	0.013	0.071	2.107	0.386
2	– 0.002	7	0	4	14.17	1.8	0.0275	0.016	0.056	1.732	0.489
3	– 0.003	6	0	3.5	16.5	1.9	0.0272	0.014	0.066	1.958	0.415
4	– 0.003	7	0	3.8	16.33	1.85	0.0281	0.015	0.065	1.863	0.433
5	– 0.004	6	0	3	16.83	1.45	0.0276	0.012	0.067	2.317	0.413
6	– 0.005	6	0	3.375	16.5	2.11	0.0276	0.013	0.066	2.060	0.421
7	0.005	6	5	4	16.5	2.2	0.0270	0.016	0.066	1.711	0.412
8	0.010	7	5	5	14.5	1.9	0.0280	0.020	0.058	1.411	0.486

Table 2 Calculated Froude's number ($\mathbf{F_{r1}}$ and $\mathbf{F_{r2}}$), and specific energy ($\mathbf{E_1}$ and $\mathbf{E_2}$) in meters of water and relative energy loss in %

S. No.	Slope	Y_2/Y_1	L_J (m)	Fr_1	Fr_2	E_1 (m)	E_2 (m)	ΔE (m)	$\frac{\Delta E}{E_1} \times 100$
1	0	5.45	2	3.70	0.29	0.26	0.19	0.07	27.62
2	– 0.002	3.54	1.8	2.76	0.41	0.19	0.15	0.04	18.69
3	– 0.003	4.71	1.9	3.34	0.33	0.22	0.17	0.05	22.63
4	– 0.003	4.30	1.85	3.05	0.34	0.21	0.17	0.04	17.40
5	– 0.004	5.61	1.45	4.27	0.32	0.30	0.18	0.12	40.59
6	– 0.005	4.89	2.11	3.58	0.33	0.24	0.17	0.07	27.32
7	0.005	4.13	2.2	2.71	0.32	0.20	0.17	0.02	12.45
8	0.010	2.90	1.9	2.01	0.41	0.17	0.16	0.01	7.84

Y_1) increases and that the length of the jump decreases with increasing adverse slopes.

5 Conclusions

Except for one observation, experimental data showed that hydraulic jumps created on positively sloping channels dissipate substantially less energy than those formed on negatively (adverse) sloping channels. It is necessary to do additional study to examine the properties of jumps on sloping floors while accounting for roughness in the channel bed.

Acknowledgements We express our gratitude for the partial funding received from La Fondation Dassault Systèmes for supporting this research project (DSF Project ID: IN-2021-2-07).

References

1. Chow VT (1959) Open channel hydraulics. McGraw-Hill, New York
2. Saghebian AM, Stanciu DD (2020) Hydraulic jump energy dissipation evaluation in rough bed channels via kernel based approach. In: Proceedings of 2nd Cilicia International Symposium in Engineering and Technology, Mersin, Turkey, pp 110–114
3. Mahtabi G, Chaplot B, Azamathulla HM, Pal M (2020) Classification of hydraulic jump in rough beds. Water 12:2249
4. Abbaspour A, Taghavianpour T, Arvanaghi H (2019) Experimental study of the hydraulic jump on reverse bed with porous screens. Appl Water Sci 91(55):1–7
5. Omid MH, Varaki ME, Narayanan R (2010) Gradually expanding hydraulic jump in a trapezoidal channel. J Hydraul Res 45(4):512–518
6. Kindsvater CE (1944) The hydraulic jump in sloping channels. Trans Am Soc Civ Eng 109(1):1107–1120
7. Rajaratnam N (1966) The hydraulic jump in sloping channels. Water Energy Int 23(2):137–149
8. Retsinis E, Papanicolaou P (2020) Numerical and experimental study of classical hydraulic jump. Water 12(6):1766. https://doi.org/10.3390/w12061766
9. Rajaratnam N (1967) Hydraulic jumps. In: Chow VT (ed) Advances in hydroscience, vol 4. Ac ademic Press, New York, NY, pp 197–280
10. RangaRaju KG (1993) Flow through open channels. Tata McGraw Hills, New Delhi, pp 196–198
11. Subramanya K (2009) Flow in open channel. Tata McGraw Hills, New Delhi
12. Mohit Kumar, Lodhi AS (2015) Hydraulic jump over sloping rough floors. ISH J Hydraulic Eng. https://doi.org/10.1080/09715010.2015.1088409

Buckling Behavior of Eccentrically Loaded Cold-Formed Steel Angle Columns

R. Malathy, O. Prashanth, K. Karthick, and M. Kasiviswanathan

Abstract The cold-formed steel (CFS) structures are increasingly being utilized in various civil engineering applications such as railway coaches, transmission towers, bridges and towers. This work focuses on the numerical analysis of axially compressed CFS equal angle sections. To capture the complex behavior of these sections, a non-linear finite element (FE) model was developed using the ABAQUS. The FE model incorporated geometric imperfections and material non-linearity, enhancing its accuracy and reliability. To validate the FE model, published experimental test data were employed for comparison. The results show good agreement between the FE model and the experimental data, confirming the model's effectiveness in capturing the behavior of axially compressed CFS equal angle sections. Additionally, a parametric study was conducted to investigate the influence of eccentricity on the strength of CFS equal angle sections. By varying this parameter within a range of values, the study aimed to identify their impact on the structural performance of the sections. The obtained data were analyzed, and a design curve was developed based on the results. This design curve serves as a valuable tool for preliminary designers, providing insights into the axial strength behavior of CFS equal angle sections under different eccentricity.

Keywords Angle · Cold-formed steel · Direct strength method · Edge stiffeners · Finite element analysis

R. Malathy · O. Prashanth · M. Kasiviswanathan (✉)
Department of Civil Engineering, Sona College of Technology, Salem, India
e-mail: kasiviswanathan.civil@sonatech.ac.in

K. Karthick
Department of Mechanical Engineering, Thiagarajar Polytechnic College, Salem, India

Y. K. Guruprasad et al. (eds.), *Recent Advances in Infrastructure Development and Management—Volume 2*, Lecture Notes in Civil Engineering 566,
https://doi.org/10.1007/978-981-96-1558-2_27

1 Introduction

The cost-effectiveness and long-lasting durability of cold-formed steel (CFS) sections have contributed to their growing prominence in the construction industry. In contrast to hot-rolled sections, cold-formed sections (CFS) present a multitude of benefits, including an increased range of cross-sectional options, a greater variety of strength-to-weight ratios, decreased weights, enhanced corrosion resistance and simplified fabrication processes. Nevertheless, the extensive implementation of CFS sections is constrained by deformation phenomena. As a result, scholars have initiated inquiries to comprehend the behavior of angle columns made of cold-formed steel.

Under eccentric and axial loading conditions, Propovic et al. [1] performed compression experiments on CFS sections at equal angle. Young [2] conducted compression experiments on columns comprised of plain angle sections made of fixed–fixed cold-formed steel. Young [3] examined CFS-lipped angle columns under fixed–fixed constraints and concentric loading. Ellobody and Young [4] created a robust and useful non-linear Finite Element Model (FEM) for CFS angle section columns. A process based on the Direct Strength Technique (DSM) was devised by Silvestre et al. [5], who investigated the behavior of equal angle short and medium CFS columns. Landesmann et al. [6] concentrated on pinned end edges of CFS short and intermediate equal angle columns.

Fasoulakis et al. [7] conducted experimental and computer-based research to explore the behavior of CFS single bolted angles under tension and compression. Their investigation specifically addressed the buckling behavior of CFS equal angle columns, considering flaws in the structural assembly [8]. Behzadi-Sofiani et al. [9] investigated CFS columns with fixed end boundaries and equal angles. Young and Ellobody [4] studied the buckling behavior of CFS angle columns with equal-lipped sides, employing Finite Element Analysis (FEA) to evaluate the accuracy of North American specification design guidelines for CFS unequal lipped angle column sections [10]. Young and Chen [11] conducted experiments on asymmetric CFS-lipped angle section columns.

Nevertheless, there is a scarcity of research regarding the application of edge stiffeners in CFS angle section columns. Although compression experiments were performed by Yan and Young [12] on CFS sections featuring complex stiffeners under fixed–fixed boundary conditions, the focus of their investigation was on channel sections as opposed to angle sections. A comparison was made between the experimental strengths and the AISI [13] and AS/NZS [14] standards. It was determined that the strengths computed using the AS/NZS [14] standard were overly conservative. In addition, Yan and Young [15] constructed a Finite Element Model (FEM) to represent CFS channel sections incorporating intricate stiffeners. Aruna et al. [16] conducted an experimental trial to assess the axial compression capability of complex edge stiffeners in CFS equal angle sections. To do so, they evaluated the applicability of the current direct strength method (DSM) by just posing experimental findings with those obtained from Finite Element Analysis (FEA) utilizing the DSM.

While numerous studies have examined the behavior of CFS angle columns, the influence of eccentricity on the buckling strength of ordinary angle columns has received notably less attention. Additionally, a parametric study was conducted to examine the impact of load eccentricity on the buckling strength of FRP columns. This was achieved through the utilization of numerical analyses conducted with the finite element software ABAQUS. The experimental results obtained from the literature were utilized to validate the finite element (FE) model, which revealed a satisfactory level of concurrence with the failure loads. The investigation utilized generated graphs to assess the influence of load eccentricity on buckle behavior.

2 Finite Element Analysis

Finite element combination (ABAQUS) was employed to evolve finite element (FE) models for cold-formed steel (CFS) disparate angle section under eccentric loading compression. Eigen value analysis and linear static analysis was carried out by using FE software ABAQUS. By using geometric imperfection and the subsequent analysis, buckling modes are identified. The accuracy of the FE models was confirmed by analyzing results from the literature.

2.1 Geometrical Model

The length and cross section of CFS angle column is shown in Fig. 1.

2.2 Material Properties

For the analysis of CFS column, CFS material properties have been used as shown in the Table 1.

2.3 Loading and Boundary Conditions

For the column, a pin-ended boundary condition was employed. To simulate this condition, displacement was allowed at the upper end of the reference point. The center of gravity (CG) of the cross-section served as the reference point. Loading was applied through a designated reference point (RP-2), employing buckle for linear analysis and static Riks for non-linear analysis. The boundary and loading conditions are illustrated in Fig. 2.

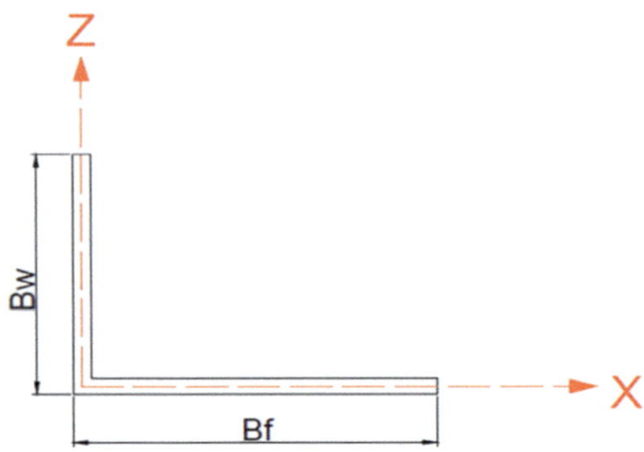

(a) Cross-section notations

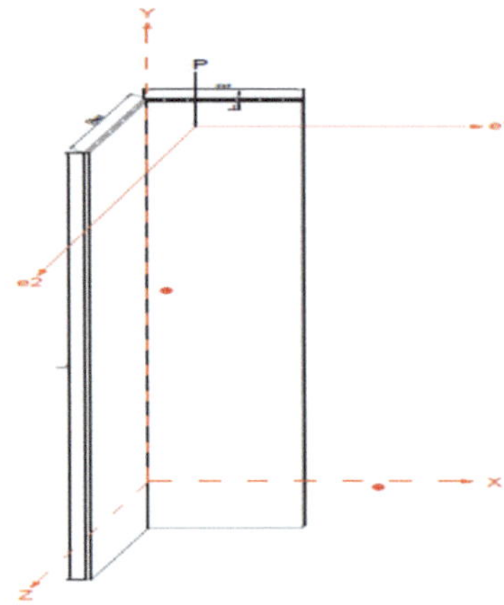

(b) Nomenclature of the specimens

Fig. 1 Geometry details

Table 1 Material properties

E (GPa)	*G* (GPa)	v	μ
200	75	0.785	0.3

Fig. 2 Boundary and loading conditions

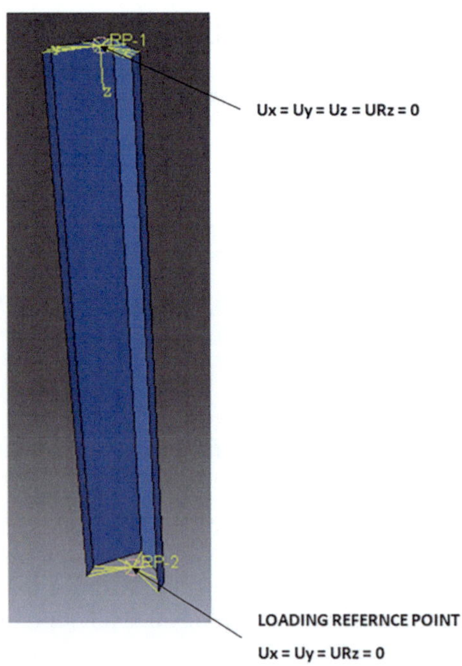

Fig. 3 FE mesh diagram

2.4 Element Type

The CFS angle column was modeled using a four-noded, doubly curved S4R shell element (edge stiffener). Figure 3 shows the meshed finite element model. To ensure a proper balance between the accuracy of the model and the computation time, a mesh size of 5 × 5 mm has been used.

2.5 Buckling Mode Shape

Static non-linear analysis was used in all cases to determine the buckling coefficient. Figure 4 depicts the buckling behavior of various sections.

2.6 Verification of Finite Element Model

When establishing constraints, material properties and mesh sizes finite element modeling requires additional care. Before the FE model can be employed in the analysis, its accuracy must be verified. Hence, results provided by En-Feng Deng et al. [17] were used to verify the precision of the present CFS column models. The comparison of the buckling findings is shown in Table 2. The experimental results of En-Feng Deng et al. [17] and the numerical model buckling strength predictions of the current work are similar. Hence, a parametric analysis with the specified sectional shape and material properties is performed in this paper.

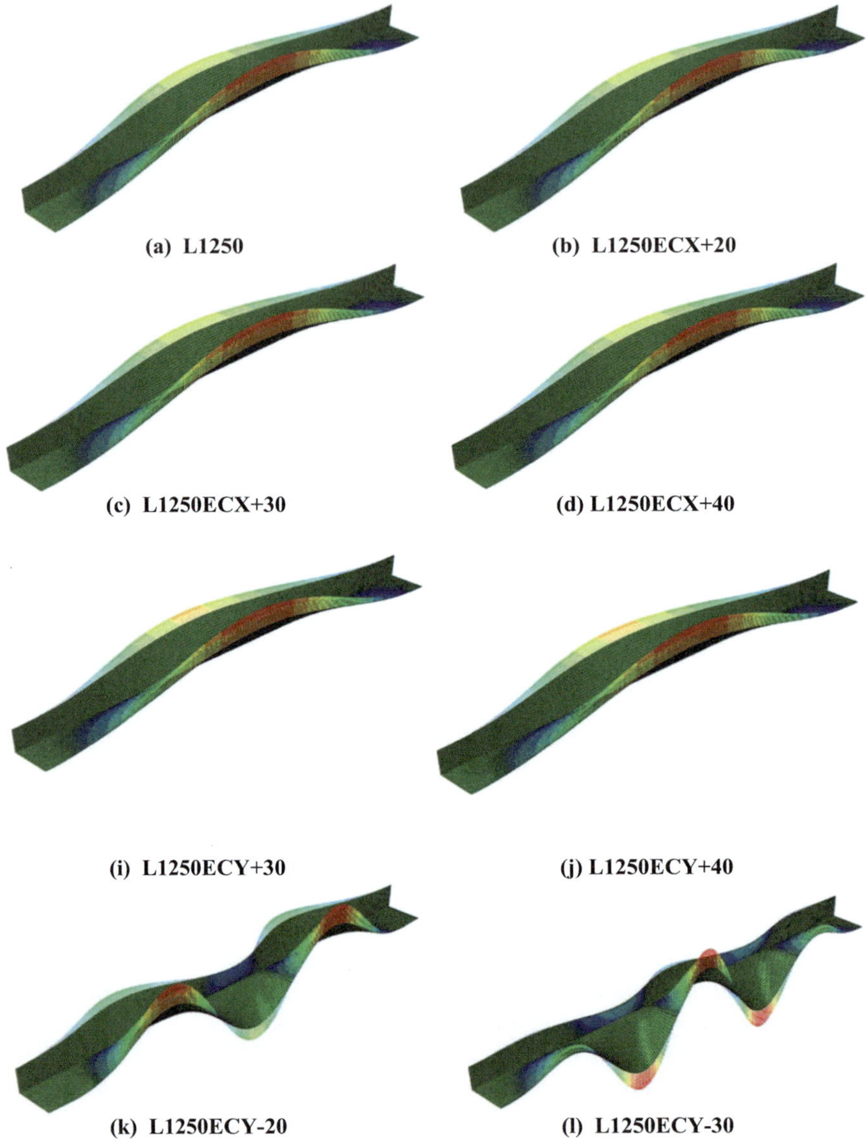

Fig. 4 Buckling mode shapes

3 Parametric Study

To assess the influence of eccentricity on axial strength of CFS angle section, parametric analysis was carried out that involved the use of 48 different FE models. The parametric analysis considered three distinct configurations of equal angle sections.

Table 2 Verification of finite element model (mm)

Series	Specimen	B_{f1}	B_{f2}	t	L	P_{FEA} (kN) a	P_{FEA} (kN) b	% difference $(a - b)/a$ * 100
L500	L500ECX+30	210.7	150.0	2.78	501.7	258.72	242.76	6.169
	L500ECX-30	210.0	149.7	2.86	496.0	200.51	215.42	– 7.436
	L500ECY+30	209.7	150.0	2.78	494.0	266.03	280.58	– 5.469
	L500ECY-30	210.0	150.0	2.84	490.7	243.85	231.69	4.987

The considered geometry as shown in Fig. 5. The width of web, width of flange and thickness of the specimens were kept constants for all specimens. Table 3 shows the dimensions and length of the specimens.

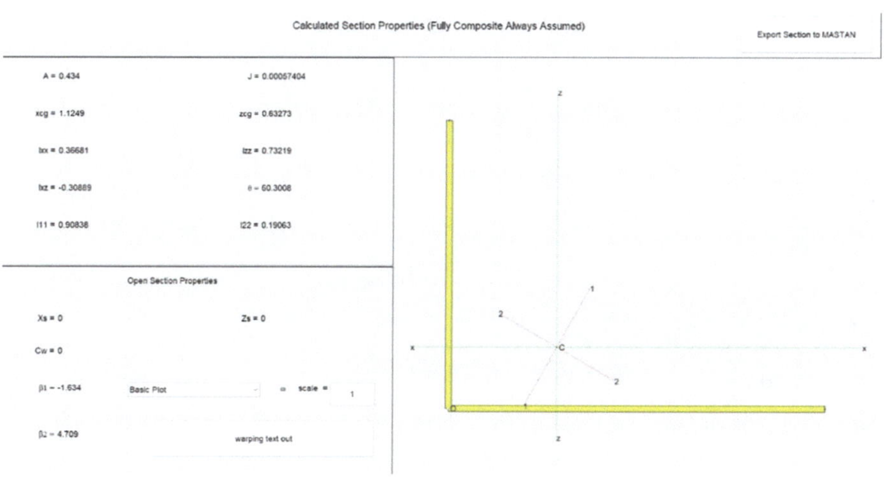

Fig. 5 Cross-sectional shape of the angle sections

Table 3 Geometry details

Parameter	Details
Eccentric direction	ECX, ECY
Thickness (mm)	1.6
Width of Flange (mm)	100
Width of Web (mm)	75
L_1, L_2, L_3 (mm)	500, 1250, 1750
Eccentricity (mm)	– 40, – 30, – 20, + 20, + 30, + 40

4 Results and Discussion

From Figs. 6 and 7, it can be observed that the buckling strength decreases as the eccentricity ($e1$ and $e2$) increases in X-direction and Z-direction, respectively.

The below graph represents the relation between length (L), eccentricity and buckling load of the specimen. Buckling load is very high in the negative eccentric direction and low in the positive eccentric direction (Fig. 8).

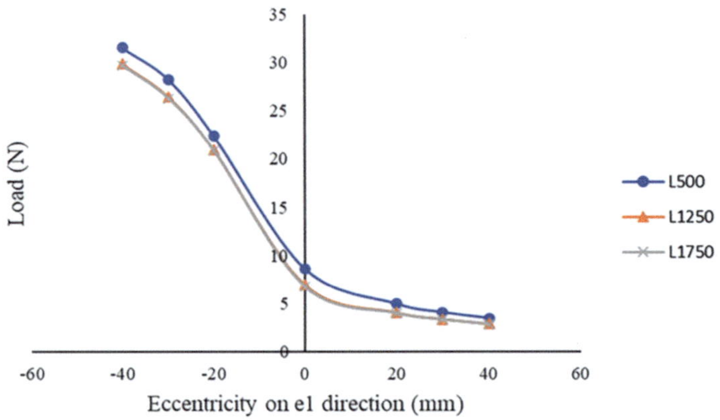

Fig. 6 FEA predicted ultimate loads versus eccentricity of the CFS specimen in X-direction

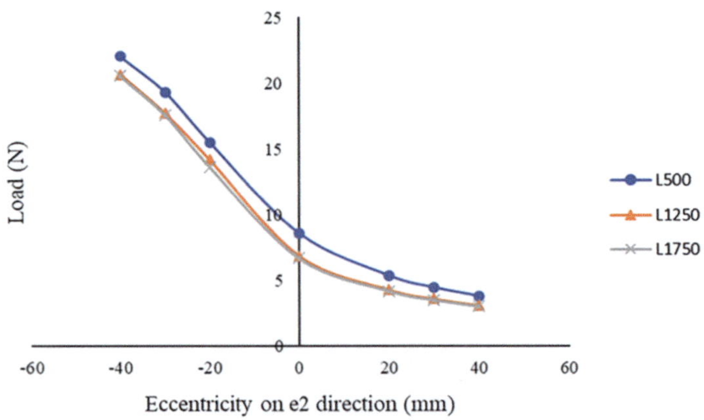

Fig. 7 Predicted ultimate loads versus eccentricity of the CFS specimen in Z-direction

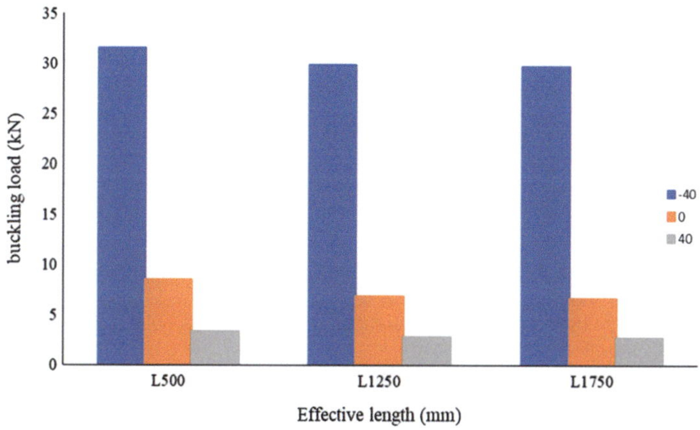

Fig. 8 Relation between length (*L*), eccentricity and buckling load of the specimen

5 Conclusion

Cold-formed steel angle sections subjected to axial and eccentric compression on both positive and negative eccentricity was studied. The following conclusions were drawn.

- Buckling strength decreases with an increase in eccentricity in the positive direction, with the rate of decrease being marginal, indicating a trend toward the lower rigidity of the column sections. Specifically, in the L500 case, the critical buckling strength decreased by approximately 60% compared to the axial load case. Similarly, for both the L1250 and L1750 cases, the critical buckling strength decreased by about 58% compared to the axial load case.
- Buckling strength increases significantly with an increase in eccentricity in the negative direction, indicating a trend toward higher rigidity of the column sections. Specifically, in the L500 case, the critical buckling strength increased by approximately 3.6 times compared to the axial load case. For the L1250 case, the critical buckling strength increased by about 4.3 times, and for the L1750 case, it increased by around 4.5 times compared to the axial load case.

Nomenclatures

L	Length of the column
B_w	Web width
B_f	Flange width
t	Thickness

K	Buckling strength
$e1, e2$	Eccentricity directions in X and Y axis

References

1. Popovic D, Hancock GJ, Rasmussen KJR (2001) Compression tests on cold-formed angles loaded parallel with a leg. J Struct Eng 127(6):600 607
2. Young B (2004) Tests and design of fixed-ended cold-formed steel plain angle columns. J Struct Eng 130:1931–1940
3. Young B (2005) Experimental investigation of cold-formed steel lipped angle concentrically loaded compression members. J Struct Eng 131(9):1390–1396
4. Ellobody E, Young B (2005) Behavior of cold-formed steel plain angle columns. J Struct Eng 131:457–466
5. Silvestre N, Dinis PB, Camotim DPB (2013) Developments on the design of cold-formed steel angles. J Struct Eng 139:680–694
6. Landesmann D, Camotim PB, Dinis RC (2017) Short-to-intermediate slender pin ended cold-formed steel equal-leg angle columns: experimental investigation, numerical simulations and DSM design. Eng Struct 132:471–493
7. Fasoulakis ZC, Raftoyiannis IG, Avraam TP (2017) Experimental and numerical study on single-bolted cold-formed angles under tension and compression. Front Built Env 3(75)
8. Fasoulakis ZC, Lignos XA, Avraam TP, Katsatsidis SP (2019) Investigation on single- bolted cold-formed steel angles with geometric imperfections under compression. J Constr Steel Res 162
9. Behzadi-Sofiani B, Gardner L, AhmerWadee M, Dinis PB, Camotim D (2021) Behaviour and design of fixed-ended steel equal-leg angle section columns. J Constr Steel Res 182
10. Young B, Ellobody E (2007) Design of cold-formed steel unequal angle compression members. Thin-Walled Struct 45(3):330–338
11. Young B, Chen J (2008) Column tests of cold-formed steel non-symmetric lipped angle sections. J Constr Steel Res 64(7–8):808–815
12. Yan J, Young B (2002) Column tests of cold-formed steel channels with complex 354 stiffeners. J Struct Eng 128(6):737–745
13. AISI (1996) North American specification for the design of cold-formed steel structural 316 members, AISI, Washington, DC
14. AS/NZS (2018) Australia/New Zealand standard: cold-formed steel structures, AS/NZS 319 4600, Sydney Australia
15. Yan J, Young B (2004) Numerical investigation of channel columns with complex 357 stiffeners part I: test verification. Thin-Walled Struct 42(6):883–893
16. Aruna G, Roy K, James BP, Suresh Kumar P (2022) Test, numerical simulation, and design of cold-formed steel equal angle with complex edge stiffener under axial compression. Pract Period Struct Des Constr 27(2)
17. Zhang JF, Wang B, Wang S, Deng EF, Zhang P (2021) Eccentric compressive distortional buckling and design of non-symmetric cold-formed angular column with complex edges. Thin Walled Struct 165

Study on Water Management by Using IoT and Machine Learning Applications

R. A. Ganapathy Raju and R. Malathy

Abstract To protect our water resources, water quality must be monitored, but conventional evaluation techniques are frequently expensive and time-consuming, relying on recurring sampling that might not provide accurate or timely data. The application of machine learning techniques to water quality monitoring is reviewed in this study, with a focus on how these techniques can enhance assessment timeliness, accuracy, and efficiency. A top priority for the world is ensuring that everyone has access to clean water. Regular monitoring of variables like pH, turbidity, and contaminant levels is crucial for determining the safety and health of water sources. This work investigates how machine learning techniques can enhance the analysis and prediction of these important water quality indicators. Rainwater harvesting is emerging as a viable solution to address water scarcity and enhance resource management in a number of regions. Traditional rainwater harvesting systems collect and use water inefficiently because they are passive and reliant on weather forecasts. This research endeavors to explore the incorporation of machine learning methodologies into rainwater harvesting systems with the ultimate objective of streamlining the water collection, storage, and distribution procedures. Predicting annual rainfall accurately is essential for agriculture, emergency preparedness, and the management of water resources. Rainfall has complex, nonlinear patterns that may be difficult for conventional forecasting techniques to fully capture because they frequently rely on historical averages and meteorological data. This paper presents a novel, data-driven approach to increase accuracy and reliability by conducting a thorough analysis of applying machine learning techniques to annual rainfall prediction.

Keywords Machine learning · Water quality monitoring · Rainwater harvesting · Water quality parameters · Rainfall prediction

R. A. Ganapathy Raju (✉)
Construction Engineering and Management, Sona College of Technology, Salem, Tamil Nadu, India
e-mail: ganamukesh05@gmail.com

R. Malathy
Department of Civil Engineering, Sona College of Technology, Salem, Tamil Nadu, India
e-mail: malathycivil@sonatech.ac.in

Y. K. Guruprasad et al. (eds.), *Recent Advances in Infrastructure Development and Management—Volume 2*, Lecture Notes in Civil Engineering 566,
https://doi.org/10.1007/978-981-96-1558-2_28

1 Introduction

1.1 Water Quality Monitoring

One of the most precious resources on Earth is water, which is necessary for life support, ecosystem maintenance, and a variety of human endeavors. Water resources are under threat from pollution, climate change, and population growth, so it's critical to protect their quality and accessibility. Effective water quality monitoring is essential for maintaining public health and environmental integrity. In this regard, the incorporation of machine learning into water quality monitoring systems presents itself as a viable path, providing cutting-edge potential to transform the evaluation, oversight, and safeguarding of our water resources. Machine learning is a subfield of artificial intelligence that enables computers to predict future occurrences and learn from data. This technology allows for the automated analysis of vast and complicated datasets from numerous sources, which has the potential to totally change the way traditional water quality monitoring is done. These sources include remote sensing, historical records, and water quality sensors, all of which add to a multifaceted and thorough understanding of aquatic systems. This abundance of data can be sorted through by machine learning models, which can identify trends, anomalies, and patterns that would be missed by human analysis.

1.2 Water Quality Parameter

An essential part of industrial processes, public health, and environmental stewardship is water quality assessment. Monitoring water qualities including pH, flow rate, turbidity, and pollutant levels is crucial to guaranteeing the sustainability and safety of water resources. Conventional techniques, which rely on chemical analysis and manual data collection, are frequently labor-intensive, time-consuming, and have a limited capacity to provide information in real-time. However, the combination of sensor technology and machine intelligence has ushered in a new age in the monitoring of water quality. We can now use data-driven solutions to tackle the issues of the twenty-first century thanks to this convergence. Machine learning is a subfield of artificial intelligence that has gained popularity due to its capacity to analyze vast volumes of data, recognize patterns, and make amazingly accurate predictions. This journal paper explores the relationship between machine learning and water quality monitoring, with particular attention to parameters such as turbidity, pH, flow rate, and contaminants. Environmental science and water resource management have found machine learning to be a valuable tool due to its many benefits for monitoring and controlling water quality parameters. Improved precision, real-time tracking, early warning systems, pattern identification, optimized sampling, automated data processing, preservation of the environment, and data integration are some of the main advantages.

1.3 Rainwater Harvesting System

Rainwater harvesting is the centuries-old practice of gathering and preserving rainwater for use in industry, agriculture, and residential settings. The importance of rainwater harvesting systems has become more apparent due to the depletion of freshwater resources and the increasing effects of climate change. In addition to offering a sustainable supply of water, these systems also reduce the risk of flooding, lessen water scarcity, and promote ecologically friendly behavior. However, the use of cutting-edge technologies, like machine learning, is becoming more and more necessary to optimize the effectiveness, performance, and dependability of rainwater harvesting systems. As a branch of artificial intelligence, machine learning has proven to be exceptionally adept in managing and interpreting massive datasets, generating forecasts, and streamlining intricate systems. When it comes to rainwater harvesting, machine learning has the power to completely change how systems are designed, run, and maintained. Rainwater collection, storage, and distribution can be made more efficient with machine learning by utilizing sensors, data analytics, and predictive models. It can adjust to shifting weather patterns, improve the quality of the water, and adjust how much water is used for different purposes. This journal article examines the relationship between machine learning and rainwater harvesting, emphasizing how artificial intelligence can revolutionize these systems' sustainability. Examined are the techniques, algorithms, and applications that enable machine learning for rainwater harvesting, with a focus on the possible advantages in terms of sustainability, resilience, and efficiency. Rainwater harvesting is a practice well known for its sustainability and good effects on the environment. Machine learning provides rainwater harvesting with a number of advantages.

1.4 Machine Learning

Algorithms are trained on data in machine learning, a kind of artificial intelligence, so they may anticipate or decide without explicit programming. It uses techniques such as supervised, unsupervised, and reinforcement learning to analyze patterns and improve performance over time is given in Fig. 1. In the context of water management, machine learning can optimize water usage, predict water demand, and enhance water quality monitoring. By analyzing data from sensors and environmental sources, it helps in efficient irrigation, leak detection, and ensuring safe drinking water through predictive maintenance and anomaly detection.

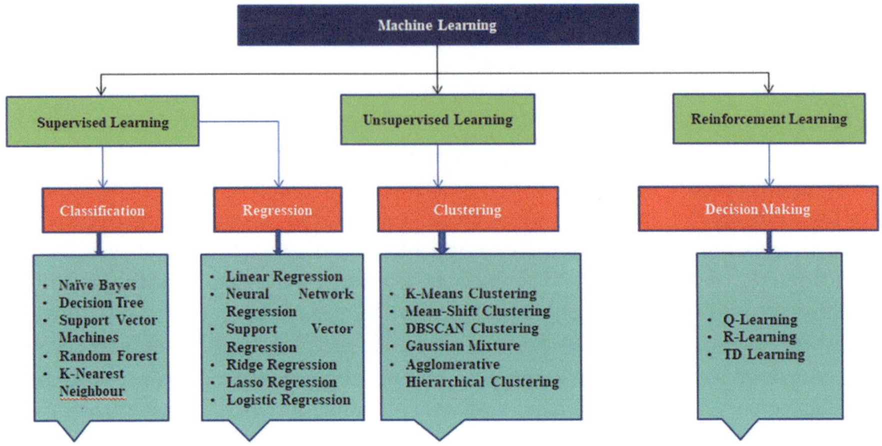

Fig. 1 Machine learning and its types

2 Related Work

2.1 Water Quality Monitoring Using Machine Learning

Unveiled a methodology for keeping an eye on rural areas' water quality. This low-cost project integrates government agencies and mobile network operators while addressing the technical and financial aspects of the system through the use of sensor cloud technology and embedded sensor systems. The system can initiate direct communication with government agencies to initiate appropriate action in response to the seriousness of water quality issues [1].

Introduced a cutting-edge system for predicting water quality in real time. Their goal is to create a sensor-based system that connects to the Internet of Things (IoT) via cloud computing architecture, making sensor data available to everyone [2]. In this paper, we discuss how advancements in communication technology have led to an increasing need for intelligent solutions in the field of water quality monitoring. This article provides a thorough overview of recent developments in smart water quality monitoring [3]. The Water Quality Class (WQC), a specific classification derived from the WQI, and the Water Quality Index (WQI), a comprehensive measure of overall water quality, are the subjects of this study's investigation of several supervised machine learning algorithms. The four input parameters that the method relies on are total dissolved solids, pH, turbidity, and temperature. Among the several methods that are employed, gradient boosting with a learning rate of 0.1 and polynomial regression with a degree of 2 are the most effective predictors of WQI, with mean absolute errors (MAE) of 1.9642 and 2.7273, respectively. Concurrently, the MLP model (3, 7) exhibits remarkable performance in WQC classification, with an accuracy of 0.8507. This proposed approach demonstrates its potential with impressive accuracy using a limited set of parameters [4].

The primary objective of this research is to use machine learning (ML) and the Internet of Things (IoT) to develop a cost-effective system for monitoring water quality in real time. The physical and chemical characteristics of water, such as its temperature, water level, moisture content, humidity, and visibility, are measured using specialized sensors. The ESP8266 central controller processes the data gathered from several sensors [5]. This paper proposes a low-cost Internet of Things, cloud computing, and machine learning monitoring system for water quality. By reducing the risk of serious illnesses like fluorosis and bone abnormalities, this system is a vital safeguard for people's health in rural areas and a contemporary alternative to conventional methods of quality assessment [6].

An intelligent water quality monitoring system is presented in this study. Over the course of a 12 h period, tests were performed on four different water sources at hourly intervals in order to validate the measurement accuracy of the system. The obtained results are consistent with what was anticipated based on previous research. This is made possible by the successful implementation of GSM technology. Additionally, classifiers will be developed using the parameter references obtained from different water sources, simplifying automated water analysis using neural network analysis [7].

The prediction of water quality using various data mining techniques and procedures in various places is evaluated in this research. It reviews current assessment techniques and provides a comparison of various water quality prediction algorithms. The analysis demonstrates how common Artificial Neural Networks are in this domain [8].

The main focus of this work is on designing a Smart Water Monitoring System (SWMS) to track water quality and usage in real time. It uses a three-tier billing framework to generate consumption bills based on water usage. In order to evaluate the quality of the drinkable water provided to clients, the Smart Water Quality meter simultaneously monitors five crucial water quality parameters: pH, temperature, turbidity, dissolved oxygen, and conductivity [9]. In this study, we evaluated artificial intelligence methods for the Tireh River in Iran, such as GMDH, SVM, and ANN, in order to predict various aspects of water quality. We collected a large dataset that included many well-known elements, including pH, SO_4, Na, Ca, Cl, Mg, HCO_3, and more. Furthermore, an examination of the applied models in comparison revealed that the GMDH and SVM models produced more dependable results than the ANN model [10].

Given the intrinsic ambiguity of assessing water quality, using a fuzzy comprehensive evaluation with a membership function offers a more logical and impartial way to evaluate water quality while recognizing its inherent imprecision. The application of the entropy weight technique, which is based on sample data, provides an unbiased depiction of each index's impact on the assessment of water quality [11]. In this study, we used data from the Kinta River to determine the most suitable classification model for categorizing water quality. The results of the experiment showed that the Lazy model with the KStar algorithm generated an amazing accuracy rate of

86.67%. These findings unequivocally demonstrate that the KStar algorithm is the most suitable choice for classifying the water quality of the Kinta River in Perak, Malaysia [12].

The research makes use of Artificial Neural Network (ANN) models that are trained using the Scaled Conjugate Gradient (SCG) on Nonlinear Autoregressive (NAR) time series. Four different ANN models that corresponded to the chosen parameters of water quality were created and carefully evaluated. Performance measures like Mean Squared Error (MSE), Root Mean Squared Error (RMSE), and Regression were used to evaluate the outcomes [13]. Understanding the limitations of traditional models for evaluating water quality, a novel method based on fuzzy neural network theory has been developed to evaluate Suzhou's surface water quality. This new model, which is called the Fuzzy Neural Network (FNN) comprehensive evaluation model, makes use of a neural network that combines competitive and Radical Basis neural networks [14].

Using information from both publicly available data and user-submitted data about the water source, we are attempting to use decision tree technology for the first time. Through adherence to the decision tree's generation process and algorithm, we have developed a decision-making tree model with the objective of forecasting the amount of chlorophyll for the upcoming day [15]. In this study, the Comprehensive Water Quality Index (CWQII) is used to evaluate the Dagu River in the Laixi neighborhood of Qingdao. The assessment's results, as shown by the assessment index, provide important new information about the particular areas of concern. This method offers benefits that are difficult to attain with other approaches, such as the ability to compare different water quality categories and the simplification of comparisons within the same category [16]. Using monthly monitoring data from 2004 to 2006 at typical sections of the Three Gorges Reservoir—including places like Huanglingmiao, Badong, Guandukou, and Xiangxi—this paper presents a Bayesian-based stochastic assessment of water quality. The following are the main conclusions: The reservoir's water quality is generally in the favorable range, which corresponds to Environmental Quality Standards for Water Quality (GB3838-2002) Class I–III [17].

In this study, we assessed artificial intelligence approaches for gauging dry desert water quality. 114 samples from six aquifers in Illizi, southeast Algeria, were used to test eight models in scenarios where a modeling approach was used in place of traditional methods. While the second scenario used parameters from sensitivity analysis, the first scenario used all chemical elements for the Water Quality Index (WQI) [18]. This study evaluated different AI techniques for predicting parameters related to desert water quality. The 114 samples that were part of our study were collected at various points in time from six different aquifers in the southeast Algerian state of Illizi. The prediction power of eight artificial intelligence models—ANN, SVM, M5P tree, RSS, RF, AR, and LWLR—was assessed using two different scenarios with varying combinations of input variables [19].

This paper explores the synergy of Machine Learning and IoT to tackle urban water challenges, contributing to the evolution of smart cities. In this framework, IoT establishes a diverse network of interconnected devices enabling seamless communication. Machine Learning imparts intelligence to assess water quality and detect pilferage, with the system delivering crucial information to end-users [20]. The investigation's main goal was to anticipate the quality of the water. To achieve this, we conducted a comprehensive comparison study of multiple Machine Learning (ML) classification models, including ANN, KNN, Rotf, NB, LR, and Ensemble Learning (RF, AdaBoostM1, Stacking, Voting, and Bagging) [21].

2.2 *Water Quality Parameters Using Machine Learning*

The creation of Adaptive Neuro-Fuzzy Inference System (ANFIS) models specifically designed for determining water safety conditions within pipe networks is presented in this paper. When the concentrations of water quality variables in the pipes rise above specified thresholds, these models are activated. These models demonstrated minimal false alarm rates of about 1% and success rates ranging from 92 to 96% during testing [22]. With applications ranging from managing water resource shortages to improving water resource allocation and water quality prediction, machine learning has become a potent tool for tackling problems related to the water environment. Existing algorithms might only be suitable for a limited number of systems due to the complexity of real-world situations in water treatment and management systems [23].

The study discovered that combining real-time sensor data from satellite and monitoring sources with experiment-based machine learning could lead to improved optimization. The partial least squares (PLS), support vector regression (SVR), and deep neural network (DNN) regression models outperformed the traditional models in terms of accuracy [24]. In this study, the parameters of four models—four regression models and four classification models—were improved using the grid search technique. The grid search-optimized GB model beat the other four classification models (RF, XGBoost, AdaBoost, and GB) that were used to predict Water Quality Classification (WQC) with an astounding accuracy of 99.5% [25].

Water quality forecasting accuracy is critical for ecosystem preservation. This model uses machine learning methods and algorithms to predict water quality based on water sample images, making it a useful tool for preliminary water quality assessment. This study's primary focus is on foundational machine learning-based water quality prediction, which evaluates water color and overall quality to determine usability or consumption suitability [26]. To identify the most effective short-term BOD prediction model, we employed the -trimmed ARIMA model. The BOD data from stations 1, 6, 8, and 10, which had notable swings in BOD values, were processed using this method. The objective was to show how well our suggested strategy worked in choosing the best model to forecast BOD for the following year during the same period of time [27]. This study looked into non-traditional input parameters for a random forest (RF) model used to predict the Water Quality Index (WQI). The recently taken into account features successfully replicated WQI, and RF demonstrated better predictive performance, capturing nonlinear features, when five machine learning models were compared [28].

2.3 Rainwater Harvesting System Using Machine Learning

In this research, the effectiveness of a smart Rainwater Harvesting (RWH) system combined with a dual water delivery system is presented and evaluated in order to address the problem of potable water shortage in locations with limited water resources. The study begins by describing the challenges associated with conventional roof-based RWH systems and then offers an architecture for a smart RWH system designed to address these issues [29].

Thc FAO guidelines, which specify ideal limits for Rainwater Harvesting (RWH) based on environmental and human factors, were followed in this study when establishing criteria for different RWH structures. For the majority of RWH structures, an annual precipitation range of 350–390 mm was found to be appropriate; steep slopes were deemed inappropriate [30]. For sustainable urban development, it is essential to address water-energy challenges. Using a Geographic Information System-Based Decision Support (GSBDS) framework, this study designed and evaluated a rainwater harvesting system in hill communities close to Taipei, assessing its hydraulic performance and economic viability [31].

The process of gathering, storing, and utilizing rainwater from rooftops. Whether on homes or other structures is referred to as rooftop rainwater harvesting. Of the alternatives available for storing excess monsoon runoff, subsurface reservoirs stand out as being especially desirable and technically feasible. These reservoirs can hold large amounts of water and take in water from a variety of surface sources [32]. The UK has been a slow adopter of Rainwater Harvesting (RWH), but faster growth is predicted. In this study, three design approaches, including continuous simulation modeling, and simple methodologies are used to evaluate the designs of two new-build RWH systems. 36% and 46% of the demand for water closets (WCs) is met by these systems. Large commercial buildings may have a higher financial viability for RWH systems than smaller residential ones [33]. An inventive Internet of Things (IoT) rainwater harvesting system is presented in this paper. We have incorporated a rainwater classification system during harvesting because we are aware that collected rainwater frequently contains impurities that necessitate expensive filtration [35] (Table 1).

3 Conclusion

In conclusion, the comprehensive review of machine learning applications in various water-related studies provides valuable insights into the diverse ways in which this technology is transforming our approach to water management and environmental science. Accurate prediction models can now be created with greater ease thanks to machine learning. Our capacity to predict important water-related parameters, such as water quality, quantity, and environmental effects, is improved by these models. The integration of machine learning with sensor networks and data analytics enables

Table 1 Related works and methods

Author name	Type of method used	Findings
Kedia [1]	Sensor cloud technology	Monitoring of water quality
Vijayakumar et al. [2]	IoT-enabled sensor system with cloud computing architecture	Monitoring of water quality
Geetha et al. [3]	Internet of Things (IoT) technology	Monitoring of water quality
Ahmed et al. [4]	Polynomial regression, Boasting mean absolute errors, Multi-layer perceptron	Water quality index (WQI)
Ashwini et al. [5]	IoT and ML technologies	Water quality monitoring
Koditala et al. [6]	Cloud computing, Machine learning (ML), and the Internet of Things	Water quality monitoring
Prasad et al. [7]	IoT, Neural network analysis, GSM technology	Water quality monitoring
Singh et al. [8]	Data mining techniques, Artificial neural networks	Water quality monitoring
Jha et al. [9]	SWMS, or Smart water monitoring system	Monitoring of water quality
Haghiab et al. [10]	Techniques for artificial intelligence, such as ANN, SVM, and GMDH	Water quality monitoring
Xing et al. [11]	FCE-EW stands for fuzzy comprehensive evaluation using the entropy weight technique	Water quality monitoring
Muhammad et al. [12]	KStar algorithm	Water quality class
Khan et al. [13]	Models like regression, mean squared error (MSE), root mean square error (RMSE), and artificial neural network (ANN) that use nonlinear autoregressive (NAR) time series	Water quality parameters
Zhu et al. [14]	Network of fuzzy neurals (FNN)	Water quality evaluation
Liu et al. [15]	Decision-making tree	Chlorophyll, Solar radiation and the change of DO (Dissolved oxygen)
Qun et al. [16]	Simple, Highly practical	Comprehensive water quality index
Li et al. [17]	Bayesian-based stochastic assessment of water quality	Environmental quality standards for water quality
Kouadri et al. [18]	Artificial intelligence models: ANN, LWLR, RSS, RF, AR, SVM, M5P tree, and MLR	Index of water quality (WQI)
Aldhyani et al. [19]	Machine learning models NARNET and LSTM; methods KNN, SVM, and Naive Bayes	Index of water quality (WQI)

(continued)

Table 1 (continued)

Author name	Type of method used	Findings
Omambia et al. [20]	IoT and machine learning	Water quality monitoring
Dritsas et al. [21]	Ensemble learning (RF, AdaBoostM1, Stacking, Voting, and Bagging) and machine learning (ML) classification models (NB, LR, ANN, kNN, and RotF)	Water quality monitoring
Mohammed et al. [22]	Neuro-fuzzy inference system	Real-time contamination event detection
Zhu et al. [23]	Machine learning techniques	Water quality
Sagan et al. [24]	Deep neural network (DNN) models, Support vector regression (SVR), and Partial least squares (PLS) regression	Water quality variables
Shams et al. [25]	Regression models: MLP, DT, SVR, GB, XGBoost, AdaBoost, and RF measures include mean squared error (MSE), mean absolute error (MAE), and median absolute error (MedAE)	Fine-tune parameters
Anand et al. [26]	TensorFlow, Convolutional neural networks (CNN), and Keras	Water quality forecasting
Treeratanajaru et al. [27]	α-trimmed ARIMA model	BOD prediction
Wong et al. [28]	Random forest (RF) and the Synthetic minority over-sampling technique (SMOTE)	Water quality index (WQI)
Judeh et al. [29]	Crowdsourcing and automation-based treatment	Smart RWH system
Darabi et al. [30]	RWH structures were established based on FAO guidelines	Annual precipitation
Chiu et al. [31]	Geographic information system-based decision support (GSBDS)	Annual precipitation
Abhinav Trivedi et al. [32]	Sub-surface reservoirs	Excess monsoon runoff
Ward et al. [33]	Continuous simulation modeling method	Rainwater classification system
RinaMaiti et al. [34]	Remote sensor network	Landslide identification
Bhuyan et al. [35]	Internet of Things (IoT)	Potable rainwater

real-time water monitoring. This capability is crucial for the early detection of water quality issues and timely decision-making in response to dynamic environmental conditions. Numerous studies reviewed in this paper focus on the application of machine learning to environmental issues such as water pollution, habitat preservation, and climate change mitigation. These developments might lead to more environmentally friendly methods of managing water resources. The review underscores the importance of interdisciplinary collaboration between experts in machine learning, hydrology, ecology, and environmental science. Such partnerships are essential for comprehensively addressing complex water-related issues. Machine learning facilitates data-driven decision-making processes. It empowers researchers, policymakers, and water resource managers to make informed choices based on empirical evidence and predictive analytics. Although machine learning has a lot of potential for studies pertaining to water, there are a number of issues that need to be resolved, including data quality, model interpretability, and computational resources. Subsequent investigations ought to concentrate on enhancing current models, exploring innovative uses, and advocating for transparency in model results. The diverse applications of machine learning in water-related studies are not limited to specific regions or contexts. They hold global significance, with the potential to benefit communities worldwide by improving access to clean water, managing water resources more effectively, and addressing environmental concerns. In summary, the reviewed studies demonstrate the growing importance of machine learning in addressing water-related challenges.

Acknowledgements The author states that he has no known affiliations or sources of funding that may have influenced the work presented in this manuscript.

Authors Contribution Mr. Ganapathy Raju R A has designed the study and prepared the manuscript and Dr. R. Malathy who helped with evaluating the manuscript.

Funding The authors confirm that there is no source of funding for this study.

Declaration

Conflict of Interest The writer states that he has no known conflicting financial interests that would have appeared to have an impact on the work that was published in this article.

References

1. Kedia N (2015) Water quality monitoring for rural areas-a sensor cloud based economical project. In: 2015 1st international conference on next generation computing technologies (NGCT). IEEE
2. Singh VK et al (2021) Development of fuzzy analytic hierarchy process based water quality model of Upper Ganga river basin, India. J Environ Manag 284:111985
3. Geetha S, Gouthami SJSW (2016) Internet of things enabled real time water quality monitoring system. Smart Water 2(1):1–19
4. Ahmed U et al (2019) Efficient water quality prediction using supervised machine learning. Water 11(11):2210
5. Ashwini K, Vedha JJ, Priya MD (2019) Intelligent model for predicting water quality. Int J Adv Res Ideas Innov Technol 5(2):70–75
6. Koditala NK, Pandey PS (2018) Water quality monitoring system using IoT and machine learning. In: 2018 International conference on research in intelligent and computing in engineering (RICE). IEEE
7. Prasad AN et al (2015) Smart water quality monitoring system. In: 2015 2nd Asia-Pacific world congress on computer science and engineering (APWC on CSE). IEEE
8. Singh P, Deep Kaur P (2017) Review on data mining techniques for prediction of water quality. Int J Adv Res Comput Sci 8(5)

9. Jha MK et al (2018) Smart water monitoring system for real-time water quality and usage monitoring. In: 2018 International conference on inventive research in computing applications (ICIRCA). IEEE
10. Haghiabi AH, Nasrolahi AH, Parsaie A (2018) Water quality prediction using machine learning methods. Water Qual Res J 53(1):3–13
11. Xing Z, Fu Q, Liu D (2011) Water quality evaluation by the fuzzy comprehensive evaluation based on EW method. In: 2011 Eighth international conference on fuzzy systems and knowledge discovery (FSKD), vol 1. IEEE
12. Muhammad SY et al (2015) Classification model for water quality using machine learning techniques. Int J Softw Eng Appl 9(6):45–52
13. Khan Y, See CS (2016) Predicting and analyzing water quality using machine learning: a comprehensive model. In: 2016 IEEE Long island systems, applications and technology conference (LISAT). IEEE
14. Zhu C, Hao Z (2009) Fuzzy neural network model and its application in water quality evaluation. In: 2009 International conference on environmental science and information application technology, vol 1. IEEE
15. Liu B et al (2021) Hydrodynamics and water quality of the Hongze Lake in response to human activities. Environ Sci Pollut Res 1–18
16. Qun M et al (2009) Application of comprehensive water quality identification index in water quality assessment of river. In: 2009 WRI global congress on intelligent systems, vol 1. IEEE
17. Li C, Wang W (2009) Assessment of the water quality near the dam area of Three Gorges Reservoir based on Bayes. In: 2009 first international conference on information science and engineering. IEEE
18. Kouadri S et al (2021) Performance of machine learning methods in predicting water quality index based on irregular data set: application on Illizi region (Algerian southeast). Appl Water Sci 11(12):190
19. Al-Yaari M, Aldhyani THH, Rushd S (2022) Prediction of arsenic removal from contaminated water using artificial neural network model. Appl Sci 12(3):999
20. Omambia A, Maake B, Wambua A (2022) Water quality monitoring using IoT and machine learning. In: 2022 IST-Africa conference (IST-Africa). IEEE
21. Dritsas E, Trigka M (2023) Efficient data-driven machine learning models for water quality prediction. Computation 11(2):16
22. Mohammed H (2018) Predictive analysis of microbial water quality using machine-learning algorithms. J Environ Res Eng Manag 74(1):7–20
23. Zhu M et al (2022) A review of the application of machine learning in water quality evaluation. Eco-Environment Health
24. Sagan V et al (2020) Monitoring inland water quality using remote sensing: potential and limitations of spectral indices, bio-optical simulations, machine learning, and cloud computing. Earth Sci Rev 205:103187
25. Shams MY et al (2023) Water quality prediction using machine learning models based on grid search method. Multimed Tools Appl 1–28
26. Anand MV et al (2023) Water quality prediction using CNN. J Phys Conf Ser 2484(1). IOP Publishing
27. Treeratanajaru W, Watcharamul S, Lipikorn R (2016) Comparison of ANN and SVM for prediction of biochemical oxygen demand in Chaophraya River. IEICE Proceedings Series 61(W2-1-4)
28. Wong WY et al (2022) Water quality index using modified random forest technique: assessing novel input features. CMES-Comput Model Eng Sci
29. Judeh T, Shahrour I, Comair F (2022) Smart rainwater harvesting for sustainable potable water supply in arid and semi-arid areas. Sustainability 14(15):9271
30. Darabi H et al (2021) A hybridized model based on neural network and swarm intelligence-grey wolf algorithm for spatial prediction of urban flood-inundation. J Hydrol 603:126854
31. Chiu Y-R, Tsai Y-L, Chiang Y-C (2015) Designing rainwater harvesting systems cost-effectively in a urban water-energy saving scheme by using a GIS-simulation based design system. Water 7(11):6285–6300

32. Gautam VK et al, Sustainable agronomic practices for improving water productivity in semi-arid areas
33. Bhuyan MGS, Yasmin MA, Remote patient monitoring in FOG computing environment using Bayesian belief network classifier algorithm
34. Ward S, Memon FA, Butler D (2010) Rainwater harvesting: model-based design evaluation. Water Sci Technol 61(1):85–96
35. RinaMaiti S, Mishra L (2018) GIS and sensor based rain water harvesting with artificial intelligence system for free landsliding. Int J Civ Eng Technol 9(6):54–66

Review on Machine Learning for Crack Analysis in Concrete Structures

M. N. A. Gulshan Taj and C. S. Sowmiya Devi

Abstract Cracks in surfaces like concrete structures, pavements, and industrial components present significant challenges concerning safety, maintenance, and structural integrity. Traditional crack detection methods typically involve manual inspections, which are time-consuming, costly, and prone to human error. This research paper introduces an innovative approach to crack detection on surfaces using advanced machine learning techniques. It investigates the use of cutting-edge machine learning algorithms, such as convolutional neural networks (CNNs) and deep learning models, to automate crack identification across various surface materials. The proposed methodology encompasses key steps: collecting data via high-resolution imaging techniques, preprocessing data to enhance image quality and reduce noise, and training deep learning models with annotated crack datasets. Existing literature highlights various methods for automatic crack detection and depth estimation using image processing techniques. This research conducts a thorough survey to pinpoint current challenges and advancements in this field. Relevant research papers on crack detection are carefully selected and reviewed, focusing on the image processing techniques used, research objectives, achieved accuracy levels, error rates, and the diversity of image datasets. This paper offers a comprehensive review of image-based crack detection techniques that integrate image processing and machine learning approaches.

Keywords Crack detection · Machine learning (ML) · Convolutional neural networks (CNN) · Deep learning (DL)

M. N. A. Gulshan Taj · C. S. Sowmiya Devi (✉)
Department of Civil Engineering, Sona College of Technology, Salem, Tamil Nadu, India
e-mail: sowmiyasaravanabhavan@gmail.com

M. N. A. Gulshan Taj
e-mail: gulshantaj.civil@sonatech.ac.in

Y. K. Guruprasad et al. (eds.), *Recent Advances in Infrastructure Development and Management—Volume 2*, Lecture Notes in Civil Engineering 566,
https://doi.org/10.1007/978-981-96-1558-2_29

1 Introduction

Identifying and monitoring cracks in surfaces such as concrete structures, pavements, and industrial components is vital for safety, preserving structural integrity, and reducing maintenance expenses. Conventional crack detection methods typically rely on manual inspections, which are labor-intensive, costly, and prone to human error, potentially undermining the accuracy and reliability of the evaluations. Recent advancements in technology have introduced more efficient alternatives, particularly in the realm of machine learning algorithms is shown in Fig. 1 [1]. This research paper proposes a novel methodology for crack detection utilizing advanced machine learning techniques. The focus is on leveraging state-of-the-art algorithms, including convolutional neural networks (CNNs) and other deep learning models, to automate the identification and analysis of cracks across a variety of surface materials. The methodology detailed in this paper encompasses several crucial steps: collecting data using high-resolution imaging techniques, preprocessing these images to enhance quality and reduce noise, and training deep learning models with meticulously annotated datasets of cracks. These steps are designed to ensure high accuracy in crack detection while minimizing the time and cost involved in traditional inspection methods. A comprehensive review of existing literature reveals numerous approaches to automatic crack detection and depth estimation, primarily employing image processing techniques. This paper conducts an extensive survey to identify and discuss the current challenges and advancements in this field. By carefully selecting and reviewing relevant research papers, the analysis focuses on the techniques used, the objectives of the research, the levels of accuracy achieved, the error rates reported, and the variety of image datasets utilized [2]. The insights gained from this survey aim to provide a thorough understanding of image-based crack detection techniques that integrate image processing and machine learning approaches, highlighting their potential to transform the field of structural health monitoring.

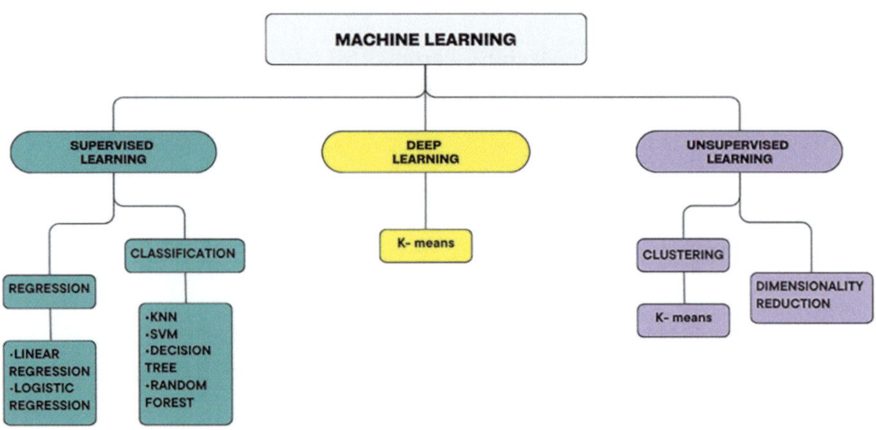

Fig. 1 Machine learning algorithms

Manual methods of crack analysis involve experts who examine the component visually and the use of specific tools to identify any deficiency in the component. Automatic crack analysis deals with using technologies to identify cracks from infrastructures. Some of the primary benefits encompass exceptional accuracy, automation, swiftness, cost-effectiveness, early anomaly detection, continuous monitoring, data integration, remote surveillance, historical data analysis, and minimized downtime. Noteworthy drawbacks involve data quality and quantity issues, intricate model complexity, data privacy and security concerns, difficulties in generalization, model robustness under varying conditions, regulatory and compliance considerations, and the need for ongoing maintenance and updates.

2 Literature Review

2.1 Structural Health Monitoring

Advancements in construction have enhanced infrastructure durability, with concrete being crucial. Despite the resilience of many nineteenth and twentieth century buildings, aging infrastructure is a global concern. The US National Academy of Engineering identifies urban infrastructure restoration as a key twenty-first century challenge. This paper explores sensor and actuator technologies for improving civil infrastructure, introducing IoT and smart structures for structural health monitoring (SHM). SHM now addresses older buildings with innovative technologies and data processing algorithms [3].

Machine learning (ML) applications in SHM for heritage buildings offer approaches for damage diagnosis and retrofitting, particularly when core sampling is impractical. ML methods using non-destructive test data and images reduce maintenance costs and aid preservation, although data availability remains a challenge. Adapting ML algorithms from concrete structures is vital for enhancing predictive capabilities. Research shows ML-based SHM accuracy, but further expansion in non-destructive testing, image processing, and modal-based identification is needed [4].

Comparative analysis evaluates a proposed network against Faster RCNN, focusing on the YOLOv2 detector for automatic crack detection in concrete bridges [5] Laiteerapong and Prendinger [6]. This algorithm also identifies other defects like corrosion and spalling. A novel crack detection approach using a deep fully convolutional network (FCN) achieves around 90% precision in detecting and assessing crack density, showing promise for concrete SHM [6].

The paper reviews crack detection methods, categorizing sensing and data analysis approaches into direct/indirect and model-based/model-free, discussing their merits and challenges as a reference for researchers in crack detection and SHM.

2.2 Machine Learning

Machine learning, a branch of artificial intelligence, allows computers to develop models from data to solve problems, unlike traditional programming, which relies on predefined rules. It is particularly useful for complex tasks where rule-based programming is impractical, provided there is enough relevant data [7]. Machine learning methods are categorized by the level of supervision: supervised, unsupervised, and reinforcement learning. In supervised learning, models are trained with labeled datasets to make predictions on new data, and it is commonly used for classification and regression tasks with algorithms such as k-Nearest Neighbors, Support Vector Machines, Logistic Regression, Linear Regression, and Neural Networks [8].

2.3 K-Nearest Neighbor

The k-nearest neighbors (KNN) algorithm is a simple supervised machine learning method used for classification and regression tasks. It is easy to implement and understand but becomes slower with larger datasets [9]. KNN predicts test data based on the most similar features in the training data, with accuracy generally improving with larger training sets. KNN operates without assumptions about data distribution, offering simplicity, flexibility, and potential high performance. KNN finds the K nearest neighbors for a given query instance in the training dataset using a distance metric, often Euclidean distance. In KNN classification tasks, the query instance is allocated the majority class label among the K nearest neighbors shown in Fig. 2 [10]. KNN averages, or weights, the target values of the K nearest neighbor to obtain the goal value for the query instance while performing regression tasks. Features can be extracted from images of concrete structures containing cracks. However, it has drawbacks, including high classification costs and computational intensity with larger datasets. As a "lazy learner," KNN can handle large datasets but struggles with high-dimensional data. It is used in recommendation systems, medical diagnosis, credit rating, handwriting recognition, financial analysis, and image recognition [11].

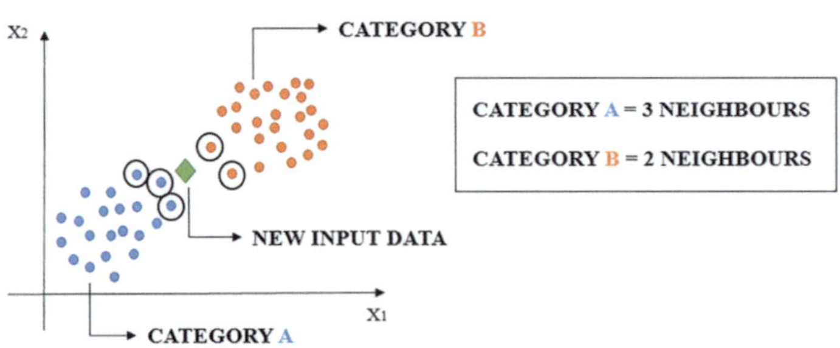

Fig. 2 K-nearest neighbors (KNN) algorithm

2.4 Support Vector Machine

Support Vector Machines (SVMs) are advanced supervised machine learning techniques similar to classical multilayer perceptron neural networks. SVMs work by maximizing the margin between a hyperplane and data classes, which helps minimize the expected generalization error. This method is highly effective for classification problems, using hyperplanes to separate high-dimensional data features and distinguish objects across classes [12]. Originally designed for binary classification, SVMs have been adapted to handle multi-class problems. They ensure that margins maximize the distance between classes while minimizing classification errors, making SVMs a powerful tool for classification tasks. The objective is to minimize classification mistakes and maximize the margin between the classes. The way SVM works is that it takes input data, maps it into a feature with high dimension space, and then determines which hyperplane best divides the data points into distinct classes. SVM finds the best hyperplane to optimize the margin between classes in classification problems [13]. An example of SVM is given in Fig. 3.

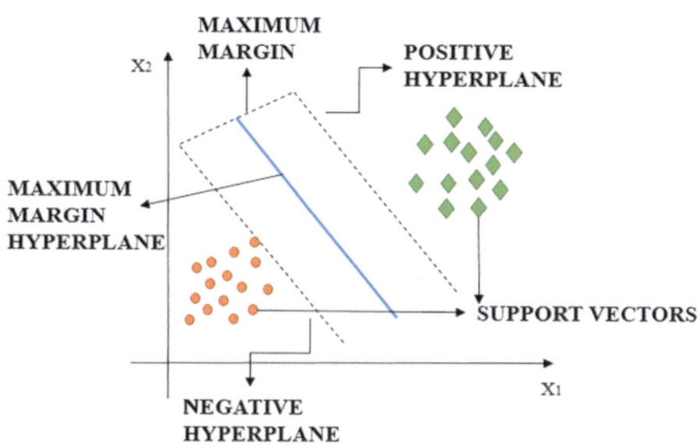

Fig. 3 Support vector machine (SVM) algorithm

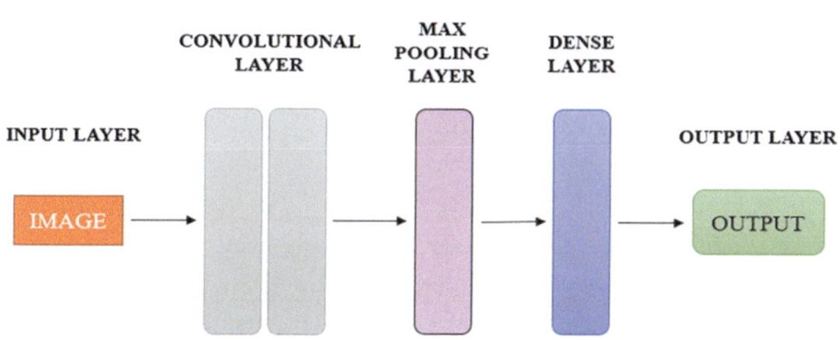

Fig. 4 Convolutional neural networks (CNNs) algorithm

2.5 Convolutional Neural Network

A convolutional neural network (CNN) is a crucial deep learning network used in computer vision and natural language processing, drawing significant attention from industry and academia. This review introduces novel concepts in 1-D, 2-D, and multi-dimensional convolutions, exploring CNN's history, models, applications, and future prospects. A CNN combines deep learning algorithms and artificial neural networks for image processing [14]. It consists of an input layer (the original image), an output layer (classified features), and a multilayer nonlinear hidden structure with convolution and sub-sampling layers for feature extraction and classification. Enhancing the convolutional and single-layer perceptron optimization improves feature extraction and classification [15]. After multiple extraction stages, the feature map is converted into a one-dimensional matrix for the fully connected layer, leading to the final output. Edges, textures, and patterns at various spatial scales are extracted from the input image using convolutional layers by the application of convolution procedures. While maintaining the most crucial information, pooling layers shrink the feature maps' spatial size [16]. The collected characteristics for tasks involving regression or classification are combined in fully connected layers at the conclusion of the network. VGG16 and other pre-trained CNN models can be adjusted for particular tasks using comparatively little datasets because they were trained on large datasets like ImageNet. CNNs automatically learn hierarchical features from raw pixel values, eliminating the need for handcrafted feature engineering [17]. Collect a dataset of images of concrete structures containing cracks, with labels indicating the presence and properties of cracks (e.g., length, width) (Fig. 4).

2.6 K-Means Clustering

The clustering algorithm operates by grouping objects with similar features into distinct clusters, effectively organizing the available data into groups where each cluster contains data with relatively indistinguishable features within a specific distance range. The K-means model is a centroid-based approach with K clusters, where the mean value serves as the center of each cluster [18].

K-means is a straightforward unsupervised learning algorithm designed to address the clustering problem. The process offers a straightforward way to categorize a given dataset into a specified number of clusters. The fundamental concept involves establishing k centers, one for each cluster, and positioning them strategically to maximize separation. The subsequent step involves associating each data point with the nearest center. When all points are assigned, the initial grouping phase is completed. Now, the task is to recalculate k new centroids, which represent the barycenter of the clusters formed in the prior step [19].

2.7 Decision Tree

The decision tree is a supervised machine learning method for classification and regression, partitioning data based on specific parameters. It's advantageous for both regression and classification, offers ease of interpretation, and efficiently handles categorical and quantitative values. In Classification Trees, the decision variable is categorical (resulting in outcomes like Yes/No), while in Regression Trees, the decision variable is continuous. However, it can face overfitting issues, which can be mitigated with Random Forest. Disadvantages include potential instability, challenges in controlling tree size, susceptibility to sampling errors, and local rather than global optima. Applications include predicting library book usage and tumor prognosis [20].

Decision Trees categorize attributes by sorting them based on their values and are primarily used for classification. Each tree is composed of nodes and branches, with each node representing attributes to be classified and each branch representing a value the node can assume [21].

3 Crack Detection by Hybrid Techniques

Cracks in concrete structures can signal serious structural problems and diminish the aesthetic appeal of the construction. The article outlines a novel image processing technique for detecting cracks in photos of concrete structures, involving three critical steps. First, the image is converted to grayscale, and Sobel's edge detection and filter are applied for initial crack detection. Second, a binary image is created by applying an appropriate threshold to distinguish between foreground and background, categorizing small areas as background. Third, the Otsu method is used to identify significant cracks after noise is reduced by Sobel's filtering. Detecting cracks can be challenging, especially if they are small or surface-level. Additionally, factors like temperature fluctuations and the hydration process can result in various crack patterns in concrete. This paper presents an innovative method for detecting crack patterns in cement using image processing techniques. It provides accurate crack detection by converting small areas to background, identifying major cracks using the Otsu method, and using filters like Sobel and the Area filter Talab [22].

Different approaches to crack detection, including deep learning algorithms and image processing techniques. These techniques included median filtering, statistical analysis, Otsu method, and thresholding in image processing. Every technique has benefits and drawbacks of its own. The results of the analysis showed that while some methods provided good accuracy at the expense of slower computational demands, others were quick but less accurate. For real-time processing to occur, accuracy and speed must be balanced. The survey also showed that accuracy and recall were generally higher with deeper neural networks [23]. The significance of crack detection for structural safety and health monitoring. To address the shortcomings of manual crack detection, they developed an intelligent model for automated crack recognition using image processing techniques. This model improves crack detection accuracy by preprocessing image thresholds with the Otsu method and incorporating the Min–Max Gray Level Discrimination (M2GLD) method. Experiments have shown that when combined with shape analysis algorithms and the Otsu method, the model is effective at detecting crack defects in digital images. This approach can benefit building management organizations and construction engineers involved in structural maintenance [24].

A study that effectively identified surface cracks in concrete. They extracted binary crack images using the OTSU method based on differential images and showed its efficacy on various concrete surfaces. They then used a connected domain de-noising algorithm, which worked well on different concrete surfaces and eliminated noise without the need to change any parameters [25]. A method for automated crack detection consists of four main steps: precise registration, pixel-level fusion, preliminary crack image generation, and refined crack extraction. The method fuses 3D point clouds and 2D images using an enhanced Otsu algorithm. It outperforms single-image and single point cloud methods in experiments, achieving an average precision of 89.0%, a recall rate of 84.8%, and an $F1$ score of 86.7%. This technique is a useful tool for automatic concrete crack detection because it has a high detection accuracy and is very versatile. The extraction algorithm used has a major influence on the detection accuracy. This method provides information on the width and depth of cracks as well as strong resistance to interference from staining, making it appropriate for widespread concrete crack detection [26]. A study by used 19 high-definition images to compare deep convolutional neural networks (DCNN) to traditional edge detection methods for concrete crack detection. They discovered that although edge detection techniques detected 53–79% of fractures, they also added noise. On the other hand, DCNNs outperformed edge detection in detecting finer cracks and achieved 99% accuracy in image labeling with shorter computation times. This study indicates that DCNN has promising potential for concrete damage detection, particularly when combined with a hybrid approach that reduces noise.

This method utilized multi-scale feature maps and anchor boxes to enhance recognition accuracy and enable detailed feature extraction. According to Thomas and Marc Maguire [27], the algorithm achieved an impressive processing speed of 96.6 frames per second (FPS) using a high-performance graphics processing unit (GPU), facilitating real-time crack detection. Compared to You Only Look Once version 1 (YOLO v1), You Only Look Once version 2 (YOLO v2), and Faster RCNN, their proposed model demonstrated superior performance in both precise and comprehensive region-level recognition [27].

To improve the detection of concrete cracks. Their hybrid approach and noise reduction techniques provide better classification performance with fewer errors. In comparison with previous models, it achieves a 10% increase in precision and accuracy when calculating crack width. These hybrid approaches perform more accurately, robustly, and precisely than existing approaches. They believe that more sophisticated segmentation techniques for image analysis will improve analysis accuracy even more in the future. Additionally, the study intends to investigate non-invasive and invasive crack detection techniques.

A study on the use of digital image analysis for concrete bridge crack detection. To precisely locate crack pixels, they introduced the gray discrimination approach after identifying limitations in the OTSU method. This new technique revealed information about the size, orientation, width, and length of cracks in addition to just detecting them. They improved on the OTSU method's crack detection results by fusing discrimination techniques. The approach may be used to identify cracks in buildings and bridges, but it needs to be adjusted for certain parameters and might miss tiny crack objects, so more testing is needed [28].

4 Crack Analysis on Concrete Structures

4.1 Crack Classification Analysis

A study used CMOS line scan cameras for classifying subway tunnel cracks, detailing image processing techniques and configurations. This method reduced misidentifications and preserved over 90% of crack length in binary images, proving effective on Beijing Subway Line 1. Neural networks play a crucial role in crack detection, involving data preparation, CNN design and training, and performance assessment. CNNs are advantageous over SVMs and boosting methods due to less preprocessing and training, identifying complex nonlinear relationships, and using binary labels for cracked surface patches. Another study tackled underwater dam crack identification using solar images and techniques like adaptive tensor voting and K-means clustering, achieving 80–86% accuracy for small cracks. An automated bridge monitoring workflow was created, using 94 annotated images to enhance classifier performance by separating mortar and brick regions before classification. The need for automated infrastructure defect detection was highlighted, with CNNs used to evaluate images and identify cracks, featuring a model with a fully connected layer, pooling layer, and feature extraction layer for classifying road and pavement cracks [29]. A CNN utilizing ResNet101 for image segmentation demonstrated the effectiveness of deep learning in analyzing cracks in concrete pavements. The use of real-world datasets enhanced the algorithm's robustness, resulting in precise and generalized fracture segmentation. Another study applied neural networks AlexNet, VGGNet16, and ResNet18 for classifying crack images, with ResNet18 and YOLOv3 models delivering the best performance. For identifying fractures in masonry buildings, deep learning techniques employing transfer learning achieved 95.3% accuracy at the patch level and a 79.6% F1 score at the pixel level by integrating Fully Convolutional Networks and Feature Pyramid Networks with pre-trained CNNs [30].

4.2 Crack Measurement Analysis

A machine vision-based system for automating crack measurement uses a single camera mounted on a truck or robot to process image sequences and estimate crack dimensions. Experimental results validate this approach's efficacy on real structures. An algorithm detects cracks in the images, marking them with red particles. These positions are recorded and fed into a measurement algorithm, which estimates the number of pixels in a cross-section to determine the crack's dimensions. Statistical analysis shows consistently low errors, ensuring accurate measurements, and this method surpasses many existing techniques in the literature [31].

A novel global method called the crack-tree approach automates crack detection by eliminating shadows from pavement images, which are typically grayscale. This study focuses on monochromatic asphalt images, introducing a surface-adjusted luminance compensation model for consistent illumination in shadowed areas. Parameters such as voting scale, edge length, path length, and the number of geodesic levels are used in this study to enhance crack detection accuracy [1].

The goal of the study to develop a comprehensive digital image processing model that accurately represents defects, in particular cracks, in concrete structures Wu and Soh [33]. This model mimics the procedure of on-site visual inspections by integrating crack quantification, change detection, neural networks, and 3D visualization. The crack quantification model takes crack tortuosity into account when evaluating crack lengths through skeleton perimeter analysis. The Fourier Transform-based change detection model does away with the requirement for conventional image registration. Crack density is incorporated into 3D visualization to produce a comprehensive representation, and neural networks are utilized to predict the depth of cracks [32].

Three software modules were created by the researchers to analyze different kinds of track defects, such as foreign objects on the track, surface cracks, and wear on the rail head from spalling. Although the paper primarily describes a software-based methodology, they also include a conceptual hardware schematic that shows how these techniques could be mounted on a rail carriage to demonstrate how they could be successfully implemented. This method enables the measurement of defects without seriously interfering with the regular operations of the railways [33]. Their suggested methodology begins with the architecture's image collection and then pre-processes the gathered images using methods like smoothing and gray scaling. The third step involves applying crucial processing algorithms to the pre-processed images so that cracks can be found in them. These images are then processed and used to detect cracks. Crack width, length, and depth are among the attributes that are extracted and evaluated in the architecture's final stage [34]. The performance of pre-trained neural networks, such as VGG19, AlexNet, ResNet101, VGG16, GoogleNet, ResNet50, and ResNet152. They were assessed in terms of network depth, training dataset size, training epochs, and how well-suited they were for different types of construction materials. Only a small number of researchers have focused on crack characterization, particularly on crack length, width, and angle. These three variables together provide a useful indication of the severity of cracks and the likelihood of structural failure. In order to make well-informed decisions about necessary actions, like replacing or repairing damaged structures, engineers rely on this information [35]. Concentrated on evaluating the length, width, and orientation of cracks. They did not, however, gauge the cracks' actual path length.

Using a CNN, we have divided the concrete images in the current study into four categories: uncracked (UC), cracks that are horizontal, vertical, and diagonal (HC, VC, and DC), and uncracked (UC). In addition, this study presents a novel method for determining crack angles, widths, and endpoint separation a measure of the separation between the crack tips by combining CNN and image processing (IP). This investigation also offers a novel approach to precisely determine the true path length of the cracks [36]. A technique for improving low-light underwater photos, with this method, we hope to get better images, less noise, no suspended particles, and precise measurement and detection of underwater concrete structure cracks. Their model performs well under difficult deep water conditions with varying illumination. By processing both the original and inverted images and combining them, it uses dual illumination estimation to produce improved underwater images. Algorithmic calculations are performed using Jupyter notebooks and MATLAB 2020. Regardless of the size, dimensions, lighting, or noise level of the input images, the suggested algorithm reliably produces accurate results. It performs exceptionally well in intricate, uneven, low-light underwater environments, accurately detecting fissures and determining their lengths in concrete constructions [37].

5 Conclusion

In summary, the literature review in this journal paper offers a comprehensive assessment of the current status of automated crack detection utilizing machine learning algorithms. The primary focus of this study was to examine and evaluate image processing-based crack detection systems, involving an extensive review of 38 research papers dedicated to this topic. The primary objectives of this review encompassed discerning whether structural elements were cracked or uncracked and, furthermore, investigating key aspects such as crack length, width, and propagation direction. It is evident that as the field progresses, the fusion of machine learning with structural health monitoring will be pivotal in ensuring the safety and sustainability of critical infrastructure. The synthesized knowledge in this survey provides a robust platform for researchers and practitioners, stimulating ongoing innovation and advancements in automated crack detection using machine learning algorithms.

In conclusion, this literature survey highlights the significant promise of machine learning for crack detection in various sectors. This survey lays a solid foundation for the research paper, providing valuable insights into the current state of machine learning-based crack detection. The journal paper serves as an essential resource for both researchers and industry experts, offering a comprehensive understanding of the latest advancements in crack detection through machine learning. It underscores the potential of these techniques to enhance the safety and reliability of critical infrastructure and manufacturing processes. With advancing technology, the adoption of machine learning for crack detection is expected to become increasingly vital for ensuring the integrity of structures and assets across various sectors.

Acknowledgements The authors would like to thank Mr.A.Naveen Kumar from Sona college of Technology for their helpful discussions.

References

1. Yao Y, Ellen Tung ST, Glisic B (2014) Crack detection and characterization techniques—an overview. Struct Control Health Monit 21(12):1387–1413
2. Habbal F, AlNuaimi A, Alshamsi M (2020) Cracks detection using artificial intelligence to enhance inspection efficiency and analyze the critical defects. J Autom Robot Constr 37
3. Song G, Wang C, Wang B (2017) Structural health monitoring (SHM) of civil structures. Appl Sci 7(8):789
4. Mishra M (2021) Machine learning techniques for structural health monitoring of heritage buildings: a state-of-the-art review and case studies. J Cult Herit 47:227–245
5. Deng J, Lu Y, Siong Lee VC (2021) Imaging-based crack detection on concrete surfaces using You Only Look Once network. Struct Health Monit 20(2):484–499
6. Rubio JJ, Kashiwa T, Laiteerapong, Prendinger H (2019) Multi-class structural damage segmentation using fully convolutional networks. Comput Ind 112:103121
7. Mahesh B (2020) Machine learning algorithms-a review. Int J Sci Res (IJSR) [Internet] 9(1):381–386
8. Baduge SK et al (2022) Artificial intelligence and smart vision for building and construction 4.0: machine and deep learning methods and applications. Autom Constr 141:104440
9. Pandey D, Niwaria K, Chourasia B (2019) Machine learning algorithms: a review. Mach Learn 6(02)
10. Ray S (2019) A quick review of machine learning algorithms. In: 2019 International conference on machine learning, big data, cloud and parallel computing (COMITCon), pp 35–39. IEEE
11. Akagic A, Buza E, Omanovic S (2017) Pothole detection: an efficient vision based method using RGB color space image segmentation. In: 2017 40th international convention on information and communication technology, electronics and microelectronics (MIPRO). IEEE
12. Osisanwo FY et al (2017) Supervised machine learning algorithms: classification and comparison. Int J Comput Trends Technol (IJCTT) 48(3):128–138
13. Bhat MM et al (2023) Automatic bridge crack detection model based on convolutional neural network and support vector machine 12(Si6):3895–3904
14. Ali L et al (2021) Performance evaluation of deep CNN-based crack detection and localization techniques for concrete structures. Sensors 21(5):1688
15. Padsumbiya M, Brahmbhatt V, Thakkar SP (2022) Automatic crack detection using convolutional neural network. J Soft Comput Civ Eng 6(3):1–17
16. Ren J, Wang Y (2022) Overview of object detection algorithms using convolutional neural networks. J Comput Commun 10(1):115–132. https://doi.org/10.4236/jcc.2022.101006
17. Li Z et al (2021) A survey of convolutional neural networks: analysis, applications, and prospects. IEEE Trans Neural Netw Learn Syst
18. Nagar R, Singh Y (2019) A literature survey on machine learning algorithms. J Emerg Technol Innov Res 6(4):471–474
19. Singh K, Mukherjee A (2022) Reliable algorithms for machine learning models: implementation research in data science

20. Sindayigaya L, Dey A (2022) Machine learning algorithms: a review. Inf Syst J ISJ-RA-3392:sp 6
21. Liu L, Wang Y, Chi W (2020) Image recognition technology based on machine learning. IEEE Access
22. Talab AMA et al (2016) Detection crack in image using Otsu method and multiple filtering in image processing techniques. Optik 127(3):1030–1033
23. Geethalakshmi SN (2018) A survey on crack detection using image processing techniques and deep learning algorithms. Int J Pure Appl Math 118(8):215–220
24. Hoang N-D, Nguyen Q-L (2018) Metaheuristic optimized edge detection for recognition of concrete wall cracks: a comparative study on the performances of roberts, prewitt, canny, and sobel algorithms. Adv Civ Eng 2018:1–16
25. Chen B et al (2019) Detect concrete cracks based on OTSU algorithm with differential image. J Eng 23:9088–9091
26. Chen X et al (2021) An automatic concrete crack-detection method fusing point clouds and images based on improved otsu's algorithm. Sensors 21(5):1581
27. Dorafshan S, Thomas RJ, Maguire M (2018) Comparison of deep convolutional neural networks and edge detectors for image-based crack detection in concrete. Constr Build Mater 186:1031–1045
28. Majidifard H, Adu-Gyamfi Y, Buttlar WG (2020) Deep machine learning approach to develop a new asphalt pavement condition index. Constr Build Mater 247:118513
29. Adam EEB, Sathesh A (2021) Construction of accurate crack identification on concrete structure using hybrid deep learning approach. J Innov Image Process (JIIP) 3(02):85–99
30. Vivekananthan V et al (2023) Concrete bridge crack detection by image processing technique by using the improved OTSU method. Mater Today Proceedings 74:1002–1007
31. Lins RG et al (2016) Autonomous robot system for inspection of defects in civil infrastructures. IEEE Syst J 12(2):1414–1422
32. Kalfarisi R, Wu ZY, Soh K (2020) Crack detection and segmentation using deep learning with 3D reality mesh model for quantitative assessment and integrated visualization. J Comput Civ Eng 34(3):04020010
33. Ai D et al (2018) Automatic pixel-level pavement crack detection using information of multi-scale neighborhoods. IEEE Access 6:24452–24463
34. Moon H, Jung H, Lee C, Park G (2017) Crack detection on pressed panel products using image processing techniques with camera system. In: PHM society Asia-pacific conference, vol 1, no 1
35. Özgenel ÇF, Gönenç Sorguç A (2018) Performance comparison of pretrained convolutional neural networks on crack detection in buildings. In: Isarc. Proceedings of the international symposium on automation and robotics in construction, vol 35. IAARC Publications
36. Sun L, Kamaliardakani M, Zhang Y (2016) Weighted neighborhood pixels segmentation method for automated detection of cracks on pavement surface images. J Comput Civ Eng 30(2):04015021
37. Zhang W, Zhang Z, Qi D, Liu Y (2014) Automatic crack detection and classification method for subway tunnel safety monitoring. Sensors 14(10):19307–19328

MIX
Papier aus verantwortungsvollen Quellen
Paper from responsible sources
FSC® C105338

If you have any concerns about our products, you can contact us on
ProductSafety@springernature.com

In case Publisher is established outside the EU, the EU authorized representative is:
Springer Nature Customer Service Center GmbH
Europaplatz 3, 69115 Heidelberg, Germany

Printed by Libri Plureos GmbH
in Hamburg, Germany